SHOREBIRDS
OF THE NORTHERN HEMISPHERE

Richard Chandler

CHRISTOPHER HELM
LONDON

Published in 2009 by Christopher Helm, an imprint of A&C Black Publishers Ltd,
36 Soho Square, London W1D 3QY

www.acblack.com

ISBN 978-1-408-10790-4

A CIP catalogue record for this book is available from the British Library

This book is produced using paper that is made from wood grown in managed sustainable forests.
It is natural, renewable and recyclable. The logging and manufacturing processes conform to the
environmental regulations of the country of origin.

Commissioning Editor: Nigel Redman

Project Editor: Jim Martin

Design by Julie Dando, Fluke Art, Cornwall

Printed in China
Lion Productions Ltd.

10 9 8 7 6 5 4 3 2 1

Cover photography: front (top) Black-tailed Godwits, (bottom) Red Knots; spine Spotted Redshank;
back (main) American Avocets, (inset left) Ruff, (inset centre) Spoon-billed Sandpiper, (inset right)
Curlew Sandpiper. All RJC.

CONTENTS

ACKNOWLEDGEMENTS

Over the years I have learnt a great deal from the classic shorebird texts: volume 3 of *The Birds of the Western Palearctic* (*BWP*) by Cramp and Simmons (1983); *Shorebirds*, by Hayman, Marchant & Prater (1986); Prater, Marchant & Vuorinen's *Guide to the Identification and Ageing of Holarctic Waders* (1977); and volume 3 of *The Handbook of the Birds of the World* (*HBW*) by del Hoyo, Elliott & Sargatal (1996). Inevitably, material from these volumes has been widely consulted. Other references are listed after the appropriate section. The distribution maps are taken from *HBW*, with kind permission.

I have received a great deal of help from many individuals, not least those photographers who have kindly supplied photographs I did not have. I am particularly indebted to Cheri Gratto-Trevor, Nobuhiro Hashimoto, Peter Kennerley, Danny Rogers, Pavel Tomkovich and Ian Wallace, all of whom endured a barrage of emails or letters, which they answered with great good humour. The staff at my local library at Oundle were endlessly helpful with inter-library loans.

As well as dealing tactfully with a demanding author, Jim Martin at A&C Black was vastly helpful in sourcing many images I would never otherwise have found! In addition, I received help in many different ways from the following; David Bakewell, Peter Basterfield, Marek Borkovski, Geoff Carey, Chung-Yu Chiang, Ian Carter, Julie Dando, Ian Dawson, Graham Etherington, David Fisher, the late Peter Grant, Jeremy Greenwood, Brian Guzzetti, Karl Ivens, Alvaro Jaramillo, Ju Yung Ki, Chris Knights, Gordon Langsbury, Wlodek Meissner, David Melville, Clive Minton, Nial Moores, Chris Kehoe, Joseph Morlan, Killian Mullarney, Ron Pittaway, Nigel Redman, Dave Richards, Roger Riddington, Alan Warmington, the late Claudia Wilds, Alexandra Wilke and Brad Winn. I apologise to anyone I may have inadvertently omitted.

But by far my greatest thanks go to my wife Eunice, for pandering to an obsessed shorebird enthusiast for half a lifetime, though I am sure that it must seem much longer than that!

Richard Chandler
February 2009

▲ A Terek Sandpiper in breeding plumage pauses briefly during feeding. South Korea, early May. RJC.

INTRODUCTION

Why shorebirds? As a group, they have everything for the birder: stunning plumages, wonderful calls and songs evocative of wild country, the most extensive migration patterns of any group of animals worldwide and, when they have lost their colourful breeding finery, they pose identification problems to challenge the most knowledgeable of birders.

This book has evolved over a period of years from *North Atlantic Shorebirds*, which was published back in 1989. The format of that small book, with individual shorebird species accounts illustrated by photographs to aid identification, was well received at the time and I make no apology for emulating a similar layout here. The objectives of this book are the same as those of the earlier title, but on a much larger scale: to bring together, for both the general birdwatcher and the shorebird enthusiast, what is known regarding the identification of shorebirds in the northern hemisphere, and to illustrate photographically the features that allow identification and ageing. Throughout the book, emphasis is placed on identification in the field; features that can only be observed with a bird in the hand are, for the most part, not considered.

Geographical area

All the shorebird species that breed in the northern hemisphere or that occur there as non-breeding birds, including vagrants, are included in this book. Though the northern hemisphere is strictly 'that portion of the earth north of the equator', this definition is impractical for a work of this kind and it is better to use natural geographic or zoogeographic boundaries. The area referred to in this book as the northern hemisphere lies north of the line shown in Figure 1. This line encompasses the North Atlantic islands as far south as the Cape Verde archipelago, runs east across Africa at about 21°N, following approximately the northern boundary of the Sahara as far as the Red Sea (which equates to the southern boundary of the Western Palearctic), runs south of the Arabian peninsula, encompasses the Indian subcontinent (including Sri Lanka), mainland south-east Asia, Taiwan and Japan (but not Sumatra, Borneo or the Philippines), then skips across the Pacific (including the islands of the North Pacific) to the Panama canal. Consequently central America and the Caribbean islands are included, but the northern portion of South America is not, and nor are the islands of Trinidad and Tobago. Of course, many of the species included in this area also occur in the southern hemisphere, particularly when non-breeding.

In all, 135 species of the world's shorebirds have occurred within the northern hemisphere as defined in this way. With one exception, all these species are covered in the species accounts. Eskimo Curlew *Numenius borealis* has not been reliably documented since 1963 and is, sadly, presumed extinct. A single, unconfirmed 19th century record of the near-sedentary Australasian Black-fronted Plover *Elseyornis melanops* in India is generally regarded as unacceptable and consequently this species is not included in the species accounts.

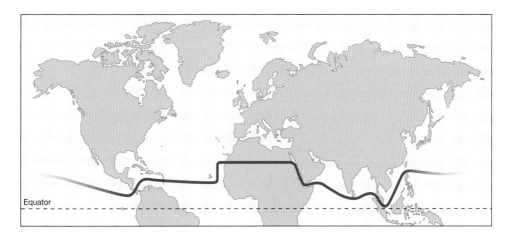

▲ **Figure. 1.** The northern hemisphere as defined in this book lies to the north of the red line.

Sequence of species

The sequence of species followed in this book is based on that used by the British Ornithologists' Union (BOU), but with the addition of those species occurring in the northern hemisphere that do not figure in the British list. In adding these further species I have been guided by the sequence used in HBW (del Hoyo *et al.* 1996), which in turn is based on Morony *et al.* (1975).

Thus the jacanas, painted-snipes and Crab Plover are treated first, followed by the oystercatchers, Ibisbill, stilts and avocets, coursers and pratincoles, plovers and lapwings and, finally, the large group of sandpipers, snipes and woodcocks. Giving the BOU sequence priority results in some differences from that used in HBW. The more significant of these differences are the placing of the small *Charadrius* plovers before, rather than after, the *Vanellus* lapwings, and placing the *Calidris* sandpipers at the beginning of the sandpipers and snipes, or Scolopacidae. Within the Scolopacidae the sequence snipes – dowitchers – woodcocks is used. Also, Stilt Sandpiper is placed in *Calidris* (rather than *Micropalama*), and the two tattlers and Willet are placed in *Tringa*, following the recent reclassifications by both the American Ornithologists' Union and the BOU.

Races

The various races (or subspecies) of shorebirds are only considered in detail if there are plumage differences that may allow them to be identified in the field. In many cases the races discussed are those generally accepted by authorities such as BWP (Cramp & Simmons 1983) and HBW. Engelmoer and Roselaar's (1998) *Geographical Variation in Waders* discusses in considerable detail the racial status of 14 northern hemisphere shorebirds. In many cases the races they propose are based on measurements rather than plumage characters, so their conclusions are not necessarily entirely appropriate here. Their authoritative analysis, however, identifies distinct populations, and the races they recognise in a number of these species are given here for completeness.

Species pairs

There are a number of pairs of shorebird species, often closely related but often with very different breeding and non-breeding distributions, that can be difficult to separate in the field. Examples are the Ringed Plover of Europe and Semipalmated Plover of the Americas, and the Red-necked Stint of eastern Asia and Little Stint of Europe and western Asia. There may be a need to identify a vagrant of one species in an area where the other species is common or, as with Red-necked and Little Stints in North America, to establish which of two vagrant species one has been lucky enough to encounter.

Species pairs are referred to as such at the start of the text for each species and, in the text for the first species of each pair, reference is made to the particular field marks that are most useful for separating one from the other. This is additional to the notes on 'Similar species', which are more wide ranging in the choice of species discussed. Where there are more than two similar species, as for example with Collared, Oriental and Black-winged Pratincoles, or when dealing with races of a single species (e.g. Red Knot or Dunlin), differences in structure and/or plumage are briefly discussed and reference is made to a table listing the essential features.

References – Introduction

Banks *et al.* (2006), Cramp & Simmons (1983), del Hoyo *et al.* (1996), Dudley *et al.* (2006), Engelmoer & Roselaar (1998), Hayman *et al.* (1986); Higgins & Davies (1996), Holden (1985), Marchant & Higgins (1993), Morony *et al.* (1975), Prater *et al.* (1977).

◄ Adult *hudsonicus* Whimbrel with a small crab caught by probing down its burrow. This one was swallowed whole. Florida, mid March. RJC.

PLUMAGES AND MOULTS

Familiarity with shorebird topography, particularly the terminology used for feather groups and plumage marks, is a valuable asset in identifying and ageing shorebirds. A basic understanding of these features is assumed in the Species Accounts that form the bulk of the book. Equally important, particularly for ageing, is a knowledge of the sequences of plumages and moults. These are discussed in detail for individual species in the Species Accounts; the general principles are presented here.

Plumage terminology

The main groups of feathers visible on a standing shorebird are shown in Figure 2, together with the terminology used for the 'bare-parts': the legs, bill and any other fleshy areas. The feather groups are also listed in Table 1 (p. 10), which draws attention to the fact that it is often convenient to lump some groups, as with the terms 'face', 'neck' and 'flight-feathers'. Another convenient portmanteau term is 'upperparts', which strictly includes mantle, scapulars, back, rump and uppertail-coverts. All of these can be seen on a flying shorebird, but only the mantle and scapulars are usually visible on a standing bird. Hence, on standing birds it is these two feather groups that are referred to collectively as upperparts.

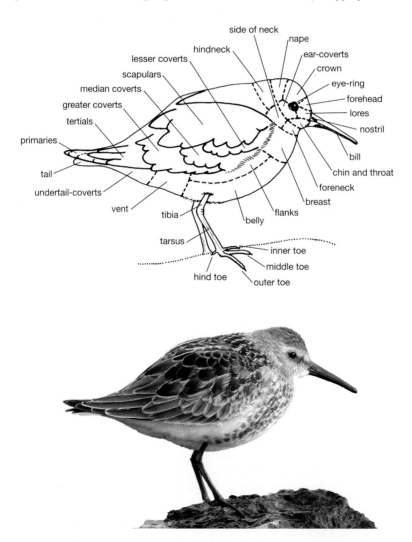

▲ **Figure 2.** Feather groups on a standing shorebird, a juvenile Dunlin. England, mid September. RJC.

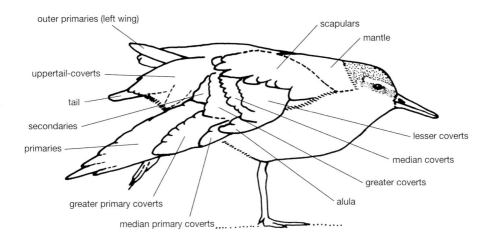

outer primaries (left wing)

scapulars

mantle

uppertail-coverts

tail

secondaries

primaries

lesser coverts

median coverts

greater coverts

greater primary coverts

alula

median primary coverts

▲ **Figure 3.** Feather groups on a shorebird when wing-stretching. This non-breeding Red Knot is completing primary moult; the outer primary is still growing. Florida, late October. RJC.

◄ **Figure 4.** The underwing of a shorebird in flight. The black axillaries at the base of the wing are diagnostic of Grey Plover. England, late February. RJC.

Table 1. Principal feather groups used when describing shorebirds.

General area	Feather groups	
Head and neck	Forehead Crown Nape	
	Chin and throat Lores Ear-coverts	'Face'
	Hindneck; sides of neck; foreneck	'Neck'
Upperparts	Mantle Scapulars Back Rump Uppertail-coverts	
Underparts	Chin and throat Breast Belly Flanks Vent Undertail-coverts	
Wings	Primaries Secondaries Tertials	'Flight feathers'
Upperwing	Greater coverts Median coverts Lesser coverts	'Innerwing-coverts'
	Greater primary coverts Median primary coverts Lesser primary coverts Alula	'Outerwing-coverts'
Underwing	Greater underwing-coverts Median underwing-coverts Lesser underwing-coverts	'Under innerwing-coverts'
	Greater under primary coverts Median under primary coverts Lesser under primary coverts	'Under outerwing-coverts'
	Axillaries	(equivalent to Scapulars)
Tail	Tail	

   The more important feather groups on a standing bird are the *mantle* (the upper portion of the back), the *scapulars* (the body feathers at the base of the wing), the various *innerwing-coverts*, and the *tertials*. As is seen in many of the images in this book, the wing-coverts are often partly obscured by the scapulars. This occurs particularly with adults, which generally have rather larger scapulars (and other body feathers) than the smaller, neater, feathers of the juveniles. The wing-coverts may also be partly covered by the breast and flank feathers, particularly in cold weather when the body feathers are fluffed up.

   The relative positions of the feather groups on the wing are well displayed on a shorebird that extends or stretches one wing sideways ('wing-stretching', Figure 3). These images show the different groups of flight feathers: primaries with their associated wing-coverts on the outerwing, and the secondaries and their coverts, together with the tertials, on the innerwing. Notice that when a shorebird stretches just one wing it twists its tail to the side, and the tertials pivot under the scapulars. Comparing the standing bird with the one wing-stretching shows that the wing-coverts visible on the folded wing are the greater, median and lesser coverts, together the innerwing-coverts. This is a result of the outerwing being folded under the innerwing, covering the outerwing-coverts. Corresponding feather groups exist on the underwing: the 'underwing-coverts' in Figure 4, with the 'axillaries' at the base of the underwing where it joins the body, corresponding to the scapulars on the upperwing. The colour and/or patterns of the axillaries are important in the specific identification of Grey and golden plovers, and in the racial identification of

Eurasian Curlews and Whimbrels. All shorebirds have eleven primaries, but the outermost is tiny, virtually one of the covert feathers. The number of secondaries varies with the size of the bird; smaller shorebirds species, for example Little and Red-necked Stints, typically have ten, whereas the largest species, such as Far Eastern Curlew, have 14. Most species have four to six tertials, the flight feathers nearest the body, which are modified secondaries that cover and protect the true secondaries on the folded wing, and which are usually clearly visible on a standing shorebird, overlying and concealing the secondaries. The patterns on the tertials are often useful when ageing a shorebird, and the length of the longest tertial relative to the underlying primary tips is helpful, for example, in separating the different golden plovers.

Plumage marks

Areas of distinctively coloured feathers or parts of feathers also provide useful field marks. These are not feather groups (as listed in Table 1), but are nonetheless invaluable for plumage descriptions. The more important plumage marks are illustrated in Figures 5 and 6. The *crown-stripe* is the pale stripe running through the centre of the crown; the *eye-stripe* is the dark stripe through the eye; the *supercilium* is the pale stripe immediately above the eye; the *wing-bar* is the white or pale bar along part or all of the length of the wing. The wing-bar is usually formed by pale tips of the greater coverts and/or adjacent white areas on the primaries. The *shoulder-patch* of some species is formed by an area of contrastingly darker lesser coverts and

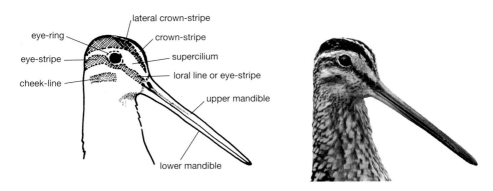

▲ **Figure 5.** Head pattern terminology. Adult Common Snipe. England, late March. RJC.

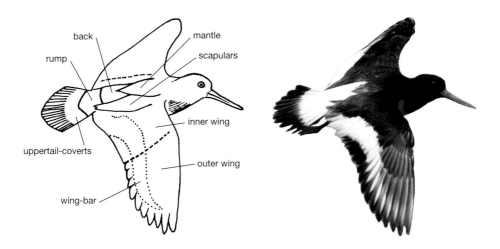

▲ **Figure 6.** Feather groups and plumage marks on a flying shorebird. Adult breeding *ostralegus* Eurasian Oystercatcher. England, late July. RJC.

◀ **Figure 7.** On juvenile Pectoral Sandpipers the feather edges align to form converging pale lines (or Vs) on both the mantle and scapulars. Canary Islands, mid September. RJC.

is often particularly obvious at the bend of the folded wing on standing birds. These dark feathers also show in flight, emphasising the wing pattern. Sanderling and Broad-billed Sandpiper are examples of species that exhibit shoulder-patches, particularly when non-breeding; however, often the dark area is covered by breast feathers.

Mantle Vs and *scapular Vs* are pale lines formed by the juxtaposition of the pale margins of the mantle and scapular feathers, and are shown particularly by most snipes, and by the juveniles of species such as Little Stint and Pectoral Sandpiper. Figure 7 shows that on juvenile Pectoral Sandpipers the feather edges align to form converging pale lines or Vs on both the mantle and scapulars.

Feather patterns and wear

Individual feathers on shorebirds may be plain (this is often the case when the bird is in non-breeding plumage), or patterned. The patterning may be simple or extremely complex. Some of the more regularly occurring feather patterns, which may be described by such terms as 'edge-spotted', or 'notched', are shown in Figure 8. All the feathers on a bird's plumage wear with time but exposed feathers are more affected, particularly those on the upperparts, the innerwing-coverts, the tertials, and the tips of

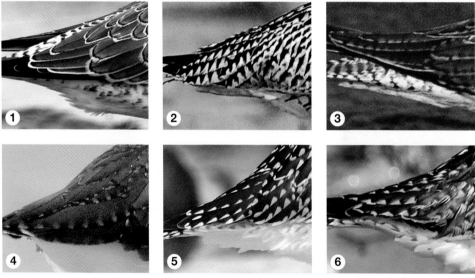

▲ **Figure 8.** Some typical shorebird feather patterns; all are tertials. **1**. Pale fringe with dark subterminal line (juvenile Red Knot); **2**. Barred (juvenile Eurasian Curlew, but basic pattern is shared by curlews of all ages); **3**. Edge-spotted (juvenile Common Redshank, but pattern is shared by several larger *Tringa* of all ages); **4**. Edge-spotted (juvenile Green Sandpiper, but pattern is shared by several smaller *Tringa* of all ages); **5**. Notched (juvenile Grey Plover); **6**. Notched (as 5, but worn and faded).

▲ **Figure 9**. Adult Green Sandpiper, showing the prominent white eye-ring and eye-lids of this species. Spain, early September. RJC.

the longest primaries. Fresh, new feathers have rounded tips and neat edges but, as time progresses, the tips at first appear frayed and slightly irregular, and eventually may become quite pointed. The sides of the feathers also wear, but this is usually less apparent except where the feather has a notched or spotted-edge pattern. In these cases the wear is greater on the paler parts of the feather, resulting in the sides of the feather having a very irregular profile (see Figure 8.6). Pale feather fringes are also often lost to wear.

Bare parts

The colours of the bare parts are also of great value for identification and ageing. This applies particularly to the bill and legs and, in some species, to the eye, where the colour of the iris and the presence, colour and size of the orbital ring may be useful. The *orbital ring* is the bare skin (which may or may not be present) immediately surrounding the eye, whereas the *eye-ring* is the ring of tiny pale feathers surrounding the eye (or the orbital ring, if the bird has one). Tiny feathers similar to those of the eye-ring also cover the eyelids, so a species with a pale eye-ring usually has pale eyelids; these can be conspicuous, especially when the bird is roosting. See Figure 9, which shows a Green Sandpiper, a species with a prominent eye-ring and white eyelids, and Figure 10, which shows the orbital ring of a Little Ringed Plover.

A coating of mud on the bill or legs can obscure the natural colours of these bare parts, and this possibility must be kept in mind. Less well-appreciated is the possibility that the bare parts may be stained, with consequent long-term discoloration. Such staining is not very common but white discoloration can occur where birds have been feeding in calcareous or other alkaline environments, and this can affect both the bill and the legs (Figure 11). Brownish iron staining also occurs.

▲ **Figure 10**. Adult male Little Ringed Plover, showing the prominent orange-yellow orbital ring of this species. South Korea, early May. RJC.

▲ **Figure 11**. A juvenile Lesser Yellowlegs with some prominent white staining on its legs. Florida, late September. RJC.

Seasonal plumage changes

Shorebirds have a succession of different plumages that change with the age of the individual and with the seasons. A number of different terminologies have been used to define the various plumages and the intervening moults; some of these are summarised in Table 2. A scheme similar to that used in *The Birds of the Western Palearctic* is used in this book. Since many northern hemisphere shorebirds spend their non-breeding season in the southern hemisphere during the austral summer, the use of the terms 'winter' and 'summer' may be confusing. Consequently, I have used the terms 'non-breeding' and 'breeding' in their place. The scheme devised by Humphrey and Parkes (1959), widely used in North America, effectively uses the corresponding terms 'basic' and 'alternate'. It is not intended here to give a complete account of the moult of shorebirds, which is well covered elsewhere, but simply to discuss those aspects that aid the ageing of the various species in the field.

Another difficulty is how one refers to the age of individual shorebirds, for which the concept of 'calendar year' is used in this book. Most shorebirds breeding in the northern hemisphere will have hatched between April and July, so the *first calendar year* covers the period from hatching to the end of December, the *second calendar year* is the next year from January to December, and so on.

Table 2. Terminology and timing of shorebird plumages and moults. The 'alternate' plumage of pre-breeding Ruffs, for example, is not included.

This book	British Birds Anon. (1985)	Humphrey & Parkes (1959)	BWP	Typical plumage or moult period	Cycle	Calendar year
Juvenile	Juvenile	Juvenile	Juvenile	June–September	1st	1st
Post-juvenile moult	Moult to first-winter or post-juvenile moult	First prebasic moult	Post-juvenile moult	August–November	1st	1st
First non-breeding	First-winter	First basic	First non-breeding	October–April	1st	1st–2nd
Moult to first-breeding	Moult to first-summer	First prealternate moult	First pre-breeding moult	February–May	1st	2nd
First-breeding	First-summer	First alternate	First breeding	February–September	1st	2nd
Moult to second non-breeding	Moult to second-winter	Second prealternate moult	First post-breeding moult	June–October	2nd	2nd
Second non-breeding	Second-winter	Second basic	Second non-breeding	August–April	2nd	2nd–3rd
Moult to adult breeding	Moult to adult summer	Prealternate moult	Adult pre-breeding moult	February–May	2nd	3rd
Adult breeding	Adult summer	Definitive alternate	Adult breeding	February–September	2nd or Adult	3rd
Moult to adult non-breeding	Moult to adult winter	Prebasic moult	Adult post-breeding	July–November	Adult	3rd
Adult non-breeding	Adult winter	Definitive basic	Adult non-breeding	October–April	Adult	3rd

The plumage cycle

On hatching, the chick has a fluffy downy plumage. This is quickly replaced, usually within the first three weeks, by the first set of true feathers, which form the *juvenile* plumage. The juvenile feathers grow as a continuation of the down, which is carried on the tips of the true feathers. The down soon falls or wears off, though occasionally, as in the thick-knees and the *Charadrius* plovers, tail-down may be retained for a week or so after the bird has otherwise attained its complete juvenile plumage; see the juvenile Killdeer shown in Figure 12. Although this newly fledged individual can probably fly, its feathers, particularly its flight feathers, are still growing, giving it a rather short rear-end; compare its overall profile with the apparently longer-bodied, full-grown, but still juvenile Killdeer (35a, p. 134). The latter bird also has a longer bill. In

◄ **Figure 12**. Juvenile Killdeer showing tail-down. Note also the neat, small feathers characteristic of all juvenile shorebirds. Texas, early August. RJC.

many shorebirds bill growth continues in otherwise full-grown juveniles, sometimes for several months; this can be particularly obvious in juvenile curlews, which have noticeably shorter bills than adults; see Eurasian Curlew (110a, p. 354).

Thus, within a few weeks of hatching (about three weeks for some of the smaller species, up to six for the larger) the bird has its first complete set of true feathers and can fly. The characteristic patterns of the juvenile feathers, particularly those of the upperparts, are often very similar, particularly in closely related species. The small *Calidris* sandpipers generally have neat pale fringes to their upperpart feathers (Figure 13), while the corresponding equally neat feathers on juvenile *Tringa* are strongly edge-spotted (Figure 14). Juvenile feathers are often quite bright and fairly strongly patterned, to such an extent that the overall plumage of juveniles of some species is remarkably similar to that of the breeding adults. Compare, for example, the juvenile and adult breeding plumages of Dunlin (82, p. 268).

The feathers of juveniles are often noticeably softer than those of the adults, and they wear more quickly. Perhaps as a consequence of this, juveniles soon – some almost immediately after fledging – commence their *post-juvenile* moult to their *first non-breeding* plumage. The post-juvenile moult is generally partial, usually involving most of the body feathers (i.e. those of head, neck, upperparts and underparts) and a variable number of the wing-coverts and tertials. The juvenile primaries, secondaries and tail-feathers are usually retained; jacanas, coursers, pratincoles and Buff-breasted Sandpiper are exceptions that replace all their feathers at the post-juvenile moult, though this moult is apparently quite lengthy in the jacanas and Buff-breasted Sandpiper.

As a result of the post-juvenile moult, the appearance of many shorebirds changes significantly. The newly acquired first non-breeding feathers are often comparatively dull and featureless, so that by the time the post-juvenile moult is complete (typically some time in November) the bird is much plainer and drabber. A typical first non-breeding Dunlin (Figure 15) shows this well.

◄ **Figure 13**. Juvenile Semipalmated Sandpiper showing the neatly pale-fringed plumage typical of many juvenile *Calidris* sandpipers. Florida, mid September. RJC.

▲ **Figure 14**. Juvenile Common Redshank showing edge-spotted plumage, particularly on the greater coverts and tertials; this is typical of many juvenile *Tringa*. Notice, too, the orange legs at this age. Wales, early August. RJC

When the post-juvenile moult is only partial, first non-breeding shorebirds can be distinguished from the otherwise very similar non-breeding adults by the unmoulted (but worn) feathers retained from the juvenile plumage. The feather groups that are usually retained are the wing-coverts and the tertials, though the number of these that remain is variable, not only between species but also between races and even individuals of the same species. The retained juvenile feathers are often differently patterned, and will certainly be more worn, compared to the fresh first non-breeding mantle and scapulars. This is illustrated by the Dunlin in Figure 15 and by a Short-billed Dowitcher in Figure 16, both of which show contrast between the worn juvenile wing-coverts and the fresh, newly acquired mantle and scapulars. In comparison, an adult non-breeding Short-billed Dowitcher at the same season (Figure 17) has fairly fresh, unworn feathers on both upperparts and wing-coverts.

Species such as Purple Sandpiper are relatively easy to age, since they retain their pale-fringed juvenile wing-coverts well into their second calendar year (80c, p. 262). Other species, however, have a much more complete post-juvenile moult, and gain upperparts and wing-coverts similar to those of the adult non-breeding bird: examples include the pratincoles, the *Gallinago* snipes, including Common and Wilson's Snipes, and the two *Actitis* sandpipers, Common and Spotted Sandpipers. With these species, ageing in

▲ **Figure 15**. First non-breeding Dunlin, showing plain featureless plumage, having lost its bright upperparts and densely black-spotted underparts. Note the retained brown-tipped juvenile greater coverts and tertials. California, early November. RJC.

▲ **Figure 16**. This first non-breeding Short-billed Dowitcher still has juvenile brown wing-coverts and a juvenile 'tiger-striped' tertial, but it has acquired several plain brown-grey non-breeding tertials. Florida, mid December. RJC.

▲ **Figure 17.** A group of roosting adult non-breeding Short-billed Dowitchers. Note the pale fringes of the newly acquired plain unworn feathers on both upperparts and wing-coverts. Florida, late October. RJC.

non-breeding plumage becomes difficult and often impossible, particularly in the field. A further category where ageing is impossible arises where the plumages are practically identical at all ages; Jack Snipe is the only example of this in the northern hemisphere.

A first non-breeding bird will moult again early in its second calendar year to attain its *first-breeding* plumage. This, too, is often a partial moult, which typically involves body feathers, some wing-coverts, and, in some species, a few (occasionally all) flight-feathers. The moult to first-breeding is thus comparable in its extent to the post-juvenile moult. There is, however, variation between species in the extent of this moult. Some species, particularly the smaller ones, gain a plumage virtually identical to that of the breeding adults. Indeed many individuals of the smaller species will breed in their second calendar year, and can only be identified as first-breeding if they retain their very worn juvenile primaries, or, in those birds that moult only their outer primaries (see below), by the contrast between new and old primaries.

Many larger species do not breed in their second calendar year and will either remain in their non-breeding areas or may move part-way towards the breeding grounds. These birds rarely attain the plumage of the adult breeding birds, but either retain very worn first non-breeding plumage or gain a mixture of adult-type breeding and non-breeding plumage; see the Short-billed Dowitchers in Figure 18. Sick or injured birds may also retain part or all of their non-breeding plumage in the breeding season and may not return to their breeding grounds.

At their next moult most species acquire their *adult non-breeding* plumage, though the oystercatchers, in particular, take two or three years to reach maturity. Subadult oystercatchers can be aged on their bare-part colours but not on their plumage, which after first non-breeding is virtually identical to that of adults.

Shorebirds are sometimes seen in adult-type non-breeding plumage as early as August. These probably represent birds that did not breed in their second calendar year, and which quickly replaced their first-breeding plumage to attain their adult non-breeding plumage before the full adults, whose moult is usually delayed until the end of their breeding season. Breeding adults do not generally attain full non-breeding plumage until October, and often later. Both first-breeding and adult breeding shorebirds have a complete moult, including their flight feathers, in attaining their adult non-breeding plumage. By this time the flight feathers of some first-breeding and second non-breeding birds are upwards of 15 months old, and they are often bleached and well-worn before they are finally shed. On average, however, second calendar-year birds also replace their primaries before the adults.

▲ **Figure 18**. First-breeding Short-billed Dowitchers in their non-breeding area. The centre and right-hand birds are typical of the range of plumages shown by first-breeding Short-billed Dowitchers, one with a large proportion of adult-type breeding plumage, the other still in almost complete non-breeding plumage. Florida, late June. RJC.

As early as the end of February in some species and populations, and as late as May in others, the adults have a body moult and gain their *breeding* plumage. In most species, the adults usually moult earlier to breeding plumage than younger individuals moult to first-breeding plumage. Differences in the timing of this moult are substantial in some species, such as Bar-tailed Godwits. In this species races *baueri* and *menzbieri*, which spend the non-breeding season in Australasia, are largely in breeding plumage by late March, when they commence their north-bound migration, whereas the nominate *lapponica*, which winters in west Africa and western Europe, does not acquire full breeding plumage until May, when the birds begin the final leg of their migration.

Breeding plumage can change noticeably during the course of the season. Many of the *Calidris* species, for example, which have pale-fringed but dark-centred upperpart feathers, lose the fringes with wear, so the overall plumage becomes darker. Feather darkening is also a result of a distinct colour change, perhaps resulting from oxidation, which can sometimes be seen in Common Redshanks and Common Greenshanks. The same process may also affect shorebirds in their greyish non-breeding plumage, some individuals appearing quite dark by the time they commence moult to breeding plumage. Feather bleaching can also occur, and this is most obvious with the ringed plovers, which in both breeding and non-breeding plumage can become noticeably paler as the feathers age.

A few species attain a breeding plumage with upperparts that consist of a mixture of breeding and non-breeding type feathers, though both feather types are acquired during the moult to adult breeding. This is the case with more southerly races or populations of species that breed relatively early, examples of which include Greater Sand Plover (of race *columbinus*), European Golden Plover, Common Redshank, Common Greenshank and Black-tailed Godwit. The effect is particularly obvious in the southern race of European Golden Plover, *apricaria*. The same may perhaps apply to Willets of the western race *inornatus* in North America.

Once they have reached adulthood, the plumage sequence of a number of shorebirds is not simply a case of seasonal alternation between non-breeding and breeding. Great Knot, Ruff, Bar-tailed Godwit, and perhaps other species with bright colours in full breeding plumage (such as Black-tailed and Hudsonian Godwits, Asian Dowitcher and Grey Phalarope), have three generations of feathers in each annual cycle. In the case of Ruff, a generation of striped feathers (the 'alternate' plumage), is gained before the birds leave Africa, from January and February. These striped feathers are replaced at migration staging areas in April and May by the large and colourful feathers of the true breeding or 'supplemental' plumage, particularly the characteristic ruff of the males. The alternate-plumage striped feathers are similar to the underpart pattern of shorebirds that have relatively dull breeding plumages, such as many *Tringa* species and the curlews. This similarity leads to the suggestions that these plumages are homologous, and that the supplemental plumage has evolved relatively recently. It is probable that female Ruffs also develop a supplemental plumage during the same time period, but the evidence is not so clear as with the males.

The 'wrong' plumage

Occasionally, shorebirds are observed in breeding plumage at a time when they would normally be in non-breeding plumage. Extreme examples include Little Stints in Australia in November and in South Georgia in December, and Red-necked Phalaropes, also in Australia, including one in October and two in December. The reason for this is not clear but it is probably associated with a lack of synchronisation of the birds' hormonal cycle.

The *Charadrius* plovers

Small plovers can have curiously worn wing-coverts, with a 'shredded' appearance (see for example 36d, p. 139). Worn coverts of this type can be shown by birds of any age, and are not necessarily an indication of first non-breeding or first-breeding plumage.

The breast-bands of juvenile and non-breeding ringed plovers are less dark than those of breeding adults, and may sometimes be incomplete, and in breeding plumage females typically have brown feathers mixed in with the black areas of the head and in the breast-band. Many individuals attain their breeding plumage (though perhaps not their full breeding bare-part colours) quite early in the year – February or earlier – and retain it until well into October. A good example of this is Ringed Plover of the largely sedentary race *hiaticula*, and also the sedentary races of Little Ringed Plover, *dubius* and *jerdoni*.

It is interesting to note that several sedentary species also have similar plumages year-round, once adulthood has been attained: Three-banded Plover, Malaysian Plover, Collared Plover and perhaps Long-billed Plover are notable examples. The bare part colours of such birds, however, may be brighter when they are breeding.

Primary moult

Most studies of the timing and feather-by-feather sequence of moult in shorebirds have concentrated on the flight feathers, particularly the primaries, which in all northern hemisphere species are shed from around August as part of the moult to adult non-breeding plumage. Less is known about the progression of body moult related to the primary moult except that, in general, the primary moult period is comparable with that of the remainder of the plumage. In flight, or when a bird lifts or stretches its wings, it is possible to see the extent of the flight feather moult, which commences with the inner primaries, typically the first three to five on both wings being shed almost simultaneously. The remaining feathers are then lost symmetrically on both wings, the primaries in sequence towards the wing-tip, and the secondaries towards the body. The Eurasian Oystercatchers in Figure 19 have just lost the inner primaries on each wing, and show the characteristic 'stepped' shape of the trailing edge resulting from the greater length of the remaining outer primaries compared with the adjacent secondaries. This wing shape is usually quite obvious in flight, even if there is no obvious gap between the primaries and secondaries.

Though the usual primary moult sequence of shorebirds is descendent, from the carpal joint outwards to the wing-tip, an exception occurs with some immatures of a number of species, which may moult just the outer primaries early in their second calendar year. Typically, two or three primaries are involved, particularly with birds occurring when non-breeding in the southern hemisphere. This partial primary-moult strategy is seen, for example, in non-breeding Sanderlings and Curlew Sandpipers in southern Africa, in non-breeding Sharp-tailed Sandpipers in Australia, and by non-breeding Least Sandpipers in South America. Species such as Dunlin and Purple Sandpiper, which are shorter distance migrants that spend the non-breeding season in the northern hemisphere relatively close to their breeding areas, do not generally moult any primaries until towards the end of their second calendar year. Figure 20 shows a first-breeding Dunlin that, most unusually for this species, has replaced its outer primaries.

▲ **Figure 19**. Eurasian Oystercatchers in flight, showing different stages of primary moult. Four of the five adults have prominent gaps mid-wing where they have lost their inner primaries. The single immature (in the centre, with white collar) is probably an older immature (red eye, broad white collar, bill dark distally) that has already completed its primary moult, having commenced earlier than the adults. England, late August. RJC.

Most of the species that have a partial primary moult early in their second calendar year also have a fairly complete body moult to breeding plumage, and breed in their second calendar year. This suggests that replacement of the outer primaries is an adaptation to enable the long-distance northward migration to the breeding grounds. This moult strategy can be of value for ageing in the field on northward migration, but a close view is needed to see the contrast between the new outer primaries and the worn inner ones.

A feature of the primary moult of adults of some shorebird species, particularly those breeding in the Arctic, is 'suspended moult'. These birds commence primary moult while on the breeding grounds. They replace some of their inner primaries, and then migrate south with completely grown new inner primaries, while still retaining their old outers. The outer primaries are then replaced on the non-breeding grounds. European Golden Plovers are one species that undergoes suspended moult; 51e (p. 181) shows an adult European Golden Plover on the breeding grounds in the early stages of moult, while 51b (p. 180) is in suspended moult, presumably still on migration, with only its outer primaries to replace. Similarly, 27f (p. 111) showes an Oriental Pratincole with suspended moult, having presumably grown the inner primaries prior to migration. The outer primaries will soon be replaced.

◀ **Figure 20**. First-breeding *hudsonia* Dunlin showing fresh outer primaries. Florida. RJC.

Hybrid shorebirds

Hybrid shorebirds are really quite rare, but they do occur, and this is a possibility to be considered when an unfamiliar bird is encountered. Typically, hybrids have plumage, size, structure and perhaps even vocal characters intermediate between their progenitors, so it is often possible to suggest their parentage. The adjectives 'apparent' and 'possible' are frequently used in the context of hybridisation since, in the wild and in the absence of molecular studies of the individual concerned, it is usually impossible to be sure of the parentage of a bird with mixed characters.

Reports of hybridisation between shorebirds seem to be limited to only a few groups: the jacanas, the oystercatchers, stilts and avocets, pratincoles, the *Pluvialis* plovers, lapwings and the *Calidris* sandpipers. It is interesting that there seems to be no suggestion of any hybridisation between any of the *Tringa* or related species.

Possible hybridisation between Northern and Wattled Jacanas in Costa Rica and Panama is discussed in the Northern Jacana account on p. 42. In this instance, natural variability of plumage and other features seems likely to explain the postulated hybridisation. Hybridisation between oystercatchers, particularly between the pied and black oystercatchers of the Americas, has long been assumed, but current research suggests that Black and American Oystercatchers, which are clearly closely related, may prove to be different colour morphs of the same species, and that intermediates are thus not true hybrids.

The golden plovers are another difficult species group. American and Pacific Golden Plovers are undoubtedly closely related. However, there are no known examples of mixed pairings of the two species, despite of intensive field research in recent years, but individuals with either mixed or uncertain characters are seen from time to time. So it is unclear if there is hybridisation between these two species. There are also reports of apparent intermediates between Pacific and European Golden Plovers from areas where the two species both occur; again more data are needed. There is also a record of an American Golden Plover paired with a Grey Plover.

Amongst the lapwings, apparent hybridisation has only been reported between Spur-winged and Blacksmith Lapwings in Africa. There are a number of reports, however, of hybridisation between various *Calidris* species; indeed there are more records of hybrid *Calidris* than for all the other genera of shorebirds combined. These include the best known shorebird hybrid, 'Cox's Sandpiper', shown by molecular studies to almost certainly be the offspring of a male Pectoral Sandpiper and a female Curlew Sandpiper. There are a surprising number of putative records of Cox's Sandpiper, the majority from south-eastern Australia, at a rate of around five a year.

Another well-known hybrid is 'Cooper's Sandpiper', the original specimen of which, taken at Long Island, New York in 1833, was thought to be a valid species when described in 1858. It has recently been re-identified as a probable Curlew Sandpiper x Sharp-tailed Sandpiper hybrid. It is apparently very similar to 'Cox's Sandpiper', as might be expected from its anticipated parentage; Cooper's Sandpiper, too, has been reported from south-eastern Australia.

There seem to be two main situations where hybridisation occurs between shorebirds in the wild. First, when isolated individuals of one species occur within the usual range of a closely related species. One example is the Spur-winged Lapwing, which periodically hybridises with Blacksmith Lapwing, a southern hemisphere African species, in central Kenya. Here the Spur-winged Lapwing is at the southernmost extreme of its range, but Blacksmith Lapwings are relatively common. Other possible examples of hybrids in this category include the several reported presumed Black-necked Stilt x American Avocet in coastal California. A Common Sandpiper seen mating with a Spotted Sandpiper in England in 1990 might have produced a hybrid offspring in just these circumstances.

The second main hybridisation situation arises where species have mating systems that result in rapid pairing. This seems to be the case with most of the *Calidris* that are believed to hybridise. For example, White-rumped Sandpiper, Pectoral Sandpiper, Sharp-tailed Sandpiper and Buff-breasted Sandpiper are all polygynous and promiscuous, and fall within this category. Other 'rapid-pairing' species include Curlew Sandpiper. These birds are polygynous, with the males attracting as many females as possible during a relatively short period before quickly leaving the breeding grounds (by late June), playing no part in incubation or caring for the young. It is not difficult to imagine that in this short period males might be confused as to which species they are mating with or that late in the season a female may find no male of her own species with which to mate and may copulate with a non-conspecific. Temminck's Stints have a 'rapid multi-clutch system', in which the female lays two or more clutches in quick succession for one or more males to incubate, incubating the last clutch herself. This is an adaptation to maximise the production of

▲ **Figure 21**. Feeding Semipalmated Sandpipers; the left-hand bird is a leucistic individual with insufficient plumage detail showing to allow it to be aged with any confidence; the right hand bird is a worn breeding adult. Alberta, Canada, early August. Gerald Romanchuk.

young in the short Arctic summer, and leads to a situation that might explain the occurrence of hybrid Little Stint x Temminck's Stint. This possible hybridisation scenario is enhanced by the rather similar 'double-clutch' breeding system of Little Stint, where the female lays two clutches, one being incubated by the male, the other by the female.

It is clear that hybridisation is unusual between monogamous species, so it is probably unsurprising that there appear to be no cases of hybrids amongst the monogamous *Tringa* species.

Aberrant plumages

Another potential source of confusion is the very occasional shorebird with an aberrant plumage. Albino, leucistic or melanistic individuals are the more usual of these uncommon forms. Of these, true albinos seem to be very rare. Leucistic birds have lost much of their normal pigmentation, so appear a dirty cream or pale brown overall, but with the usual plumage pattern showing faintly. Birds with an excess of melanin are sooty black or very dark brown, either overall or restricted to a particular area of the plumage. Examples of the former include dark morphs of Common and Wilson's Snipes, the so-called 'Sabine's Snipe', while the Eurasian Stone-curlew in Figure 22 is an example of the latter, with a restricted area of melanistic feathering in the folded wing.

Shorebirds with these types of aberrant plumages can pose an identification challenge, particularly those that are uniformly coloured. Comparison of the size and structure of the aberrant bird with nearby normal birds is usually helpful. Individuals that are only partially or irregularly mis-coloured are usually fairly straightforward to identify.

▲ **Figure 22**. First non-breeding Eurasian Stone-curlew with dark, melanistic feathering in its folded wing. England, late September. RJC.

References – Plumages and moults

Anon. (1981, 1985), Bamford *et al.* (2005), Battley *et al.* (2006), Betts (1973), Campbell & Lack (1985), Chandler & Marchant (2001), Cox (1989, 1990), Hale (1980), Humphrey & Parkes, (1959), Jonsson (1996), Jukema & Piersma (2000), Lawrence (1993), Pearson (1983), Piersma & Jukema (1993), Rogers (2006), Sangster (1996), Skewes *et al.* (2004), Taldenkov (2006), Vinicombe (1988).

SHOREBIRD BEHAVIOUR

What is that shorebird doing? In terms of a shorebird's daily routine, its behaviour can readily be categorised under three headings: feeding, roosting and plumage care. The following is intended to provide an illustrated introduction to these topics.

Feeding behaviour

The majority of shorebirds are diurnal feeders, at least by preference, though some, particularly stone-curlews, snipes and woodcocks, are largely nocturnal. Additionally, many of the birds that are dependent on coastal mudflats have to feed at low tide, and will do so at night, particularly around full moon.

The feeding behaviour of shorebirds is closely associated with their bill sizes and shapes, which provide a clear indication of their feeding methods. A rather sweeping generalisation is that short-billed species are 'pickers', most usually taking food items from the ground surface (Figure 25), such as the plovers, while those with long bills (such as godwits and curlews) feed by probing deeply into soft sediments. Those with medium-length straight bills are generalists, able to feed by picking or probing, though some also use more specialist methods. Examples of generalists are many of the *Calidris* and *Tringa* species that, in addition to picking or probing, may use other techniques, such as chasing and catching small fish in shallow water, as do Spotted Redshanks and Common Greenshanks (Figure 26).

▲ **Figure 23**. A Willet feeds by picking, walking along the tide-line. This generalist feeder is able to pick or probe. Adult non-breeding of the western subspecies, California, early November. RJC.

In addition, there are a number of specialist feeders, which have bills adapted to either their favoured food or their primary feeding technique. Species with particularly specialised bills include the avocets, Crab Plover, Wilson's Plover and Spoon-billed Sandpiper. It is rash, however, to generalise, as all shorebirds will take foods other than those for which they are best adapted if the opportunity arises, or if conditions

▲ **Figure 24**. Adult non-breeding Surfbird feeding by picking on a mussel bed exposed at low tide. Also note the feet, which are adapted for a good grip on wet rocks. California, early November. RJC.

▲ **Figure 25**. Grey Plovers feed by picking; this adult has extracted a ragworm from its hole in the mud, having located the worm by its movement. Florida, early April. RJC.

do not allow their usual feeding method. There are many examples of exceptions to these generalisations, an important one being that many species will feed on insects (Figure 27) or on berries from low-growing plants (Figure 28), particularly in the relatively short period when they are on their breeding grounds.

A brief discussion of the usual feeding method is given for each individual species in the relevant Species Accounts.

◀ **Figure 26**. Common Greenshank, a generalist feeder, with a small fish. Adult breeding, South Korea, late April. RJC.

◀ **Figure 27**. Adult male breeding *islandica* Black-tailed Godwit feeding on insects. Iceland, late June. RJC.

◀ **Figure 28**. Adult *islandicus* Whimbrel feeding on berries. Iceland, late June. RJC.

▲ **Figure 29**. Adult breeding Ringed Plover foot-paddling on a silty sand substrate. Note that the plover has chosen an area where the sand surface is wet; it is vibrating its front foot, which has liquefied the sand. In the two seconds between the two photos, both feet have settled below the surface as a result of liquefaction and, although difficult to see, the small puddle in front of the bird has increased in size. England, late February. RJC.

Feeding techniques

Short-billed pickers, which include the plovers and lapwings, take prey from the surface or from very shallow depths, hunting visually. They all share the 'run-stop-peck' feeding technique, alternately standing motionless scanning for food, and then abruptly moving a few steps before bending forward to pick items from the ground or water surface. The smaller plovers tend to be more active in their movements. On wet ground all plovers and lapwings use the 'foot-trembling' technique, in which the leading foot is vibrated rapidly for several seconds (Figure 29). On wet sand or silt this has the effect of liquefying the sediment and either causing food items to float to the surface, or perhaps disturbing prey which gives itself away by moving and is subsequently caught. When this method is used on wet but vegetated surfaces the roots inhibit liquefaction but the technique still disturbs prey (Figure 30).

A variant of this technique is used by many *Calidris* species. One of their feeding methods is 'stitching', where the bill is held at an angle with the tip just below the wet surface and probed rapidly and continuously into the substrate (Figure 31). This feeding method is not well understood but it may enable feeding on micro-organisms, the efficiency of which is perhaps improved if the sediment is liquefied by the rapid bill movement. Stitching may also disturb larger food items, which are then more easily encountered by the rapidly probing bill-tip. An interesting further exploitation of the liquefaction phenomenon is exhibited by

◀ **Figure 30**. Northern Lapwing foot-paddling with the front foot on wet grass, hoping to disturb invertebrates. Breeding adult male, England, late March. RJC.

◄ **Figure 31**. Non-breeding Western Sandpipers 'stitching', making rapid, continuous probing movements, pausing only briefly to swallow a food item or to move to recommence feeding a few steps away. These two may be using their brush-tipped tongues to graze on the biofilm. California, mid November. RJC.

Red Knots, which have been shown to be able to detect cockles at depths of several centimetres. Apparently using the stitching technique to liquify the sand, they generate a wave of water pressure in the pores of the sediment, which reflects back from the cockle. This reflected pressure wave is then registered by the sensitive Herbst's corpuscles at the bill tip, enabling the knot to detect the cockle.

Some shorebirds probably feed on the biofilm that grows on the surface of coastal mudflats. This mat of organic material lightly reinforced by a polymer matrix can be up to two millimetres thick. The forming and breaking of polymer chains gives rise to 'thixotropy', whereby disturbance of the film reduces viscosity. It is perhaps this phenomenon that is being exploited by a recently described feeding mechanism used by Western Sandpipers and Dunlins when feeding on the biofilm. These two species (and perhaps other *Calidris* sandpipers) have tongues with tips that act as a 'brush' of fine, hairlike filaments; the stitching technique breaks down the polymer matrix of the biofilm, which is then mopped up by the brush-tip of the tongue.

Longer-billed shorebirds that primarily feed by probing in soft sediments do so to depths determined by bill length, enabling different species to harvest prey from different depths (and thus reducing interspecific competition). They locate their prey by touch, using their sensitive bill-tips. The longer-billed species usually have longer legs, enabling them to feed in deeper waters. Species with straight bills, such the snipes, dowitchers, the woodcocks and the godwits, probe vertically, often continuously, sewing-machine style, until a prey item is encountered. Dowitchers and godwits in particular will frequently feed when wading (Figure 32).

◄ **Figure 32**. First non-breeding Black-tailed Godwit of the race *islandica* probing while wading. England, early February. RJC.

▲ **Figure 33**. A Eurasian Curlew probes deeply, rotating its decurved bill; it is probably searching for ragworms. Adult non-breeding, England, early March. RJC.

The various curlew species use their curved bills to good effect, twisting their bills once inserted, increasing the likelihood of encountering prey (Figure 33). The curved bill is also of value in probing for crabs in their burrows.

Other specialist feeders

Stilts and avocets The delicate, pointed bill of the stilts has evolved for picking food, principally aquatic insects, from the water surface or just below, or, less frequently, from vegetation or the ground. In contrast, the upturned bill of the avocet is flattened and rather broader at the base, though pointed at the tip. Internally the bill has a finely lamellar structure that serves to filter small organisms from the water as the bird feeds, scything its bill from side-to-side in water or liquid mud.

Pratincoles Though all shorebirds will feed on insects, only the pratincoles specialise in doing so. With forked tails, narrow crescentic wings and short bills with a huge gape, they resemble large swallows, and feed aerially, often in flocks. On the ground they may be opportunistic and feed by picking, rather like plovers.

Oystercatchers Oystercatchers specialise in feeding on shellfish, particularly bivalves such as cockles and mussels (Figure 34). Immatures particularly feed by probing for worms or small shellfish, as it takes them some time to master the technique of opening bivalve shells. Once learnt, the method involves forcibly opening the shellfish by using the bill-tip as a hammer or chisel. The hard outer case of the bill, which is composed of keratin, grows continuously (like the teeth of rodents), at a rate more-or-less matched by the wear of the tip. Consequently the shape of the bill tip differs, particularly in the two pied oystercatchers,

◀ **Figure 34**. Adult American Oystercatcher feeding on a bivalve that it has just opened. Florida, mid December. RJC.

◀ **Figure 35**. Ruddy Turnstone turning stones in search of food. Breeding adult, Korea, early May. RJC.

such that birds feeding by probing have relatively pointed bills and those feeding on molluscs, or by picking from a hard stratum, have a chisel-tipped bill. A change of feeding environment, as occurs when Eurasian Oystercatchers move inland to breed and then feed in pastures by probing rather than by chiselling molluscs, results in less wear and a lengthening of the bill.

Spoon-billed Sandpipers These charismatic shorebirds regularly use a particularly rapid form of stitching, probing to depths of five millimetres, typically on wet mud at the edge of small puddles. As with the Western Sandpipers and Dunlins referred to above (p. 26), they seem likely to be exploiting the thixotropic properties of the biofilm, using their specialised bills, which are about 12mm wide, to maximise the area of the surface that they graze for food. Their alternative technique of sweeping the bill sideways when wading in shallow water appears to be used less frequently.

Turnstones As their name suggests, both Ruddy and Black Turnstones frequently feed by turning stones, weed and other debris with their bills as they search for food (Figure 35), though they will also pick and probe as the occasion demands.

Crabs as shorebird prey

Crabs are a food item taken by many shorebirds. They are either caught on the surface after a short run, as with the Crab Plover and the true plovers, or by deep probing, as with the curlews. Crab Plover and Wilson's Plover, though very different in size, are both crab specialists, and have evolved short, sturdy bills to deal with them, as have the two large *Esacus* thick-knees.

Crabs are dealt with in a similar manner by all the shorebirds that feed on them. If the crab is small compared to the size of the bird it is swallowed whole (Figure 36). If it is relatively large it is picked up by a leg and shaken until the leg comes off, which is swallowed if large enough to be worthwhile or otherwise discarded (Figure 37). The process is repeated until only the body remains, which is then swallowed whole. Discarded legs may be pilfered by other species, particularly turnstones.

▼ **Figure 36**. A Kentish Plover with a tiny crab, which it caught as it emerged from its burrow and swallowed whole. Korea, early May. RJC

▼ **Figure 37**. A Wilson's Plover dismembering a small fiddler crab, holding it by one of its remaining legs. When the legs have been removed the body is swallowed whole. Florida, late August. RJC.

◀ **Figure 38**. An adult Long-billed Curlew bringing up a pellet of indigestible material. Florida, early September. RJC

A diet of molluscs or crabs results in shorebirds ingesting shell material that has no nutritional value, which is later brought up as a pellet (Figure 38). The birds try to avoid consuming more inedible material than they need and, if close to water, will often wash larger items before they swallow them (Figure 39).

▲ **Figure 39**. A Eurasian Curlew with a small ragworm, which it washes before swallowing. England, early March. RJC.

Other aspects of feeding in shorebirds

Research on Red-necked Phalaropes has shown that once a food item has been captured it is transported up the bill between the mandibles by surface tensional forces in the water droplet surrounding the item. The bird opens its bill, causing the surface tension to pull the water droplet, together with the food item, up to its mouth. Figure 40 shows a Grey Phalarope doing this. The mechanism appears to need a narrow bill to work most effectively; other narrow-billed shorebirds, such as stilts and avocets, probably transport food items up the bill in a similar way.

Many shorebirds have a flexible tip to the upper mandible, remarkably so in some species. This phenomenon, rhynchokinesis, is occasionally seen when shorebirds stretch forward, usually when yawning or head-scratching, actions which tighten the sinews that control the bill-tip movement, causing the bill-tip to turn upwards (Figure 41). The reason for the bill-tip flexibility is to enable prey to be captured and manipulated more easily, particularly in the ground, though to do this the bird will generally turn the bill-tip down, as shown in Figure 42.

▲ **Figure 40.** First non-breeding Grey Phalarope with a prey item in a water droplet in its bill. All phalaropes feed by picking. Red-necked and Grey Phalaropes do so when swimming. Wilson's Phalarope often feeds while swimming, but also does so at the water's edge or when wading. As the mandibles are opened, the food item is automatically pulled into the mouth by the surface tension forces in the surrounding water droplet. England, early October. RJC.

◄ **Figure 41.** Non-breeding adult Long-billed Dowitcher head-scratching, showing rhynchokinesis. Most shorebird species head-scratch in this manner, 'directly', or 'underwing'. California, mid November. RJC.

◄ **Figure 42.** First non-breeding Marbled Godwit manipulating prey, showing the flexibility of the upper mandible. California, early November. RJC.

▲ **Figure 43**. Semipalmated Plovers disputing feeding territory; adult breeding left, first-breeding/second non-breeding right. Florida, late August. RJC.

Feeding territories and cooperative feeding Many species have individual feeding territories that they will defend against congeners (Figure 43). These territories may only be temporary, particularly if the birds are using the site as a migration stop-over, or they may be maintained for days or weeks. Species such as Common Redshanks spread out to maintain sufficient distance from other individuals so as to avoid frequent disturbance of prey, which may reduce their foraging success. In contrast to territorial feeding, a few species will feed cooperatively where their numbers are sufficiently large, presumably in pursuit of shoaling prey such as small fish. They do this when either wading or swimming; examples include avocets, particularly American Avocets, and Spotted Redshanks, both of which appear to drive shoals of small fish in front of them (Figure 44).

▲ **Figure 44**. American Avocets feeding in a group in shallow water, probably chasing a shoal of small fish. Florida, mid April. RJC.

◀ **Figure 45**. An adult non-breeding Common Redshank swimming. This species rarely feeds while swimming, but often swims for short distances when the water depth is too great for wading. England, late November. RJC.

Feeding while swimming Many shorebirds will swim. The phalaropes, particularly the two smaller species, almost always feed while swimming, while avocets and stilts will regularly do so. Other species feed when swimming, but only occasionally. Curlew Sandpipers, for example, have been seen in Australia with Red-necked Phalaropes, swimming and feeding on insects. Other species will swim briefly when the water depth forces them to do so, and when shallower water is not far away. Many *Tringa* do this (Figure 45) though they rarely seem to feed while swimming, preferring to wade in shallower water.

Resting

Shorebirds need to rest, and they will do so when they do not have to feed, or when they cannot do so owing to inclement weather or when high tide renders their feeding grounds inaccessible. When roosting – that is sleeping – they conserve energy by minimising heat loss, turning their heads and tucking their bills under their scapulars, and standing on one leg with the other tucked under the breast feathers (Figure 46). They sleep in this pose, eyes shut but blinking fairly regularly; if even slightly disturbed they will open their eyes.

▲ **Figure 46**. Sanderlings roosting at high tide. The two left-hand front birds and the last but one on the right are first non-breeding, the others are non-breeding adults. England, late February. RJC.

◀ **Figure 47**. Short-billed Dowitcher loafing, resting on its tarsi. Adult breeding, Florida, late June; at this date this is either a failed breeder or a bird that for some reason has not returned to the breeding grounds. RJC.

Another form of resting is loafing. Loafing birds simply stand around idly, perhaps preening, or just simply digesting their last meal. Solitary species and breeding birds roost or loaf singly but, particularly on the coast, it is usual to see groups or flocks of either loafing or roosting shorebirds, or birds doing both in the same flock. Some loafing birds may rest on their tarsi, particularly in warmer climates. Thick-knees, which roost during the day, are particularly known for doing this but other species also do so occasionally (Figure 47).

Plumage care

Feather maintenance is important for all birds. Bathing (Figure 48) is usually followed by preening (Figure 49), which is usually carried out while loafing, before roosting, or between feeding bouts. Most of the body and wing feathers are preened with the bill; if the bird is standing in water the bill-tip is often dipped in the water to aid the process (Figure 50). The preen gland, situated just above the tail (Figure 51), is squeezed gently to provide natural oil for waterproofing, which is then systematically worked into the plumage. The areas of the head and neck that the bird cannot reach with its bill are preened by scratching with the feet. The whole process serves to keep the plumage in good order, and helps to remove parasites.

▲ **Figure 48**. A Marbled Godwit bathing. Florida, mid October. RJC.

▲ **Figure 49**. Adult Long-billed Curlew preening. Florida, mid September. RJC.

▲ **Figure 50**. Adult Long-billed Curlew dipping its bill-tip into water while preening. Adult, California, early November. RJC.

▲ **Figure 51**. Adult Long-billed Curlew using its preen gland. The gland releases oils that the bird then works into its plumage. Florida, mid September. RJC.

All birds preen the head by scratching, which they do in one of two ways: directly (or 'underwing'), and indirectly (or 'overwing'). Many shorebirds do the former, as shown in Figure 41, but the closely related genera of oystercatchers, stilts and avocets, and also the plovers, head-scratch by the rather ungainly overwing method, as seen in Figure 52.

▲ **Figure 52**. An adult Pied Avocet head-scratching over the wing, as do all stilts and avocets. Compare with the head-scratching dowitcher in Figure 41, which shows 'underwing' scratching. England, early April. RJC.

References – Shorebird behaviour

Campbell & Lack (1985), Elner *et al.* (2005), Ferns & Siman (1994), Minton (2000), Piersma *et al.* (1998), Robinson *et al.* (1997), Rogers *et al.* (2008), Rubega & Obst (1993).

INTRODUCTION TO THE SPECIES ACCOUNTS

Each account follows a standard format, as follows.

Species number / Species name / *Scientific name*

All species are numbered, from 1 to 134, and the corresponding photographs 1a, 1b, etc. This numbering system is used throughout the text for cross-referencing purposes. There is a brief introduction to the species, along with status if endangered and any commonly encountered alternative names.

Identification A discussion of main identification criteria. Dimensions are **L** = overall length from bill-tip to end of tail, **WS** = wingspan, in centimetres (and inches). It is difficult to give a measurement that relates with any precision to the overall dimensions of a shorebird seen in the field. The figures given enable the size of different species to be compared, but they are not in any sense absolute measurements. Overall size is indicated by the bird's length, from bill-tip to tail-tip, when placed on a flat surface, and is referred to here as the 'overall length'. The figures represent average sizes for the species concerned; individuals may vary in size from that suggested. The overall length is the basis of the size category used, ranging from very small to very large, as shown in Table 3 below. The wingspans enable comparisons to be made of the apparent size of birds in flight; where possible these measurements are taken from freshly dead birds, with their wings spread using reasonable, not exceptional force. It is found that this measurement is about three times the usual measurement of the wing-length (from the bend of the wing to the tip of the longest primary) using the flattened straightened-wing technique. The average of this ratio for shorebirds as a group is about 3.1, and for species where wing-span measurements are not available, the wing-length is multiplied by 3.1 to give the wingspan. It should be noted that this method yields wingspans that are considerably less than those quoted in *BWP* (Cramp & Simmons 1983).

   If one of a species pair, the other member of the pair is given, along with a cross reference. The size categories of relative bill and leg lengths in the species description are obtained by expressing the bill or tarsus length as a ratio of the overall average length of the species. These categories are defined in Table 3. Some discretion has been used with the leg length classifications, since the proportional leg length should also include the length of the tibia (for which measurements are not readily available), not just that of the tarsus. For example, the Dotterel's tarsus length indicates 'long' legs, whereas they are better described as being of 'medium-length'; similarly the Willet has 'long', not 'medium-length' legs. Moreover, the apparent leg length of shorebirds is affected by the ambient temperature, feathers being held close to the body in hot weather, when the legs appear long, and *vice versa*.

Ageing Descriptions of plumages which enable the different ages and sexes to be distinguished in the field. These are subdivided as juvenile, first non-breeding, adult non-breeding, first breeding and adult breeding where appropriate. Age of first breeding is given where known.

Call Brief description of most usual call (or calls); song is not discussed.

Table 3. Descriptions used for shorebird size, bill length and leg length.

Size	Very small	Small	Medium	Large	Very large
Overall length (cm)	<15.9	16–21.9	22–32.9	33–49.9	>50
Bill length	Very short	Short	Medium	Long	Very long
Bill/overall length ratio	<0.08	0.08–0.12	0.13–0.17	0.18–0.22	>0.22
Leg length	Very short	Short	Medium	Long	Very long
Tarsus/overall length ratio	–	<0.13	0.13–0.17	0.18–0.23	>0.23

Status, habitat and distribution As these apply in the northern hemisphere, both breeding and non-breeding. The map shows world distribution.

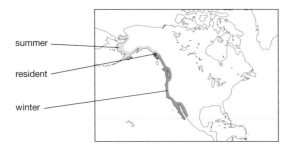

Racial variation Discussion of morphological and plumage differences between the races occurring in the northern hemisphere that are observable in the field. Details of any records of hybridisation are given where relevant.

Similar species Summary of the main distinguishing features from species of similar appearance.

References Source material and further reading on identification, ageing, sexing and racial differences.

Photographic captions The captions summarise the main points of interest of the photograph. For full details of identification and ageing, reference should be made to the text for each species. The geographical location of each photograph is given since it may be helpful in attempting to decide the race to which the individual may belong. Similarly, it is valuable to know the date on which the photograph was taken; consequently, so far as possible, the month is given, qualified by 'early' (1st to 10th day of the month), 'mid' (11th to 20th), and 'late' (on or after 21st).

For plumage terminology it may be helpful to refer to Figures 2 to 6 on pages 8–11, as well as to the general discussion on pages 8–22.

▲ Adult Long-billed Curlew defending its feeding territory. California, early November. RJC.

JACANAS

There are four species, in three genera, in the northern hemisphere: Pheasant-tailed Jacana *Hydrophasianus chirurgus*, Bronze-winged Jacana *Metopidius indicus*, Northern Jacana *Jacana spinosa* and Wattled Jacana *J. jacana* (represented in the northern hemisphere by race *hypomelaena*); the family Jacanidae contains eight species (in seven genera) worldwide. Though jacanas are largely sedentary, they will move in response to local rainfall or drought. Pheasant-tailed Jacana is the most prone to wander and there are a number of records of vagrants at quite substantial distances from their usual areas.

Jacanas are rail- or crake-like tropical freshwater shorebirds, resembling moorhens or coots in general size and structure, and also in the possession of a fleshy forehead-shield. The resemblance is emphasised by the similarity of their preferred freshwater habitat. Jacanas have extraordinarily long toes and claws, which enable them to spread their weight over floating vegetation, giving rise to their popular name of 'lily-trotters'.

All northern hemisphere jacanas have a distinctive juvenile plumage that is replaced by an equally distinctive adult one. Worldwide, the only exception to this is the small African species, the Lesser Jacana *Microparra capensis*, which has a juvenile-type plumage throughout its life. Unlike any other shorebirds, the post-juvenile moult of jacanas starts quite soon after fledging and is rather prolonged, with juveniles commencing both body and wing moult at about two months, plumage then being more-or-less continuously replaced for the next ten to twelve months until adult plumage is attained.

The adult plumages of the various jacanas remain the same year-round, with the exception of Pheasant-tailed Jacana, which has both breeding and non-breeding plumages. Both sexes of all jacanas have similar plumages, though females tend to be about 10% larger and, at least when breeding, have somewhat larger forehead shields or wattles. The Pheasant-tailed Jacana is again an exception in being the only jacana not to have a forehead shield. In common with a number of lapwing species, the jacanas discussed here all have a carpal spur at the bend of the wing, the purpose of which is uncertain. A characteristic of Northern and Wattled Jacanas and of Pheasant-tailed Jacana is that they have strikingly conspicuous pale primaries and secondaries of which the observer is unaware until they either raise their wings, which they do frequently as a contact or territorial signal, or when they fly.

Though Wattled Jacana has been reported to hybridise with Northern Jacana in Costa Rica and western Panama, it seems likely that this suggestion is based on a misunderstanding of the range of variation of plumage and forehead-shield colours of these two species. This is discussed in the species accounts.

With the exception of the monogamous Lesser Jacana, jacanas are polyandrous; females mate with as many as four males, each of which has its own sub-territory within that of the female. In Northern and Wattled Jacanas each of the males cares for the eggs and young more-or-less simultaneously, but the breeding process appears to be more extended in the other species. It seems likely that individual adults 'rest' between breeding attempts, and that resting birds, particularly the females, probably have forehead- shields and wattles that are smaller than those of actively breeding individuals.

1. PHEASANT-TAILED JACANA
Hydrophasianus chirurgus

The only jacana with a distinctive breeding plumage, at which time it acquires a long tail.

Identification L up to 58cm (23") with tail in breeding plumage, 31cm (12") when non-breeding; **WS** 67cm (26"). The only jacana with a distinctive breeding plumage, when it acquires the long tail which gives it its English and scientific names. When non-breeding medium-sized, with short bill and medium-length blue-grey legs, and long toes and claws. In breeding plumage has a very long, black, drooping tail. Non-breeding resembles juvenile, with black on forehead and crown running down nape to mantle. Prominent black line from bill through eye and down side of neck merges with dark brown breast-band; remainder of hindneck pale yellow-buff. Wing-coverts barred dark brown and whitish, extensive white of remainder of wing not showing until bird flies. *In flight* wings above and below white, with black outer primaries and black tips to outer primary coverts, primaries and outer secondaries. When breeding long tail extends well beyond feet; when non-breeding wing-coverts brown, but leading edge of wing white, and wings still appear pale; feet and legs extend beyond tail. Sexes similar, with female larger than male; plumages vary seasonally and with age. Small whitish spur at bend of wing; lacks forehead shield of other northern hemisphere jacanas. *Feeds* by picking on floating vegetation, on aquatic invertebrates.

Juvenile Forehead, crown, entire hindneck, and mantle rufous-brown. Dark-brown line from bill through eye and down neck-side joins across breast, forming ill-defined brownish band. Sides of neck pale buff. Reddish-buff fringes to upperparts and wing-coverts. Underparts white; with age, buffish neck and upper breast become whiter and breast band more prominent. Iris pale yellow.

Non-breeding Similar to juvenile, but more contrasted, with crown and hindneck brown-black, sides of neck brighter yellow buff and breast-band black. Wing-coverts whitish or pale brown, with dark brown barring. Legs greenish or greyish; bill bluish-grey, sometimes with yellowish base. Iris pale yellow.

First-breeding Age when first breeds not known; may attain only partial breeding plumage.

Adult breeding Head and foreneck white; hindneck bright orange-buff, narrowly outlined black down side of neck, with black patch on nape; upperparts dull brown, white wing-coverts prominent on folded wing; underparts and very long, flexible tail black. Primaries narrowly elongate with small 'spoon' at tip, showing (given good view) on folded wing. Legs and bill grey-blue; iris brown.

Call In breeding season a loud *me-e-ou* or *me-onp*; groups when non-breeding give a nasal *tewn, tewn* in alarm.

Status, habitat and distribution Fairly common on larger open wetlands with surface vegetation. Pakistan, India, Nepal, Bangladesh, Sri Lanka, south-east Asia, south China and Taiwan. Will move within its breeding range in response to floods or drought; when non-breeding higher-altitude and more northerly populations move to lower altitudes

◀ **1a. Juvenile**. Sri Lanka, mid October. Head and neck less contrasting than adult non-breeding; dark fringes on upperparts and tertials. Ray Tipper.

and may migrate south to Thailand, Malay peninsula, Sumatra and the Philippines, and west to Oman and Yemen; non-breeding birds sometimes gather in small flocks. Occasional on Japan north to Honshu. Vagrant to Socotra, Java, Bali, South Korea and Western Australia.

Racial variation None.

Similar species Breeding adult with long tail is unmistakable. Non-breeding and juvenile separated from Bronze-winged Jacana of similar age (2) by dark lines on neck-sides and white wings in flight.

References Ali & Ripley (1983), Marchant & Higgins (1993).

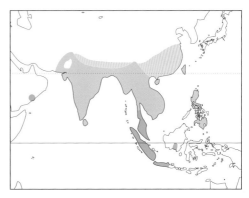

▲ **1b. Adult non-breeding**. India, early January. Much brighter than juvenile, with plain upperparts and brown barring on wing-coverts. RJC.

▼ **1d. Adult breeding in flight**. China, late July. The white wings are unexpected until the bird takes off; note the 'spoons' at the tips of the longest primaries, which can also be seen on the folded wings of 1c. Adult non-breeding shows some brown on the wing-coverts. Yuan Xiao.

▼ **1c. Adult breeding**. India, May. Unmistakable in this plumage. Jugal Tiwari.

2. BRONZE-WINGED JACANA
Metopidius indicus

A medium-sized south Asian jacana with a blue forehead-shield in adult plumage.

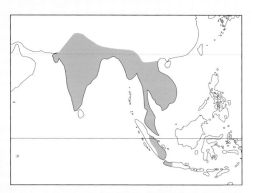

Identification L 30cm (12"); **WS** 54cm (21"). Medium-sized but rather short-tailed (folded wings extend beyond tail), with short pale yellow bill, long pale grey legs, and very long toes and claws. *In flight* from above green-bronze wing-coverts and mantle contrast (in good light) with black flight feathers; legs and feet extend well beyond tail. Sexes have similar plumage year-round, but female is larger than the male; juvenile separable. Has a small red or blue forehead-shield that extends onto the base of the upper mandible, and a dark spur at bend of wing. Iris is dark brown at all ages. *Feeds* while walking on floating vegetation, by picking aquatic invertebrates.

Juvenile/immature Crown brown, darker rufous on forehead and behind eye. Tiny white area just above eye; white sides to head. Bill dull pale yellow, with bluish-grey upper mandible and tiny blue forehead-shield. Blackish-brown hindneck and mantle; scapulars and wing-coverts dull brown, fringed paler when fresh. Neck and upper breast rusty-buff, remainder of underparts white.

Adult Head, neck and breast glossy black, with prominent pure white supercilium extending from above eye to hindcrown where the two almost join. Bill yellow, small grey-blue forehead-shield extends to tiny red area at base of upper mandible.

Mantle and wing-coverts have green-bronze sheen. Underparts black; tail, including uppertail- and undertail-coverts, chestnut. Underwing black.

Call A shrill *seek-seek-seek*; also a short harsh grunt.

Status, habitat and distribution Common in lowland wetlands in India (but does not occur in Sri Lanka), Bangladesh, Nepal, south-east Asia, west Java and Sumatra. Sedentary.

Racial variation None.

Similar species Pheasant-tailed Jacana (1), for which see.

Reference Ali & Ripley (1983).

▼ **2a. Juvenile**. India, early January. Much duller than the adult. RJC.

▲ **2b. Adult.** India, early January. Has similar plumage year-round once adult. Statements in the literature that the forehead-shield becomes red when breeding do not seem to be correct, the red being limited to the tiny wattle at the base of the bill, as here. René Pop.

▼ **2c. Adult in flight.** Eastern India, late December. Note the extraordinarily long toes and claws, typical of all jacanas. Sumit Sen.

3. NORTHERN JACANA
Jacana spinosa

One of two closely related small jacanas occurring in Central America. Alternative name: American Jacana.

Identification L 20cm (8"); **WS** 36cm (14"). A small jacana with a short bright yellow bill and, in the adult, a prominent fleshy bright yellow forehead shield, usually three-lobed on the upper edge, with the central lobe longest. Long-necked, with long, pale greenish-grey legs and extremely long toes and claws. *In flight* has short rounded wings with greenish-yellow primaries and secondaries that contrast dramatically with chestnut body, wing-coverts and tail; legs and feet extend well beyond short tail. Frequently lifts and spreads wings to expose brightly coloured primaries and secondaries. Sexes have similar plumage year-round, but female is larger than male (averaging 10% on most dimensions), with proportionally larger forehead-shield; juvenile separable. Bright yellow spur at bend of wing. *Feeds* by picking on floating vegetation, mainly on small insects and other invertebrates.

Juvenile/immature Crown and hindneck brown, becoming black with age; both small forehead-shield and bill dull yellow, dusky when young. Broad off-white supercilia meet at nape. Upperparts and wing-coverts dull mid-brown, slightly darker fringes showing at close range. Underparts off-white, with buffish wash on upper breast; brown flanks are only seen when wing-lifting. Primaries dull greenish-yellow, with extensive dark brown tips and outer-web margins to outer primaries. Iris pale dull green, yellow carpal spur shorter than adult; legs pale greenish grey. Post-juvenile moult (both body and wing) is protracted, starting with some body feathers at about two months, though primaries are replaced before body moult is complete, at 10–14 months.

Adult Head, neck, shoulders and upper breast black; back, wing-coverts and flanks rusty-chestnut, underparts (apart from flanks) entirely black. Primaries and secondaries bright greenish-yellow, with tips and outer margins to outer primaries black, narrower and more sharply defined than on juvenile. Forehead-shield extends to base of upper mandible, but base of shield usually separated by narrow red band from off-white or pale blue area at base of bill. Occasionally the red area may be more extensive, and the bluish area at bill base may have poorly developed rictal wattles (compare Wattled Jacana). Narrow yellow-green orbital ring visible at close range, iris dark. Long, sharp yellow carpal spurs; legs pale greenish grey. Breeding females have larger forehead-shield than males, extending further up forehead and laterally almost to the eye.

Call Noisy, with loud, harsh *check* or *kak* calls, similar to Wattled Jacana.

Status, habitat and distribution Common in lowland wetlands. Usually sedentary. Lowland southern Mexico to western Panama; also Cuba, Jamaica and Hispaniola. Individuals wander sporadically to southern Texas, particularly the lower Rio Grande valley, where it formerly bred, and less frequently to Arizona; there are possible records from Florida.

Racial variation and hybridisation The various races differ only marginally in size and are not separable in the field. Reports of hybrid Northern x Wattled Jacanas in Costa Rica and western Panama are likely to be individuals of the black race of Wattled Jacana, *hypomelaena*, some of which may show some chestnut on the body or wings; or they may perhaps be Northern Jacanas with a larger than usual red area at the base of the forehead-shield.

Similar species Wattled Jacana (4) of race *hypomelaena* overlaps with Northern Jacana in Costa Rica and Panama; this race has a largely or completely black body and, like all Wattled Jacanas, has a red two-lobed shield and rictal wattles (Northern Jacana has a three-lobed yellow shield and chestnut body).

References Bent (1929), Betts (1973), Jenni & Mace (1999).

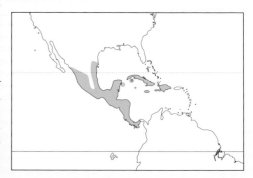

◄ **3a. Immature.** Costa Rica, mid November. Identified to species by small yellow forehead-shield (the only jacana with a yellow shield); aged by neatly fringed juvenile upperparts, white face and underparts, but acquiring black adult feathering on face and neck. RJC.

◄ **3b. Immature.** Costa Rica, mid November. Similar age to 3a; both Northern and Wattled Jacanas frequently raise their wings, perhaps to maintain contact with nearby individuals. Note the small yellow carpal spur, and the contrast between the outer six (broadly fringed juvenile) primaries and inner four (newly grown narrowly fringed adult) primaries. RJC.

◄ **3c. Adult male.** Costa Rica, mid November. All primaries adult-type with narrow black fringes at tips, black head and chestnut underparts, large forehead shield, and large carpal spur. RJC.

▲ **3d. Adult female.** Costa Rica, mid November. When direct comparison is possible, the female is obviously larger than the male. RJC.

◀ **3e. Juvenile.** Costa Rica, mid November. A young individual, with pale green iris, tiny forehead shield, and greenish tinge to upper mandible. RJC.

▶ **3f. Immature.** Costa Rica, mid November. Similar age to 3a and 3b, with scattering of black adult feathers, all-yellow bill, more extensive yellow shield with small pink area below, bluish-white wattle on bill, darker eye than 3e, and suggestion of an orbital ring. RJC.

◀ **3g. Adult male.** Costa Rica, mid November. Brighter colours than 3f, with all-black head and neck, narrow yellow-green orbital ring and dark iris. RJC.

▶ **3h. Adult female.** Costa Rica, mid November. Same bird as 3d. As 3g, but yellow shield is larger, extending higher on crown and closer to the eyes. RJC.

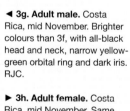

4. WATTLED JACANA
Jacana jacana

A small jacana with a red forehead-shield that extends down as small wattles on either side of the bill. Widespread in South America, it reaches its northern limit in western Panama.

Identification L 20cm (8"); **WS** 40cm (16"). Small, with short yellow bill and fleshy red forehead-shield that extends down to base and sides (as rictal wattles) of upper mandible. Adult of race *hypomelaena* black (sometimes with some chestnut on rear body), with greenish-yellow flight feathers. Long-necked and short-tailed, with long, pale greenish-grey legs and extremely long toes and claws. Bright yellow carpal spur at bend of wing. Frequently lifts and spreads wings, exposing bright greenish-yellow primaries and secondaries. *In flight* has short rounded wings; both above and below greenish-yellow primaries and secondaries contrast with otherwise all-black plumage. Legs and feet extend well beyond short tail. Female larger than male, with proportionally larger forehead shield and (sometimes) slightly longer rictal wattles; sexes similar year-round, juvenile separable. *Feeds* by picking, often while walking on floating vegetation, mainly on small insects and other invertebrates.

Juvenile/immature Crown, hindneck and mantle brown, gradually replaced by black with age. Small red forehead-shield and tiny rictal wattles; bill dull yellow, slightly dusky when young. Broad off-white supercilia meet at nape. Upperparts and wing-coverts dull mid-brown, slightly darker fringes showing at close range. Underparts off-white, with (sometimes) buffish wash on upper breast; brown flanks soon become black, as do axillaries and underwing-coverts. Primaries dull greenish-yellow, outer ones have extensive dark-brown tips and outer-web margins. Iris pale dull green, yellow carpal spur shorter than adult; legs pale greenish grey. Probably has same protracted post-juvenile moult as Northern Jacana (which see).

Adult Head, neck, upper and lower body and wing-coverts are entirely black, or (not infrequently) with small areas of chestnut admixed. The primaries and secondaries are bright greenish-yellow, the outer primaries have black tips and outer margins, narrower and more sharply defined than in the juvenile. The red bi-lobed forehead-shield extends to the upper mandible, with small lobate wattles at the sides of the bill base. Narrow yellow orbital ring is visible at close range; iris dark. Has long, sharp, yellow carpal spurs, and pale greenish-grey legs.

Females have a larger forehead-shield than the males, extending further up the forehead, laterally almost to the eye, and longer rictal wattles often extending below the bill.

Call Noisy, with loud, harsh *check* or *kak* calls, similar to those of Northern Jacana.

Status, habitat and distribution Common in freshwater wetlands from western Panama to South America, where it is widely distributed. There is a record (presumably of *hypomelaena*) from southern Costa Rica.

Racial variation and hybridisation Race *hypomelaena* (described above) occurs in Panama and northern Colombia; adults of this race are largely black. Juvenile *hypomelaena* is initially probably indistinguishable from the nominate *jacana* (the adults of which are rusty chestnut), but quite soon the crown, hindneck, mantle and flanks become black (only the crown and hindneck become black in the nominate). Adults of other races of Wattled Jacana (particularly the nominate race, which is widespread) have a rusty chestnut back, wing-coverts and flanks, similar to Northern Jacana, though with a red forehead shield. For discussion of possible hybrid Northern x Wattled Jacana, see Northern Jacana.

Similar species Northern Jacana (3), which see.

References Jenni & Mace (1999), Stiles & Skutch (1989).

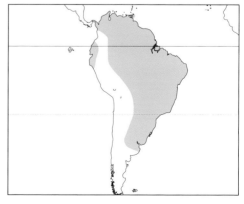

▲ **4a. Juvenile *hypomelaena*.** Colombia, mid April. This race reaches north to west Panama, just inside the region covered. Juveniles are very similar to the nominate race (4b) but quickly acquire black where the nominate is brown. The relatively young age of this bird is shown by the absence of a forehead shield. RJC.

▲ **4b. Juvenile *jacana*.** Venezuela, mid November. Very similar to *hypomelaena* at this age. RJC.

▼ **4c. Adult *hypomelaena*.** Colombia, mid April. Adults of this race may be completely black, as here, or show variable amounts of chestnut on the upperparts. As in Northern Jacana, the sex can only safely be assigned by comparing body size and the extent of forehead-shield and wattle with other nearby individuals. RJC.

▲ **4d. Adult *hypomelaena*.** Colombia, mid April. A different bird from 4c, showing some chestnut on the upperparts. RJC.

▲ **4e. Adult male *jacana*.** Venezuela, mid November. More southerly races of this widespread South American species have a chestnut (not black) body, as here. RJC.

▼ **4f. Adult *jacana* in flight.** Venezuela, mid November. The flight pattern of all races is similar (and is also similar to Northern Jacana), though *hypomelaena* has a black (or largely black) body. RJC.

PAINTED-SNIPES AND CRAB PLOVER

There is one species of painted-snipe in the northern hemisphere – Greater Painted-snipe *Rostratula benghalensis*; three species (in two genera) worldwide form the family Rostratulidae. Australian Painted-snipe *R. australis* has recently been split from Greater Painted-snipe on the basis of differences in morphology, plumage, and mitochondrial DNA. It has only been recorded once away from Australia (in New Zealand in 1986), and seems unlikely to occur in the northern hemisphere. The Crab Plover *Dromas ardeola* is the sole representative of its genus and family, the Dromadidae; it occurs in both hemispheres, but only in the Indian Ocean.

Painted-snipe are more closely related to the jacanas than to the true snipes, which they superficially resemble. Like jacanas they are freshwater shorebirds that move haphazardly, rather than migrate, in response to changing wetland conditions. They are polyandrous, with the female laying several successive clutches, each of which is incubated and the young subsequently cared for by a different male. As is often the case with polyandrous species, the female is larger and brighter; the juvenile and adult male plumages are very similar and quite cryptic in comparison to the female.

The closest relatives of the Crab Plover are probably the pratincoles and coursers. As suggested by its name, crabs are its favoured prey, for which a particularly heavy bill has evolved. It is an unusual shorebird that breeds colonially and, uniquely, for a shorebird, digs a nest burrow in which it lays a single white egg. Once the chick hatches it remains in its nest burrow for some time, where it is fed by its parents; the juvenile begs for food for several months after fledging. The details of this bird's breeding biology are not well known but the observation that as many as ten birds may attend a single nest burrow suggests that the species may not be entirely monogamous, and it may be a cooperative breeder, with nest helpers. If the latter is the case then it is likely that the nest helpers are failed breeders or immature individuals. Breeding colonies of up to 2,500 birds have been reported.

5. GREATER PAINTED-SNIPE
Rostratula benghalensis

A colourful, snipe-like shorebird, widespread thoughout Africa and Asia.

Identification L 25cm (9.5"); **WS** 43cm (17"). A medium-sized, short-tailed, snipe-like shorebird with a long brownish or greyish dark-tipped and slightly decurved bill, and long pale yellow or greenish legs. In all plumages has white or pale area around eye, more extensive behind, a pale crown stripe, and a 'harness' around the body formed by pale lines running from the upper breast around the folded wing. Underparts largely white. *In flight* from above dark, with rounded wings having a broad paler trailing edge, the 'harness' showing as pale lines at base of wings and around breast; legs and feet extend beyond tail, and are often allowed to dangle. Sexually dimorphic, with juveniles and adult males similar, differing from slightly larger and brighter adult female. No seasonal variation. *Feeds* by snipe-like probing in mud, appearing hunched, round-shouldered; largely crepuscular.

Juvenile Very similar to adult male. Differs mainly in pale buff (not white) area around eye, upperparts grey-brown with broad, bright buff tips; throat white, upper breast washed brown, these areas much paler than adult. Bill all-dark, at least on young individuals.

Adult male Male probably breeds in second calendar year. Area around eye off-white, mantle grey, pale 'harness' outlined black, wing-coverts tipped with large paired creamy spots; hindneck and breast grey-brown with black lower border.

First non-breeding female/adult female Female probably breeds in third calendar year. Mantle and scapulars dark greenish bronze with black markings, wing-coverts finely barred black; pure white around eye, hindneck and breast maroon, streaked paler on throat, upper breast and neck-side. First

non-breeding identifiable so long as some juvenile head pattern or wing-coverts are retained.

Call Far-carrying resonant calls (like blowing across top of empty bottle) given by female in breeding season. Otherwise a croaked *kek* when flushed.

Status, habitat and distribution Widespread in tropical freshwater wetlands, from Egypt and sub-Saharan Africa east through Middle East, Indian subcontinent, south-east Asia, China and Japan, rare Indonesia. Australian Painted-snipe now generally considered a separate species. Largely sedentary, but may move in response to drought or flood.

Racial variation None.

Similar species Australian Painted-snipe, which is endemic to Australia, seems unlikely to occur in northern hemisphere; it has very similar plumages and is of a similar size, but it has longer wings (WS 47cm, 18.5"), a slightly shorter bill, and shorter (medium-length) legs. The head and neck of female Australian are mainly dark brown (not streaked paler as in Greater), with a small rufous patch on

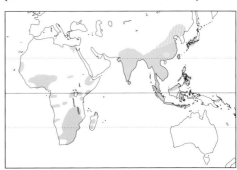

▼ **5a. Adult male (left) and four young juveniles**. Taiwan, April. The different sizes of the juveniles, which are not yet fully grown and still show traces of down, suggests they may be from different broods. Note the similarity of the adult male and the juveniles, though the male has a bolder head pattern. Ming-Li Pan.

the hindneck, cream (not pure white) around eye; males differ less from Greater Painted-snipe, and they may be difficult to separate.

References Ali & Ripley (1983), Baker *et al.* (2007), Lane & Rogers (2000), Marchant & Higgins (1993), Ueki (1986).

▲ **5b. Immature female**. India, late December. Extensive pale flecking on face and throat and pale buff (not white) around eye suggest that this is an immature female. RJC.

▲ **5c. Adult female**. Kenya, early September. More colourful than the immature female (5b), with chestnut collar and sides of neck, though the chestnut area can be brighter and less patchy than this. RJC.

▼ **5d. Adult male**. Kenya, early September. Basic pattern similar to female, but with large pale creamy spots on wing-coverts. RJC.

▼ **5e. Adult female in flight**. The Gambia, mid February. The broad pale trailing edge to the wing and feet (often dangling) extending beyond the tail are diagnostic in flight. James Lidster.

6. CRAB PLOVER
Dromas ardeola

A unique black-and-white bird with a distinctive heavy black bill.

Identification L 40cm (15.5"); **WS** 66cm (26"). Has a unique combination of black-and-white plumage and a medium-length, heavy black bill. Large, mainly white, with a long neck; very long blue-grey legs. Adults have mantle and flight feathers black, immatures grey. *In flight* from above black-and-white; mantle, primaries and secondaries, together with primary and secondary greater coverts, black; remainder white, including V up back; primary shafts also white. Legs and feet extend well beyond tail. Sexes similar in size and plumage, no seasonal variation; juvenile separable. Partial webbing between all three toes. *Feeds* mainly on crabs, which it catches in a plover-like manner after a short run.

Juvenile Mantle grey, scapulars and wing-coverts light brownish-grey; primaries and secondaries dark grey. Rear crown pale grey, streaked black.

First non-breeding Hindneck and mantle become blacker, but can be aged well into second calendar year by retained pale brownish-grey scapulars and wing-coverts. Older non-breeders perhaps also have grey scapulars and wing-coverts.

Adult Probably does not breed until at least third calendar year, perhaps later. Head all-white, or, more usually, with small blackish area in front and behind eye; some have greyish or blackish streaking of variable extent on the hindcrown. Base of hindneck and mantle black, back and rump white.

Calls Various, including far carrying *ha-how* and *crow-ow-ow*.

Status, habitat and distribution An entirely coastal species, usually gregarious. Locally common on mudflats around the Indian Ocean, but needs sandy beaches to excavate nest burrow. Breeds colonially, at the southern end of the Arabian peninsula, Eritrea and Somalia; when non-breeding is widespread around the western Indian Ocean and East Africa south to north-eastern South Africa (small numbers) and Madagascar, east to north-west India, northern Sri Lanka, many of the Indian Ocean islands, western Thailand; vagrant to peninsular Malaysia, the Maldives and the eastern Mediterranean.

Racial variation None.

Similar species None. Long legs, smart black-and-white plumage and shortish heavy black bill are distinctive.

References Anderson & Baldock (2001), Bhuva & Soni (1998), Bouwman (1987), Chiozzi & De Marchi (2003), Hockey & Douie (1995), Urfi (2002).

▼ **6a. Juvenile.** Kenya, early August. Pale grey above where adult is white. Steve Garvie.

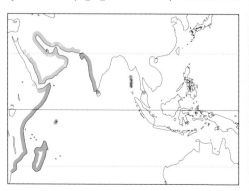

◄ **6b. Adult**. Kenya, late March. Adults of both sexes are similar year-round. RJC.

▼ **6c. Adults in flight**. Western India, mid December. As unmistakable in flight as on the ground! Niranjan Sant.

◄ **6d. Adults in flight**. Oman, early May. Three age classes in this flock: adult (with black mantle and primaries, not in wing moult), older immatures (black mantle, in wing moult), and younger immatures (grey mantle and primaries, limited wing moult). Hanne and Jens Eriksen.

OYSTERCATCHERS

There are three species in the northern hemisphere: Eurasian Oystercatcher *Haematopus ostralegus*, American Oystercatcher *H. palliatus* and American Black Oystercatcher *H. bachmani*. The Canarian Black Oystercatcher *H. (moquini) meadewaldoi* is presumed extinct (probably from 1913); excluding this species there are 10 species (in one genus) worldwide, forming the family Haematopodidae.

Oystercatchers are stockily built, noisy shorebirds with black- (sometimes brown) and-white ('pied') or all-black plumage, long orange-red bills, short legs and slightly webbed feet. Young are fed by their parents for up to two to three months after hatching. They are widespread, largely coastal species, distributed from the Arctic to the tropics; the American Oystercatcher is the counterpart of the Eurasian Oystercatcher in the Americas. The American Black Oystercatcher is restricted to the Pacific coast of North America and is confined to rocky shores, as are the other of the world's black oystercatchers. All three northern hemisphere species tend to move south during the non-breeding season, the Eurasian Oystercatcher being the most migratory of the three.

Oystercatchers have distinct juvenile and adult plumages; apart from the Eurasian Oystercatcher, the adult plumage of all oystercatchers is similar year-round. Eurasian Oystercatcher differs in having a white foreneck half-collar in immature (but not juvenile) and adult non-breeding plumage (though there are exceptions to this; careful scanning of any large group of Eurasian Oystercatchers outside the breeding season will usually reveal one or two adults lacking a white collar, apparently still in breeding plumage).

It is thought that the most likely reason for the Eurasian Oystercatcher's white collar is that it enables breeding birds to distinguish non-breeders from other breeders, so minimising territorial disputes. The possession of a good feeding area adjacent to the nest site is important to Eurasian Oystercatchers, and much time – sometimes years – is spent acquiring and then defending such a territory. In coastal breeding areas there are often many non-breeding individuals, and much valuable time can be saved if territorial squabbles can be limited to breeding rivals, with disputes being quickly ritualised once neighbouring breeding pairs have established their territorial boundaries.

It is interesting that the eastern race of Eurasian Oystercatcher, *H. o. osculans* (which it might be appropriate to call 'Korean' Oystercatcher) does not gain the non-breeding white collar, though some immatures show a few flecks of white on the throat. This feature, together with a somewhat different wing pattern, a considerably longer bill, and its geographical isolation from the other two races, raises questions regarding its taxonomic status.

Adult bare-part coloration in all oystercatchers is usually not fully developed until the fourth calendar year, and breeding is probably delayed at least until then. This is certainly the case with all three northern hemisphere species, and it represents a longer period of immaturity than that of most, if not all, other shorebirds. Juveniles have an extensive dark tip to the bill, typically half its length, with a dull orange base; the iris is dark, the narrow dull orange or yellow orbital ring is incomplete, and the legs are dull brownish (American Black Oystercatcher) or pale grey (American and Eurasian Oystercatchers). These colours do not change much as the bird acquires its first non-breeding plumage. Thereafter, three further age classes

▲ **Heads of adult male (left) and adult female (right) American Black Oystercatchers.** The male has no iris flecking, while the female has dark flecking on the iris below and just forward of the pupil, giving the (erroneous) impression that the pupil is oval. The bills are of different length, with that of the male slightly shorter and deeper centrally. California, early November. RJC.

can be identified: younger immatures, older immatures and adult. Very often these age classes follow an annual cycle, corresponding to first-breeding/second non-breeding, second breeding/third non-breeding and adult, but there is variability, with some maturing more quickly and the true age of all birds may not correspond to their apparent age class. See Table 4, which shows the typical features that enable the age classes of the northern hemisphere oystercatchers to be identified.

Research on American Black Oystercatchers shows that adults can be sexed by their eye pattern, most females having more extensive dark flecking in their irides than males. This flecking occurs at the front lower edge of the pupil, making the latter appear as a slightly blurred black oval. Males virtually lack eye flecks and so show clear, round pupils. Flecking of this type occurs in all oystercatcher species although it is most easily seen in the two American species, both of which have yellow irides. It appears that this feature may be used to aid the sexing of all oystercatchers but this has yet to be confirmed for all but the American Black Oystercatcher.

Table 4. Typical variation of bare-part colours of northern hemisphere oystercatchers with age; a similar sequence is probably followed by all oystercatchers. There is much variation, however, and the age-class details shown should only be regarded as indicative. The presence or absence and extent of the white foreneck collar further aids the ageing of Eurasian Oystercatchers. Key: (1) American Black Oystercatcher, (2) American Oystercatcher, (3) Eurasian Oystercatcher.

Age class	Bare-part colours
Juvenile (first/second calendar year; juvenile and 1st non-breeding plumages)	Distal ½ to two-thirds of dull orange bill dark grey; iris brown; narrow and incomplete dull orange orbital ring; legs pinkish grey
Younger immature (second calendar year; 1st breeding and 2nd non-breeding plumages)	Distal ½ or less of bill greyish; iris dull yellow (1, 2) or dull red (3); narrow dull orange orbital ring; legs pinkish grey
Older immature (third calendar year; 2nd breeding and 3rd non-breeding plumages)	Distal ridge of upper mandible and bill tip greyish; iris brighter yellow (1, 2) or red (3); complete orbital ring orange; legs pale pink
Adult (fourth calendar year and older; adult plumage)	Bill bright orange (distal ridge of upper mandible may show greyish when non-breeding); iris bright yellow (1, 2) or bright red (3); thick fleshy bright orange orbital ring; legs pale pink (1), pale pinkish yellow (2) or bright pink (3)

7. AMERICAN BLACK OYSTERCATCHER
Haematopus bachmani

The only extant all-black oystercatcher in the northern hemisphere.

Identification L 44cm (17"); **WS** 80cm (31.5"). Large, with a medium-length, straight, laterally-compressed orange-red bill with a blackish or yellowish tip, and medium-length pale pink (adult) or greyish (immature) legs. Black, with brownish-black mantle and wing-coverts. See Table 4 (p. 54) for a summary of variation of bare-part colours with age. *In flight* from above all-black, underwing silvery-grey; feet do not reach end of tail. Female averages marginally larger, and has slightly longer, less deep bill; no seasonal variation, but juvenile and immatures separable. Some webbing between outer toes, virtually none between inner toes; no hind toe. *Feeds* by prising open shellfish, and by probing to moderate depths.

Juvenile Mantle and wing-coverts blackish-brown, with buff tips when fresh. Bill brownish-orange at base, blackish distally. Iris dark, narrow orbital ring orange. Legs pinkish grey.

Immature Buff tipped mantle and wing-coverts either wear, or are replaced, so ageing probably only possible from bare-part colours. As in other oystercatchers there are two immature age classes: younger birds with dark iris, poorly developed dull orange-red orbital ring, and extensively dark-tipped orange-red bill; older birds with orange-yellow iris, moderately developed red orbital ring, and bill orange-red with some dusky on upper mandible and at tip. Legs greyish-pink, becoming pale pink with age (see Table 4).

Adult non-breeding/adult breeding Narrow white fringes to underparts when in fresh plumage. Yellow iris, prominent bright red orbital ring, bill bright orange-red with yellow tip; legs pale pink.

Call Similar to other oystercatchers; loud, sharp *keee*, often repeated.

Status, habitat and distribution Uncommon along American Pacific rocky coasts, from the western Aleutians, Alaska and Canada south to Baja California. Largely sedentary.

Racial variation and hybridisation No racial variation; occasionally hybridises with American Oystercatcher (which see).

Similar species With the extinction of the Canarian Black Oystercatcher *Haematopus (moquini) meadewaldoi* there is no other black oystercatcher in the northern hemisphere, though African Black Oystercatcher *H. moquini* (iris red, legs deep pink) might wander up the west African coast from the southern hemisphere.

References Andres & Falxa (1995), Collar (1994), Guzetti *et al.* (2008), Webster (1942).

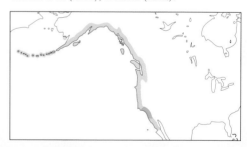

◄ **7a. Juvenile being fed by adult**. British Columbia, July. Juvenile has buff spots on the feather fringes of both upperparts and wing-coverts, plus a dark iris and an extensive dusky tip to the bill. Young of all oystercatcher species are fed by their parents for anything up to three months. Jim Martin.

▲ **7b. Juvenile age class (first non-breeding)**. California, early November. About three months older than 7b; upperparts newly moulted, wing-coverts slightly faded, iris dull pale yellow, orange orbital ring developing. Relatively short bill suggests a male; at this age the iris flecking cannot be used for sexing. RJC.

▼ **7c. Younger immature age class**. California, mid April. Extensive dusky coloration on bill tip shows that this is probably a first-breeding individual; iris, bill and legs are brighter than 7c. Shortish bill suggests a male. RJC.

▲ **7d. Adult male non-breeding**. California, early November. Similar plumage year-round, but fresh underpart feathers have white fringes, and bare-parts are duller than in the breeding season. RJC.

▲ **7e. Adult female breeding**. California, mid April. Iris flecking and longish bill show this to be a female; note bright bill and slightly pinkish legs typical of breeding season. RJC.

▲ **7f. Juvenile/first non-breeding American Black Oystercatcher in flight**. California, early November. This bird still retains the majority of its juvenile plumage. RJC

8. AMERICAN OYSTERCATCHER
Haematopus palliatus

A large 'pied' species of the Americas with a longer
bill and browner plumage than its Eurasian counterpart.

Identification L 42cm (16.5"); **WS** 76cm (30").
Large, with long, straight, laterally-compressed
blackish- (immature) or yellowish-tipped (adult)
orange bill and medium-length pale-pink legs.
Head and neck black, upperparts and wing-coverts
(except greater coverts) brownish-black; greater-
coverts and underparts white. See Table 4 (p. 54)
for a summary of variation of bare-part colours with
age. *In flight* from above black, with white wing-bar
on secondaries extending to inner primaries (but
see race *frazari* below), and white uppertail-coverts;
feet just reach end of tail. Sexes of similar size, male
has slightly shorter bill; adult plumage similar year-
round, juvenile and immature separable. Some webbing between outer toes, virtually none between
inner toes; no hind toe. *Feeds* on shellfish and also
by probing to moderate depths.

Juvenile Head, neck and upperparts brownish-
black, scalloped with closely spotted buff fringes.
Dark-tipped, dull orange, pointed bill; legs pale
pinkish-grey, iris brown, inconspicuous orbital ring.

First non-breeding Much as juvenile, though
bare-part colours are brighter. Upperparts have
buff fringes following post-juvenile moult but the
scalloped appearance is reduced by wear; retains
some juvenile wing-coverts. Legs pale pinkish
yellow.

Adult non-breeding/adult breeding Head and
neck black; upperparts brownish-black, mantle and
scapulars fringed paler when fresh, becoming more
uniform with wear. Bright orange, yellow-tipped
bill; legs flesh, becoming pale yellowish pink when
breeding. Iris yellow, orbital ring orange-red, legs
pale pinkish-yellow. Male has blacker upperparts
and brighter bare-parts.

Calls A strident, piping *kleep* and *kip, kip*.

Status, habitat and distribution Breeds coastally in
eastern United States from New England (with isolated records north to Nova Scotia) south to Florida,
the Gulf Coast, the West Indies and Baja California,
south through Central America to South America.
Largely sedentary; in eastern North America
non-breeding birds occur from Delaware south.
Nominate race *palliatus* occurs throughout most
of North America, with *frazari* in south California
and western Mexico; other South American races
unlikely to occur in northern hemisphere.

Racial variation and hybridisation Nominate race
is described above. Race *frazari* differs in having irregular, mottled lower border to black foreneck (but
some *palliatus* also show this) and, in flight, white
wing-bar on secondaries does not extend to inner
primaries. Hybrids between American Oystercatcher
and American Black Oystercatcher are similar to

◄ **8a. Juvenile age class**. New
York, mid July. A recently fledged
juvenile, still with tail-down, a dark
iris, pale tips to all upperparts and
wing-coverts, and with a rather
short bill that will continue to grow
for a while yet. Lloyd Spitalnik.

American Oystercatcher but have variable amounts of black on breast and belly. It is possible that *frazari*, too, represents a hybrid form.

Similar species Eurasian Oystercatcher (9) has a proportionally less deep, considerably shorter brighter orange bill, is black, not brown, on upperparts, and has a white half-collar on foreneck when non-breeding; in flight white rump of Eurasian Oystercatcher extends in a V up the back. The all-black, near-sedentary American Black Oystercatcher (7) does not occur in eastern north America.

Reference Nol & Humphrey (1994).

▲ **8b. Juvenile age class (first non-breeding).** Florida, mid September. Replaced upperparts, but has very worn wing-coverts, iris is still largely dark, and has extensive dusky tip to bill. The grey or dark-grey upperparts and wing-coverts contrast with the black head and neck in all plumages. RJC.

◄ **8c. Younger immature age class (first-breeding/second non-breeding).** Florida, mid September. Ringed as a chick (in Virginia) in August the previous year. Appears to be an older immature age class on bare-part colours, but ring proves otherwise. RJC.

▲ **8d. Adult male non-breeding *palliatus***. Florida, mid December. Lack of iris flecking and shorter bill indicate a male; note the thin white fringes of the more recently acquired wing-coverts. RJC.

▼ **8e. Adult female breeding *palliatus***. Florida, early May. Extensive iris flecking and longer bill indicate a female; the white upperpart fringes have worn away, otherwise very similar to adult non-breeding. RJC.

◄ **8f. Adult female breeding** *frazari*. California, early April. The irregular lower border to the black foreneck is distinctive of this race. Wes Fritz.

◄ **8g. Adult female non-breeding** *palliatus* **in flight**. Florida, early January. The white wing-bar extends onto the inner primaries in this race. RJC.

◄ **8h. Adult** *frazari* **in flight**. California, late September. The white wing-bar is restricted to the secondaries. Kyle E. Pias.

9. EURASIAN OYSTERCATCHER
Haematopus ostralegus

A conspicuous and familiar black-and-white Eurasian shorebird.

Identification L 43cm (17"); **WS** 76cm (30"). Large, with medium to long (depending on sex and race), straight, laterally-compressed orange or blackish-tipped orange bill and short pink legs. Head, neck, upperparts and wing-coverts (except greater coverts) black or brownish-black; greater coverts and under-parts white. See Table 4 (p. 54) for a summary of variation of bare-part colours with age. *In flight* from above, black with broad white wing-bar (latter differing slightly between races), and white rump extending in V up back. Black tail-band is narrower in juvenile and first non-breeding plumages; feet do not reach end of tail. Sexes of similar size but adult male has slightly shorter, deeper bill; plumages differ seasonally and with age. Bill pointed on individuals that feed by probing, chisel-tipped in shellfish feeders, but may change with variations in diet. Some webbing between outer toes, virtually none between inner toes; no hind toe. *Feeds* on shellfish, also by probing to moderate depths.

Juvenile Head and neck dull sooty-black, no white half-collar on foreneck; upperparts and wing-coverts brownish-black, indistinctly scalloped with narrow paler fringes. Blackish tipped, orange-yellow pointed bill; legs greyish-pink. Iris brown, inconspicuous orbital ring.

First non-breeding Upperparts sooty-black, irregular white half-collar on foreneck; bare parts much as juvenile. Retains some juvenile wing-coverts; best aged by contrast of black upperparts and browner wing-coverts, and juvenile-like iris and bare-part colours.

First-breeding to third non-breeding A few may breed in their third calendar year but most do not until their fourth or even later. As adult non-breeding, but has wider white foreneck collar than first non-breeding. Bare-part colours are transitional between first non-breeding and adult non-breeding, bill gradually becoming all orange, iris all red, orbital-ring redder and more fleshy, and legs all pink. Average differences may allow younger immatures to be distinguished from older immatures, but there is much overlap.

Adult non-breeding Upperparts black; most have irregular white foreneck collar, narrower than that of immatures, usually acquired from August and largely lost by end February, though a small proportion (about 5%) apparently do not acquire a white collar. Bill orange; iris red, orbital ring orange, legs pink.

Adult breeding As adult non-breeding, but loses white half-collar; bare parts brighter, orbital ring more fleshy. Male separable in breeding pairs by deeper, shorter bill, and blacker upperparts, female shows dark smudge on iris by pupil.

Calls A strident, piping *kleep*, and *kip kip*.

Status, habitat and distribution Common in Europe, less so further east. Nominate race *ostralegus* generally breeds near the coast (but sometimes inland), from Iceland, Britain and Ireland, Scandinavia and north-west Russia south to France and locally around north-west Mediterranean; non-breeding birds found from Britain and Ireland, Denmark, south to Africa; vagrant to South Africa (perhaps of race *longipes*), Greenland, Spitsbergen and eastern Canada (presumably race *ostralegus*). Race *longipes* breeds around the Adriatic, north-east Mediterranean, Russia, east to western Siberia, non-breeding birds found coastally around east Africa, the Arabian peninsula, east to India; race *osculans* breeds from Kamchatka to north-east China and Korea, non-breeding birds found coastally from Japan, South Korea to south China.

Racial variation Race *longipes* has a longer bill on average (with nasal groove longer than half bill length; less than half bill length in *ostralegus*), has longer legs and typically has browner upperparts, scapulars and wing-coverts than both *ostralegus* and *osculans*; black upper breast may extend below bend of wing when standing (does not reach bend of wing in *ostralegus*), but care is needed to separate from *ostralegus* in the field as intergrades occur. Race *osculans* has a longer bill than either *ostralegus* or *longipes* (with nasal groove longer than half bill length), has longer legs and shows

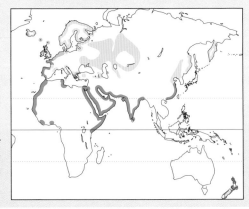

more white on closed wing when standing; in flight shows no white on outer five primaries (no white on outer two in *ostralegus*), and usually outer two or three primary shafts are black (white in other races), white inner wing-bar broader, extending to median coverts. Immature and non-breeding *osculans* do not have a white half-collar but some immatures show a trace of white on the throat where the white collar would be.

Similar species Confusion possible only with American Oystercatcher (8), which has a proportionally longer, deeper, less orange bill, a yellow iris and no white collar, and (in flight) lacks white V up back.

References Barter (2002), Dare & Mercer (1974), Heppleston (1982), Hockey & Douie (1995), Lambeck *et al.* (1995), Rusticali *et al.* (2002), Shirihai (1996).

▲ **9a. Juvenile *ostralegus***. Eastern England, mid September. Lacks white foreneck collar, has dark iris, minimal orbital ring, distal half of bill is dusky and legs are pinkish grey. RJC.

▼ **9b. Juvenile age class (first non-breeding) *ostralegus***. Eastern England, late November. Has replaced juvenile upperparts and acquired white foreneck collar, but still has a dark iris, a minimal orbital ring, the distal half of bill is dusky, and the legs are pinkish grey. RJC.

▼ **9c. Younger immature age class (first-breeding/ second non-breeding) *ostralegus***. Eastern England, late July. Broad white foreneck collar; iris is dull red, with a largely complete dull-orange orbital ring; the distal half of the bill is dusky and the legs are pinkish grey. RJC.

◄ **9d. Older immature age class (second non-breeding)** *ostralegus*. Eastern England, mid February. White foreneck collar, iris dull red, largely complete dull orange orbital ring, distal third of bill dusky and legs pink. The dull iris and very worn outer primaries show that this is not an adult bird, as an adult would normally have completed primary moult by mid February. RJC.

◄ **9e. Adult non-breeding** *ostralegus*. Eastern England, mid February. White foreneck collar, iris is completely red with complete orange orbital ring; bill is orange, duskier at tip, and has pink legs. The iris flecking suggests that this is a female, though the bill is rather short. RJC.

◄ **9f. Adult female non-breeding** *ostralegus*. Eastern England, mid December. Unusually in December, lacks white foreneck collar; iris red, orange orbital ring, bill orange, but slightly dusky at tip, a non-breeding characteristic. The iris flecking and rather long bill show that this is a female; from the ring this bird is known to have been at least 22 years old when photographed. Perhaps older birds do not acquire the non-breeding white collar. RJC.

◀ **9g. Adult male breeding** *ostralegus*. Eastern England, early April. No foreneck collar; iris red with orange orbital ring, bill orange, no duskiness at tip, pink legs. The very restricted iris flecking and rather short bill show this is a male. RJC.

◀ **9h. Adult female breeding** *ostralegus*. Eastern England, early August. No foreneck collar; iris red, orange orbital ring, bill orange, no duskiness at tip, pink legs. The iris flecking and rather long, less deep bill show this is a female; note the worn and rather faded plumage compared to 9g in April. RJC.

◀ **9i. Adult female breeding** *ostralegus* **in flight**. Eastern England, late July. Note the white on the outer web of all but the outer two primaries, characteristic of this race. RJC.

▲ **9j. Immature *longipes***. Oman, early January. Broad white collars as with *ostralegus*. Hanne and Jens Eriksen.

▼ **9k. Adult breeding *longipes* in flight**. Samara River, Russia, early July. Similar flight pattern to nominate *ostralegus*. Igor Karyakin.

▼ **9l. Adult breeding *longipes* (probably a male)**. Western Russia, mid June. Note longer bill (nasal groove > half bill length) and rather browner upperparts than in the nominate; the extent of black on breast is apparently variable in this race, and is not particularly extensive in this individual. Alexander Mischenko.

◀ **9m. Younger immature age class (first-breeding) female** *osculans*. South Korea, early May. Female *osculans* has the longest bill of any oystercatcher. Note faded (brownish) juvenile primaries, and hint of white on throat where the other two races at this age would have a broad white collar. Most immature *osculans* have no white at all on the throat; compare with *ostralegus* of the same age class (9c). RJC.

▶ **9n. Adult male breeding** *osculans*. South Korea, early May. Has a greater amount of white in folded wing than the other two races. RJC.

◀ **9o. Adult *osculans* in flight**. South Korea, mid May. This race has no white on the outer five primaries, though there is some white on the primary shafts. Compare with *ostralegus* (9i), which has whiter outer primary shafts. RJC.

IBISBILL, STILTS AND AVOCETS

The Ibisbill *Ibidorhyncha struthersii* is the only species in its family, the Ibidorhynchidae. It is restricted to mountain areas of central Asia. It is a striking species, whose relatively large size and decurved bill give it a superficial resemblance to an ibis, though it is most closely related to the oystercatchers, stilts and avocets. It is found on the pebble or boulder-strewn valley floors of mountain rivers, usually singly or in pairs, though when non-breeding may occur in small flocks. Ibisbills can be very difficult to locate among the boulders, where they spend considerable time standing or sitting motionless.

Stilts and avocets form the family Recurvirostridae. There are two stilt species in the northern hemisphere, Black-winged Stilt *Himantopus himantopus* and Black-necked Stilt *H. mexicanus*, with four species (in two genera) worldwide. There are also two avocet species in the northern hemisphere, Pied Avocet *Recurvirostra avocetta* and American Avocet *R. americana*, with four species (in one genus) worldwide.

Stilts and avocets are large shorebirds, strikingly black-and-white with extremely long legs, and they are obviously closely related. Though very similar, the two northern hemisphere stilts differ in size and, particularly, in head pattern. The North American species, Black-necked Stilt (which is rather smaller), has a consistent head and hindneck pattern, greyish in juveniles and black in adults. By contrast, the Eurasian Black-winged Stilt has an extremely variable grey and/or black head and neck pattern, with two individuals rarely the same, and there is a tendency for younger and non-breeding birds to show more grey and less black. Occasionally, individuals may be found with head and neck patterns recalling those of Black-necked Stilt, or the equally consistent head-pattern of the Australian race of Black-winged Stilt ('White-headed Stilt'), *leucocephalus*. Careful examination, however, usually yields some variation from the head and neck patterns of the latter two taxa.

10. IBISBILL
Ibidorhyncha struthersii

A striking sedentary bird or altitudinal migrant of the Central Asian mountains with a long, decurved reddish bill.

Identification L 40cm (15.5"); **WS** 74cm (29"). Large, with a long, decurved brown or red bill, and short greyish to red legs. Face and crown black, narrowly bordered white; remainder of head and neck pale blue-grey. Upperparts and wing-coverts pale grey-brown; black breast-band bordered white above, underparts entirely white. *In flight* from above, grey with elongate central white oval on primaries, tail appears rather short, with narrow black terminal band, broken at centre; feet do not reach tail. Sexes similar, though marginally shorter bill of male may allow breeding pairs to be separated; some seasonal variation, juvenile is distinctive. Partial webbing between outer and middle toes. *Feeds* largely on insects and crustaceans in short vegetation, in a plover-like manner; also in rivers and streams by picking from surface and by submerging head to probe among gravel and boulders. Frequently wades belly-deep, sometimes in surprisingly fast-flowing water.

Juvenile Crown dark-grey; face plain, mottled white, lacking black of adult. Head and neck grey, bordered on breast by dark brown breast-band.

Upperparts and wing-coverts greyish-brown, narrowly fringed bright buff when fresh. Bill a dull brownish-red, becoming red with age.

First non-breeding/adult non-breeding Very similar to adult breeding but with white streaking on black face at base of bill. Legs pale greyish-purple. Immature soon becomes practically inseparable from adult but sometimes shows a few remaining browner feathers on upperparts and wing-coverts, retains a largely white face for several months, and may have darker base to red bill.

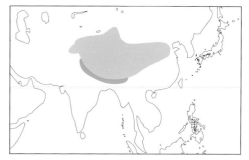

▲ **10a. Juvenile**. Qinghai, China, early August. Lacks dark face of adult. John and Jemi Holmes.

Adult breeding Age when first breeds not known. At close range, black face has faint white streaking around bill base. Bill deep red; legs variable, bright purple-red, greyish-purple or grey.

Calls *Sissi-sip, sissi-sissi-sip* on flying, recalling Common Sandpiper, and a noisy, Common Greenshank-like *klew-klew*, particularly when with young.

Status, habitat and distribution Uncommon in gravel and boulder-strewn mountain valley floors in central Asia, from eastern Kazakhstan to northeast China, south to north India, moving to lower altitudes in colder weather when non-breeding.

Racial variation None.

Similar species None.

▲ **10b. Adult non-breeding**. India, early January. Very similar to adult breeding, but has more extensive white streaking on face. RJC.

▼ **10c. Adult breeding**. Qinghai, China, early August. Less white at the base of the bill than in non-breeding. John and Jemi Holmes.

▼ **10d. Adult breeding**. Qinghai, China, early August. Wing-stretching, showing flight pattern. John and Jemi Holmes.

11. BLACK-WINGED STILT
Himantopus himantopus

An elegant, long-legged and long-necked shorebird, widespread throughout many of the warmer areas of the Old World.

Identification L 37cm (15"); **WS** 69cm (27"). Large, elegant and long-necked, with a medium-length, needle-fine straight black bill, whitish head and neck with variable amounts of black or grey, black upperparts and wings; very long red legs. *In flight* has uniform black upper and underwings (but see juvenile, below), white uppertail-coverts, and white rump extending in V up back, pale grey tail; legs and feet project well beyond tail. Sexes similar in plumage (but see below), with no seasonal variation, but male typically 10% larger than female; juvenile separable. Restricted webbing between outer and middle toes only; no hind toe. *Feeds* usually by wading, picking from the water surface and by sweeping bill from side to side.

Juvenile Crown and hindneck grey, upperparts and wing-coverts greyish-black, scalloped with buff fringes, underparts white; male has greenish sheen to wing-coverts. White tips to inner primaries and secondaries show as narrow trailing edge in flight. Legs pinkish-grey; reddish base to lower mandible.

First non-breeding/first-breeding A few breed in their second calendar year. As juvenile, but upperparts uniformly blackish-brown, and retains many juvenile wing-coverts; white trailing edge to wing in flight becomes less conspicuous with wear. Head pattern has much diffuse grey; legs red.

Adult non-breeding/adult breeding Head pattern variable, in part related to both season and sex, may be all-white (usually breeding males), or with differing amounts of black or grey on head and nape, black occasionally extensive running from crown to lower nape; much overlap between sexes. Mantle and scapulars black in male, very dark bronze-brown in female; wing-coverts black. Bill completely black; legs red, brighter when breeding.

Call A noisy, resonant *kek*.

Status, habitat and distribution Common in suitable wetlands with shallow water. Breeds from southern Europe, Africa, Russia, Middle East, south and south-east Asia, China, to Japan; northerly populations move south when non-breeding. Vagrant north to Britain, Ireland, Sweden and the Aleutian Islands.

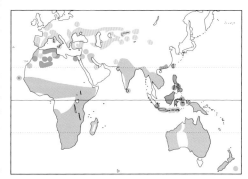

◄ **11a. Juvenile**. Spain, early September. A pale, rather grey version of the adult, with neatly fringed upperparts and wing-coverts. The greenish metallic gloss of the wing-coverts, contrasting slightly with the upperparts, suggests this may be a male. RJC.

Racial variation and hybridisation No racial variation. Breeding of Black-winged x (escaped) Black-necked Stilt reported in the Netherlands, but plumage of offspring not recorded.

Similar species Separation from Black-necked Stilt (12) involves head and neck patterns. Black-winged juvenile has greyish crown and hindneck retained through first non-breeding; adult non-breeding/adult breeding has amounts of black on crown and hindneck varying from none to extensive in both sexes. Black-necked in all plumages has crown, nape, ear-coverts and hindneck to mantle black, with small white area above and behind eye. Occasionally Black-winged has head pattern recalling Black-necked, but it does not have white area above eye and black on lower nape does not reach mantle. Individuals resembling Australian race of Black-winged, *leucocephalus*, also occur; this race averages slightly smaller, with proportionally shorter legs, slightly longer bill and broader wings.

References Barter (2002), Meininger (1993), Xeira (1987), Zeillemaker *et al.* (1985).

▲ **11b. Presumed first-breeding**. Israel, early April. The rather pale upperparts are faded first non-breeding plumage, with a few recently replaced darker feathers. This and the following images show something of the considerable range of head patterns of this species. RJC.

▼ **11c. Adult male breeding**. Spain, early May. A typical white-headed male. RJC.

▼ **11d. Adult male breeding**. Cyprus, mid April. A head pattern with some similarities to Black-necked Stilt. RJC.

▲ **11e. Adult female breeding**. Cyprus, late April. The completely white head is usually a feature of males, but the bronze-brown upperparts show this to be a female. RJC.

◄ **11f. Adult female breeding**. Cyprus, mid April. A more usual female head pattern. RJC.

▼ **11g. Non-breeding Black-winged Stilts in flight**. The Gambia, mid January. Adult front, first non-breeding rear; note the white trailing edge of the younger bird. RJC.

12. BLACK-NECKED STILT
Himantopus mexicanus

An elegant black-and-white shorebird, the New World counterpart of the Old World Black-winged Stilt.

Identification L 36cm (14"); **WS** 66cm (26"). Very similar to Black-winged Stilt. Large, elegant and long-necked, with long needle-fine straight black bill; white, with head and neck black from fore-crown down hindneck, small but conspicuous white area above eye; black upperparts and wings, and very long red legs. *In flight* has uniform black upper and underwings (but see juvenile below) with white uppertail-coverts, and white rump extending in V up back, tail pale grey; legs and feet project well beyond tail, tail-tip only reaching top of tarsus. Sexes similar in plumage year-round, male typically 10% larger; juvenile separable. Restricted webbing between outer and middle toes only; no hind toe. *Feeds* usually by wading, picking from the water surface and by sweeping bill from side-to-side.

▼ **12a. Juvenile *mexicanus*.** Florida, mid July. Dark grey where adult is black, including head and neck pattern, but with extensive pale fringes to upperparts. Jay Paredes.

Juvenile Crown and hindneck black, upperparts and wing-coverts greyish-black, scalloped with buff fringes; underparts white. White tips to inner primaries and secondaries show as narrow trailing edge in flight. Male has greenish sheen to wing-coverts. Legs pinkish-grey, reddish base to lower mandible, iris brown.

First non-breeding/first-breeding A few perhaps breed in their second calendar year. Plumage as adult, though black of head often flecked white and upperparts uniformly blackish-brown, not black; retains many, if not most, juvenile wing-coverts, male still with greenish sheen. White trailing edge of wing in flight becomes less conspicuous with wear. Legs red.

Adult non-breeding/adult breeding Head pattern similar in both sexes, having black from crown down hindneck to mantle, and on sides of head, with small but conspicuous white patch above and behind eye; remainder of head and neck white. Mantle and scapulars black in male, very dark bronze-brown in female; wing-coverts black. Bill completely black; legs and iris red, brighter when breeding, when male sometimes has pale pink flush to breast.

Call A noisy, resonant *kek*.

Status, habitat and distribution Nominate race *mexicanus* breeds on coastal and inland wetlands in eastern North America south from Delaware, in interior North America discontinuously from Oregon south to Mexico and Central America, and on the Pacific coast in California; also West Indies. Southern populations are sedentary; northern birds move south when non-breeding to Florida, California and Central and South America. Has occurred as far north as Washington, Alberta, Saskatchewan and Newfoundland. 'Hawaiian' Stilt, race *knudseni,* breeds in coastal wetland on most of the Hawaiian islands, where it is largely sedentary, moving only in response to seasonal drying of its habitat.

Racial variation and hybridisation Adult 'Hawaiian' Stilt *knudseni* differs from nominate *mexicanus* in being larger, and having more black on head and neck (particularly in the male), white spot above eye variable, usually smaller or absent; juveniles of the two races do not differ significantly. Hybridisation has been recorded between Black-necked Stilt and American Avocet, both in captivity and in the wild. Descriptions are incomplete but hybrid has slightly upturned bill; compared to Black-necked Stilt has restricted grey, not black, on crown and hindneck; mantle and scapulars dark grey, (folded) wing black, legs grey, with webbing on feet intermediate between Black-necked Stilt (restricted) and American Avocet (complete). Flight pattern of hybrid not reported.

Similar species Black-winged Stilt (11), which see.

References Meininger (1993), Morlan *et al.* (2004), Principe (1977), Robinson *et al.* (1999).

◄ **12b. First non-breeding** ***mexicanus***. California, early November. Diffuse head and neck patterns; both sexes have brown upperparts. RJC.

▼ **12c. Adult female breeding** ***mexicanus***. Florida, mid May. Female has bronze-brown upperparts, as in Black-winged Stilt. RJC.

◄ **12d. Adult male non-breeding** *mexicanus*. California, early November. Male has all-black upperparts. RJC.

▼ **12e. Adult (right) and first non-breeding (left)** *mexicanus* **Black-necked Stilts in flight**. California, mid September. The narrow white trailing edge of the wing is only shown in juvenile and first non-breeding plumage, but it gradually wears and can eventually be very difficult to see. Note that the immature is generally browner than the adult. Peter Basterfield.

▲ **12f. Adult** *mexicanus* **in flight**. California, late July. As with Black-winged Stilt, adult Black-necked Stilts have entirely black wings. This individual is undergoing wing moult, but it still retains its old outer five primaries. Donald Metzner.

▲ **12g. Adult** *knudseni*. Hawaii, late January. Female left, male right. Slightly larger than the nominate, with longer bills and legs. Also more black on head, and restricted white above the eye, particularly in the male. Wes Fritz.

13. PIED AVOCET
Recurvirostra avocetta

A graceful black-and-white wader with a slender, upturned bill, occurring across much of Europe, Asia and Africa.

Identification L 43cm (17"); **WS** 67cm (26.5"). Large, black-and-white, with long upturned slender black bill, long neck and long bluish-grey legs; forehead, crown, hindneck, upper scapulars, greater coverts and outer primaries black; remainder white. *In flight* from above, wings strikingly patterned black-and-white, rump and tail white; below, only primaries are black, legs and feet extend some way beyond tail. Males marginally larger; female tends to have a shorter, more upcurved bill, which often enables pairs seen together to be sexed. Plumage similar year-round, juvenile separable. Feet webbed. *Feeds* while wading, continuously scything bill from side-to-side, and often swims, when probes mud well below water surface; will also pick from surface of mud or water.

Juvenile Pattern as adult but with brownish markings on white mantle, wing-coverts and tail, and pale fringes on sooty-brown tertials, which together give an overall 'dirty' appearance. Black areas duller and browner than in adults.

First non-breeding As juvenile, but mantle white; some juvenile wing-coverts are retained, upper scapulars fringed white.

Adult non-breeding/adult breeding Most individuals probably first breed in their third calendar year. Entirely black-and-white, except for brownish tertials. Whitish oval area formed by lower scapulars and wing-coverts pale grey when fresh, bleaching to white. Some individuals have yellow-tinged legs.

Call A mellow *kluit.*

Status, habitat and distribution Fairly comon on coasts or coastal wetlands with areas of alkaline or saline shallow water. Breeds locally from southern

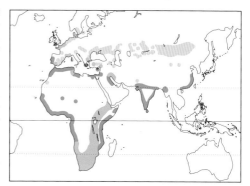

▲ **13a. Juvenile**. Eastern England, late July. In complete juvenile plumage, including brownish upperparts, greyish crown and hindneck, and a short, incompletely grown bill. RJC.

Sweden to the Mediterranean, and across central Russia to Mongolia and north-east China. Non-breeding birds occur from the coasts of southern England and southern North Sea south to the Mediterranean, north and equatorial Africa, and locally east to southern China; rarely Korea and Japan. Vagrant Philippines, Indochina.

Racial variation None.

Similar species Sharply upturned bill and striking pied plumage distinguishes Pied Avocet from all but American Avocet (14), with which it is unlikely to occur.

References Barter (2002), Salvig (1995).

▲ **13b. First non-breeding**. Eastern England, early September. Has gained adult-type head-pattern and white upperparts but still has juvenile brownish wing-coverts and one or two juvenile tertials. RJC.

▼ **13c. Adult**. Eastern England, late July. Females (which this individual probably is) typically have more strongly upturned bills, but there is much overlap between sexes, unlike American Avocets, which can usually be sexed confidently on their consistently different bill shape. Note the near-white scapulars compared to 13e. RJC.

▲ **13d. Adults**. Eastern England, early April. The female (right) is soliciting copulation. Note the smaller, less bulky female with a subtly more upturned bill than the male. RJC.

▼ **13e. Adult**. Eastern England, early April. Has pale grey inner-wing, which contrasts with the white secondaries; these feathers bleach and can be almost completely white by July, as seen in 13c. RJC.

14. AMERICAN AVOCET
Recurvirostra americana

A slender, long-legged North American avocet, with a striking breeding plumage.

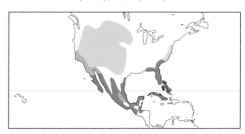

Identification L 45cm (17.5"); **WS** 68cm (27"). Large, with long, upturned, slender, black bill, long neck, black-and-white body with (depending on age and season) greyish or rust-brown head and neck, and long bluish-grey legs. Upperparts and wings mainly black or brownish, with white mantle and white band extending across wing-coverts and lower scapulars; underparts white. *In flight* from above largely black wing with white bar on inner-wing, body white, with black stripe formed by upper scapulars, rump and tail white; below only primaries are black; legs and feet extend some way beyond tail. Male marginally larger, with longer, straighter bill than female; plumage varies seasonally, juvenile separable. Feet webbed. *Feeds* usually when wading (often swims), scything bill from side-to-side, but may pick items from mud or surface of water. Will also feed in active but coordinated groups, when either wading or swimming, apparently hunting small fish.

Juvenile Head and neck largely pale grey, hind-neck pale cinnamon; mantle and wing-coverts have contrasting tips when fresh: white feathers with dark tips, dark feathers with pale tips. White tips to inner five primaries.

First non-breeding/first-br/adult non-breeding Most probably breed first in their third calendar year. Head and neck pale grey, upper scapulars sooty black. First non-breeding probably replaces some scapulars, but retains (as may first-breeding) worn, faded juvenile wing-coverts. Primaries of first non-breeding/first-breeding browner and more worn than those of adults (which are black), some still with pale tips to inner primaries.

Adult breeding Head and neck become rusty-brown, brighter in male.

Call A far-carrying *kleek*.

Status, habitat and distribution Fairly common, often in flocks, on fresh, saline and alkaline shallow waters. Breeds in west and central North America, from British Columbia to southern California and Texas south to Central America. Migrates south from California and on east coast from North Carolina to Florida, the Gulf, and south to Mexico and central America. Vagrant to Alaska.

Racial variation and hybridisation No races recognised. Hybrid American Avocet x Black-necked Stilt is described under Black-necked Stilt (12).

Similar species Upturned bill and pied plumage eliminate all but Pied Avocet (13), which see.

References Morlan *et al.* (2004) Principe (1977), Robinson *et al.* (1997), Weir (1997).

▼ **14a. Juvenile/first non-breeding female**. California, mid November. Plumage relatively fresh for this time of year. This bird retains faded juvenile wing-coverts but it is replacing some upperpart feathers; note pale tips to the (juvenile) primaries. Bill is more or less completely grown at this stage; the degree of bill curvature indicates a female. RJC.

◄ **14b. Juvenile/first non-breeding male**. California, early November. More worn and faded than 14a, this individual is in plumage more typical for early November. Longer, near-straight bill indicates a male. RJC.

▼ **14c. Adult non-breeding American Avocets**. California, early November. Male front with longer, straighter bill; female to rear. RJC.

◄ **14d. Adult male non-breeding**. Florida, late March. Just starting to acquire adult breeding plumage of cinnamon head and neck. RJC.

▲ **14e. Adult female breeding**. Texas, late April. RJC.

◀ **14f. Adult non-breeding in flight**. California, early November. RJC.

THICK-KNEES

The thick-knees or Burhinidae include six northern hemisphere species in two genera: Eurasian Stone-curlew *Burhinus oedicnemus*, Senegal Thick-knee *B. senegalensis*, Spotted Thick-knee *B. capensis*, Double-striped Thick-knee *B. bistriatus*, Great Thick-knee *Esacus recurvirostris* and Beach Thick-knee *E. magnirostris*. There are nine species (in two genera) worldwide.

Thick-knees (known as stone-curlews in Europe or dikkops in southern Africa) are a group of large, bulky, generally brown shorebirds with proportionally large heads, large eyes and short bills. The two *Esacus* species are very large, with particularly heavy bills and strongly black-and-white patterned heads. All the thick-knees are species of relatively open, rather dry habitats, the precise nature of the habitat varying with species. They are largely nocturnal and during the breeding season they can be quite noisy at night, giving plaintive, far-carrying calls. During the daytime they spend considerable periods standing in a hunched posture or sometimes squatting on their tarsi, their cryptic plumage resulting in them being easily overlooked. The majority of northern hemisphere species occur in the more southerly areas of Europe and Asia; Eurasian Stone-curlew occurs in more temperate areas and is the only species that is migratory. Double-striped Thick-knee occurs in central America and has wandered to Texas.

All species have very similar plumages at all ages, and the differences between the sexes are very minor. Their flight patterns, though rather complex with white areas on the black flight feathers, do not differ greatly between the various species. They all lack a hind toe.

15. EURASIAN STONE-CURLEW
Burhinus oedicnemus

A Eurasian and north African species, and the only migrant thick-knee. Alternative name: Northern Thick-knee.

Identification L 42cm (16.5"); **WS** 79cm (31"). Forms a species pair with Senegal Thick-knee (16). Large, predominantly pale brown, with large yellow eye, short black-tipped yellow bill (black sometimes extending along cutting edges) and long, pale yellow legs. Narrow whitish band across lesser coverts on folded wing is best distinction from Senegal Thick-knee. Generally looks hunched, large-headed, large eyed and short-necked. *In flight* above shows black primaries with two white flashes (similar to all *Burhinus* thick-knees), black primary coverts and secondaries and two white bars across brown inner-wing-coverts; underwing white, with fairly broad black trailing edge; feet do not project beyond tail. Sexes of similar size, with similar plumages year-round; juvenile separable with good views. Small webs between all toes. *Feeds* mainly at night on earth-worms and other invertebrates, and occasionally on small vertebrates, stalking heron-like before stabbing at prey on the ground or on low vegetation.

Juvenile Similar to adult, but distinguished by plain face with white ear-coverts, diffuse pale supercilium, lacking dark 'moustache' of adult. No dark lower border to buffy-white lesser-covert bar which is wider than in adult. Neck and upper breast finely streaked dark brown. Often retains tail-down after fledging.

▼ **15a. Juvenile *oedicnemus*.** Eastern England, late September. Aged by facial pattern with very little white above eye, white area on ear-coverts, and slightly wider white wing-covert bar than adult, lacking dark borders. RJC.

First non-breeding Most moult head and body, so from about September head pattern as adult, with dark moustache and buff, streaked brown, ear-coverts. Neck and breast streaking coarser than juvenile.

Adult non-breeding/adult breeding Some breed in second calendar year, remainder in third. White forehead, broad white supercilium restricted to area above eye, white from forehead below eye to ear-coverts; dark moustache, from the lower mandible to side of neck. Neck and breast with coarse dark brown streaks. Male has more contrasting lower dark border to white lesser-covert bar and may have darker moustache and yellower legs than female.

Call A series of fluty whistles, *cur-lee, cur-lee*, with rising inflection, particularly at night during the breeding season.

Status, habitat and distribution Uncommon on dry, open, frequently stony terrain, often inland. Nominate *oedicnemus* breeds from east and southern England south to Spain, around the northern Mediterranean, east to the Caucasus; *distinctus* and *insularum* in the Canaries; *saharae* in North Africa and the Middle East to Iran; *harterti* from the Caucasus to Kazakhstan and Pakistan; *indicus* in India and Sri Lanka to Indochina. Southern populations are largely sedentary; more northerly populations move south when non-breeding to the Mediterranean, North Africa, Africa just south of the Sahara and the Arabian peninsula. Vagrant to Ireland, Scotland and north to Norway, Sweden and Finland.

Racial variation Much variation between and within races; the nominate is typically rather darker, *harterti* greyer and *saharae* more sandy, but these extreme examples are doubtfully distinguishable in the field. Smaller and darker *indicus* has less yellow on bill, longer legs and paler wing-panel, and may be a separate species.

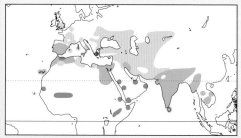

Similar species None in Europe; Senegal Thick-knee (16) overlaps in the Nile Valley. It is smaller, usually has more black on bill, has broad, narrowly black-bordered, pale panel in folded wing, lacking narrow white band of Eurasian Stone-curlew; in flight shows more white in upperwing with more extensive white primary flashes. Occasionally Eurasian Stone-curlews with worn and bleached wing-coverts may give an impression of Senegal Thick-knee's wide pale wing-panel. Spotted Thick-knee (17) of Africa and Arabia is boldly spotted on upperparts and wing-coverts, while Great (19) and Beach Thick-knees (20) of Asia are both much larger, with striking black-and-white facial patterns and heavy bills.

References Green & Bowden (1986), Rasmussen & Anderton (2005), Shirihai (1994).

▲ **15b. Adult non-breeding** *oedicnemus*. Eastern England, late September. This individual is completing its moult to non-breeding, and its new feathers are more buff than when breeding, by which time some fading will have occurred. RJC.

▼ **15c. Adult male breeding** *oedicnemus*. Eastern England, mid July. Wear and fading result in a paler bird than when freshly moulted (as in 15b). The broad black borders to the wing-covert bar indicates a male. The ring showed this bird to be nine years old. RJC.

▼ **15d. Adult female breeding** *oedicnemus*. Eastern England, mid July. Has less black margins to the wing-covert bar than the male. RJC.

▲ **15e. Adult *oedicnemus* with chick**. Female on nest and male standing; same individuals as 15c and 15d. RJC.

▲ **15f. Adult non-breeding *indicus***. India, early January. Slightly smaller than *oedicnemus* but otherwise very similar. RJC.

▲ **15g. Eurasian Stone-curlews in flight**. Eastern England, late September. RJC.

16. SENEGAL THICK-KNEE
Burhinus senegalensis

The smallest of the thick-knees, this African species reaches the northern hemisphere only in the Nile Valley.

Identification L 34cm (13.5"); **WS** 68cm (27"). Forms a species pair with Eurasian Stone-curlew (15). Large, generally pale brown, heavily streaked above and on throat and upper breast; large yellow eye, short black bill (yellow at sides of base of upper mandible) and long straw-coloured legs. Broad, pale, sandy panel in folded wing, narrowly bordered black above and below, is main distinction from Eurasian Stone-curlew. *In flight* the upperwing exhibits black primaries with two white flashes, black primary coverts and secondaries and a broad pale secondary covert panel; the underwing is white, with a fairly broad black trailing edge; the feet do not project beyond tail. Sexes of similar size, with similar plumages irrespective of age; small webs between the toes. *Feeds* mainly at night in typical thick-knee manner.

Juvenile Inseparable from adult unless tail-down present.

First non-breeding/adult non-br/adult breeding Similar plumages year-round (described above). No known differences between the sexes.

Calls Similar to Eurasian Stone-curlew.

Status, habitat and distribution Fairly common along rivers and lake shores; widespread and largely resident (but moving in response to rain or drought) in this habitat in Africa north of the Equator, from Senegal east to Kenya and north along the Nile Valley to the Mediterranean. Often near villages in the Nile Valley; nests on rooftops in Cairo.

Racial variation Nominate race *senegalensis* in west Africa, race *inornatus* in eastern Africa from Kenya north to Egypt; *inornatus* is typically darker, but there is much overlap with the nominate.

Similar species Overlaps in Egypt with Eurasian Stone-curlew (15), which see, but the habitats of the two species differ.

References Snow & Perrins (1998), Shirihai (1994).

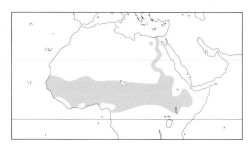

▼ **16a. Senegal Thick-knee of race *senegalensis*.** The Gambia, early January. Smaller than Eurasian Stone-curlew and lacking narrow white wing-covert bar. There are no known characters by which adult and juvenile Senegal Thick-knee may be separated. Race *inornatus* occurs in the northern hemisphere, but this is probably indistinguishable from *senegalensis*. RJC.

▼ **16b. Senegal Thick-knee of race *senegalensis* in flight**. Cameroon, early May. Differs from Eurasian Stone-curlew by larger white flashes on primaries, and lack of pale covert-bar on inner-wing. Gilles Monnoyeur.

17. SPOTTED THICK-KNEE
Burhinus capensis

A boldly spotted African thick-knee that reaches the region in southern Arabia. Alternative name: Spotted Dikkop.

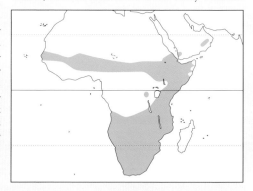

Identification L 43cm (17"); **WS** 72cm (28"). Large, with large yellow eye; short, black, yellow-based bill and long yellow legs. Immediately distinguished from other thick-knees by boldly spotted upperparts. Sandy brown above, with dark brown-centred, pale-fringed upperparts and wing-coverts. Underparts off-white, with long fine dark streaks. No wing-bar, either on folded wing or in flight. *In flight* from above generally dark, black primaries and secondaries; tail barred, unlike Eurasian Stone-curlew or Senegal Thick-knee. White primary flashes smaller than other *Burhinus* thick-knees. Below, wings show dark trailing edge with a central line formed by dark tips to underwing-coverts, broader on primary coverts; toes do not extend beyond tail. Sexes of similar size; small webs between all toes. *Feeds* in similar manner to other thick-knees.

Juvenile As adult, but wing-coverts more barred, less strongly dark-centred.

First non-breeding/adult non-br/adult breeding Similar plumages year-round. No differences known between the sexes.

Call Usually nocturnal; whistled *ti, ti, ti, tee-tee-tee, ti, ti, ti,* rising, then falling in volume.

Status, habitat and distribution Fairly common

▼ **17a. Adult *dodsoni*.** Oman, mid January. A widespread African species, of which this race occurs in our region; slightly more barred above than other races, but the differences are slight. Hanne and Jens Eriksen.

in dry, open or lightly wooded areas, sometimes on playing fields or other urban open areas; local in southern parts of the Arabian peninsula (Oman, and perhaps Yemen), but more widespread in Africa south of Sahara. Largely sedentary, probably making only local movements.

Racial variation Three largely similar races in Africa, including the nominate. In our region, *dodsoni* occurs in southern Arabia and adjacent parts of Somalia. Race *dodsoni* is slightly more barred and streaked on the mantle and scapulars than the other races, but it has equally spotted wing-coverts.

Similar species None. The boldly spotted upperparts separate Spotted Thick-knee from other species of thick-knees.

Reference Hockey & Douie (1995).

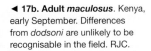

◄ **17b. Adult *maculosus*.** Kenya, early September. Differences from *dodsoni* are unlikely to be recognisable in the field. RJC.

◄ **17c. Adult *maculosus*.** Kenya, mid March. White on the primaries is less extensive than on other *Burhinus* thick-knees. This individual is in wing moult, with the outer four primaries still to be replaced. RJC.

18. DOUBLE-STRIPED THICK-KNEE
Burhinus bistriatus

A long-legged Central and South American thick-knee with a distinctive face-pattern.

Identification L 46cm (18"); **WS** 79cm (31"). Large, with large head, large yellow eye; short, black, yellow-based bill and very long pale yellow legs; often has a more upright and long-necked stance than other *Burhinus* thick-knees. Broad white supercilium, with broad dark line above; remainder of head, neck and breast pale brown. Upperparts and wing-coverts brown, with broad buffish to off-white lateral fringes. *In flight* from above uniform brown, with black primaries and secondaries, primaries having typical white *Burhinus* flashes. Underwing largely white with dark trailing edge; feet extend well beyond tail. Sexes of similar size, with similar plumage; juveniles perhaps separable with close views. Small webs between all toes. *Feeds* in typical thick-knee manner.

Juvenile Very similar to adult but upperparts and wing-coverts have broader, paler, buff fringes, giving an paler appearance overall than adult.

First non-br/adult non-br/adult breeding Described under Identification. Similar plumages year-round. Male may show darker crown than the female but otherwise no known differences between the sexes.

Calls Largely nocturnal; loud, repeated sequences *dit-dit, dit-dit, dit dit dit churr* and accelerating *prip prip prip pip pip pipipipi.*

Status, habitat and distribution Widely distributed in dry open grassland from southern Mexico to Costa Rica (nominate *bistriatus*) and northern Colombia (*pediacus*); Venezuela, Guyana to northern Brazil (*vocifer*); Hispaniola (*dominicensis*). Largely sedentary; vagrant to Texas.

Racial variation Slight and variable; nominate and *vocifer* are very similar, the former with a greyer breast, the latter browner. Race *pediacus* is paler overall; *dominicensis* is smaller.

Similar species None in North and Central America, where the similar Peruvian Thick-knee *B. superciliaris* is unlikely to occur. Peruvian Thick-knee is plainer and more uniform, with paler upperparts that lack Double-striped Thick-knee's broad pale fringes.

References Arvin (1992), Hilty & Brown (1986).

▼ **18a. Adult male** *vocifer*. Venezuela, mid November. Darker crown separates from female, at least in paired birds; 18b is the female of this pair. Race *vocifer* is probably inseparable in the field from the nominate, which is the race that occurs in our region. RJC.

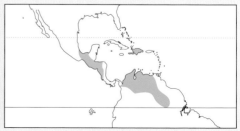

▲ **18b. Adult female *vocifer*.** Venezuela, mid November. RJC.

▼ **18c. Adult *vocifer* in flight.** Venezuela, mid November. Flight pattern similar to other *Burhinus* thick-knees. RJC.

19. GREAT THICK-KNEE
Esacus recurvirostris

A large Asian thick-knee with a striking head pattern and heavy bill. Alternative name: Great Stone Plover.

Identification L 52cm (20.5"); **WS** 82cm (32"). Very large, with yellow eyes; medium-length, heavy, black bill with small yellow patches at base of both mandibles (the upturned shape – though not size – recalls the turnstones); crown medium-brown, sides of head patterned black-and-white. Upperparts sandy buff, wing-coverts paler grey-buff with black bar bordered white in upper part of folded wing; throat, foreneck and underparts white, breast buff. Medium-length dull yellow or greenish-yellow legs. *In flight* short-tailed; body and tail uniform sandy buff, primaries and secondaries black with prominent white primary patches (particularly on inner primaries); broad, pale, grey-buff inner-wing panel with narrow white and black lesser-covert bars. Underwing white, with black trailing edge and black primary-covert crescent; toes just extend beyond tail. Sexes are of similar size with similar plumage, and there is no seasonal variation; juvenile may be separable at close range. Small webs between all toes. *Feeds* as other thick-knees by picking, mainly at night, sometimes using bill to turn pebbles; food includes crustaceans, particularly crabs, and other small animals.

Juvenile As adult, but initially with buff fringes to upperparts and wing-coverts.

First non-breeding/adult non-br/adult breeding Similar plumages year-round. No known differences between sexes, but some (perhaps males) have more black in face pattern.

Calls Loud, harsh single notes, and a wailing *kree-kree-kree-kre-kre-kre*.

Status, habitat and distribution Locally common

▲ **19a. Adult**. India, late December. The amount of black in the head pattern varies, with this individual and 19b showing the normal range of variation. It is not known if this has any age or sexual significance. RJC.

in Indian subcontinent, less common elsewhere, particularly to west and east ends of its range. Largely a species of the margins of larger rivers, lakes and reservoirs, but may also be coastal. Oman, southeast Iran, Pakistan, India and Sri Lanka, Bangladesh, east through Burma and, northern Thailand to northern Vietnam and Hainan, China. Sedentary.

Racial variation None.

Similar species Beach Thick-knee (20), has similar black-and-white facial pattern but with more black above supercilium, a heavier, less upturned bill, and more prominent black bar on folded wing; different flight pattern with all-white inner primaries, and grey, not black, secondaries.

References Ali & Ripley (1983), Evans (2001), Étchécopar & Hüe (1978).

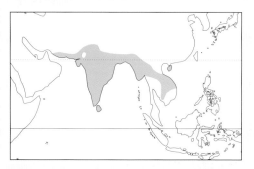

▲ **19b. Adult**. India, early January. An individual with less black on head. RJC.

▼ **19c. In flight**. Oman, late January. Crabs, such as the one in the background, form a major component of the diet of this species. Hanne and Jens Eriksen.

20 BEACH THICK-KNEE
Esacus magnirostris

This coastal species is the largest of the thick-knees. Near Threatened.

Identification L 55cm (21.5"); **WS** 85cm (33.5"). Very large, with yellow eyes; medium-length, heavy, black bill with small yellow patches at base of both mandibles; forehead and crown brown, sides of head strikingly patterned black-and-white. Upperparts are medium brown, wing-coverts pale grey with broad white-bordered black bar in upper part of folded wing; throat, foreneck and underparts white, breast pale buff. Medium-length pale yellow legs. *In flight* short-tailed; body and tail uniform sandy buff, outer primaries black with prominent white patches, inner primaries entirely white; inner-wing pale grey with narrow white and black lesser-covert bars. The underwing is white, with a black trailing edge; toes just extend beyond tail. Sexes are of similar size with similar plumages, and there is no seasonal variation; the juvenile may be separable at close range. Small webs between all three toes. **Feeds** as other thick-knees by picking, mainly at night, on crustaceans, particularly crabs.

Juvenile As adult but upperparts, wing-coverts (darker than adult) and tertials all initially pale fringed. Longer scapulars and tertials have dark submarginal lines.

First non-breeding/adult non-br/adult breeding Similar plumages year-round. No known differences between sexes. First non-breeding may show retained juvenile coverts and tertials.

Call A mournful, bi-syllabic *weeloo*.

Status, habitat and distribution Uncommon. Coastal, on sandy beaches; in northern hemisphere only in Burma and peninsular Thailand and Malaysia. Elsewhere range extends from the Andaman Islands (not Nicobars), Indonesia, Philippines, south to Australia and the south-west Pacific. Sedentary.

Racial variation None, but slight geographical variation.

Similar species Great Thick-knee (19), which see.

References Ali & Ripley (1983), Marchant & Higgins (1993).

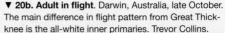

◄ **20a. Adult**. Queensland, Australia, early September. RJC.

▼ **20b. Adult in flight**. Darwin, Australia, late October. The main difference in flight pattern from Great Thick-knee is the all-white inner primaries. Trevor Collins.

EGYPTIAN PLOVER AND COURSERS

The Egyptian Plover *Pluvianus aegyptius*, an African species, is the only member of its genus worldwide; it occurs in both hemispheres. Four species (in two genera) of coursers occur in the northern hemisphere: Cream-coloured Courser *Cursorius cursor*; Temminck's Courser *C. temminckii*; Indian Courser *C. coromandelicus*; and Jerdon's Courser *Rhinoptilus bitorquatus*. There are nine species (in two genera) worldwide. Coursers are placed with the pratincoles in the family Glareolidae.

The taxonomic position of Egyptian Plover is uncertain, but it is perhaps best regarded as an aberrant small courser. As with the coursers, it has a distinctive juvenile plumage, though adults of both sexes are similar year-round. Egyptian Plover, a riverine species, is mainly sedentary, but it may make irregular movements in response to changing river levels, and is perhaps a seasonal short-distance migrant in some areas of Africa. It nests on sand bars, and is unusual in burying its eggs to a depth of a few millimetres (using the bill) during the hotter part of the day, and then both sheltering the eggs from the sun and wetting the sand above the eggs with water transported in the breast feathers. The eggs hatch underground; as with the eggs, the chicks may also be buried if danger threatens. There seems to be no evidence that the Egyptian Plover (sometimes known as the 'Crocodile Bird') actually picks food from between the teeth of crocodiles.

Coursers are small to medium-sized, with a distinctively slender, small headed and long-necked appearance; *Cursorius* coursers have long off-white legs. They are shorebirds of hot climates and arid open or lightly wooded habitats, have short, slightly decurved bills, are not sexually dimorphic, and often prefer to run rather than fly. They are not truly migratory apart from the more northerly populations of Cream-coloured Courser and perhaps some Temminck's Coursers, but they are nomadic, particularly those species favouring more arid areas.

As with Egyptian Plover, coursers have a distinctive juvenile plumage, and an adult plumage that is the same year-round. Coursers, but not Egyptian Plover, have a complete post-juvenile moult. *Rhinoptilus* coursers differ from *Cursorius* coursers in their generally larger size and more patterned head, neck and breast, white uppertail-coverts (Indian Courser is the only *Cursorius* courser to have white uppertail-coverts), more variably coloured legs, and large eyes indicative of their generally nocturnal behaviour.

The critically endangered Jerdon's Courser, long believed extinct, was rediscovered in 1986 having not previously been sighted since 1900.

21. EGYPTIAN PLOVER
Pluvianus aegyptius

A striking and unmistakable vagrant to the region.

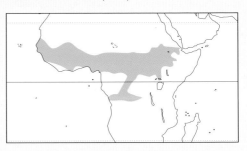

Identification L 20cm (8"); **WS** 42cm (16.5"). Unique combination of black, blue-grey and buff allows immediate identification. A distinctive small shorebird, with short neck and horizontal carriage, short black bill and medium-length grey legs. Black forehead, crown and nape to lower back, with bold white supercilium from bill meeting at nape; black mask around eye extends forward to bill and back to black of nape and breast-band. Scapulars and wing-coverts bluish grey; chin and throat white, remainder of underparts pinkish-buff. *In flight* from above black of head extends narrowly down back; scapulars, coverts and tertials blue-grey, white outerwing narrows across secondaries, with prominent diagonal black band from near base of primaries to tips of inner secondaries. Tail blue-grey with white terminal band. Underwing white with black diagonal band as on upperwing. Adults similar year-round, juvenile separable; lacks hind toe. *Feeds* on insects and invertebrates, running actively, picking, sometimes probing and turning stones.

Juvenile Black on head, back and breast-band has extensive brown admixed, resulting in similar overall pattern to adult but with reduced contrast, particularly breast-band. Upperpart feathers with brownish fringes. Lesser and median coverts brownish. Soon becomes indistinguishable from adult in field.

Adult Described in Identification. Male and female similar.

Call A rapid *cherk cherk cherk* usually repeated at least three times.

Status, habitat and distribution Fairly common on large lowland rivers with sand banks. In small flocks when not breeding. Occurs from Senegal and Gambia east to Ethiopia and Uganda; formerly Egypt, where now extinct. Makes local movements in response to changing water conditions; occasional longer movements. Vagrant to Canary Islands, Spain, Libya, Jordan, possibly Israel.

Racial variation None.

Similar species None.

Reference Zielinski (1992).

▼ **21a. Adult**. The Gambia, late November. An unmistakable species. Steve Garvie.

▼ **21b. Adult in flight**. The Gambia, mid December. As distinctive in flight as it is on the ground. James Lidster.

22. CREAM-COLOURED COURSER
Cursorius cursor

The palest of the coursers, widespread
from North Africa to central Asia.

Identification L 23cm (9"); **WS** 51cm (20").
Medium-sized with an upright stance, longish neck;
short, black, decurved bill with pale base to lower
mandible and very long off-white legs. Strikingly
patterned head with rusty-brown forehead, grey
rear crown and nape, bold white supercilia meeting at nape, bordered below by black line from
eye to nape. Neck, upperparts, throat and breast
sandy-buff; belly and undertail-coverts white. *In
flight* from above sandy-buff body, inner wings
(with narrow white trailing edge) and tail contrast
with black primaries and primary coverts; underwing black, with brown leading edge to inner wing,
feet extend beyond tail. Sexes similar year-round,
juvenile separable; lacks hind toe. *Feeds* largely on
insects, often making short runs to catch prey, then
picking from surface, plover-like; occasionally digs
to shallow depths.

Juvenile As adult, but head pattern diffuse, upperparts with narrow subterminal bars giving lightly
scaled effect. Primaries broadly fringed pale buff,
though the fringes wear quickly.

First non-breeding Body and scapulars quickly
replaced, but extent of wing-covert, tertial and
primary moult variable (probably related to date of
hatching), but may be complete; where incomplete
can be aged by retained juvenile feathers and contrast between adult-type black inner primaries and
dark-brown, pale-fringed, juvenile outer primaries.

Adult Described under Identification. Breeds in
second calender year.

Call A whistled, repeated *quit*, usually given in flight.

Status, habitat and distribution Uncommon, in
dry open desert, sometimes with sparse vegetation.
From Cape Verde Islands and Canary Islands, east
through north Africa to Sudan, Socotra, Arabia to
Afghanistan, Pakistan and north-west India. Bred
southern Spain in 2001. More northerly populations
of nominate race migrate to sub-Saharan Africa

▼ **22a. Juvenile/first non-breeding**. Southwest England, October. Still has most of its juvenile coverts and tertials,
but has replaced much of the juvenile upperparts. Simon Stirrup.

and Arabia. Vagrant to much of Europe, including Ireland, Britain, northern Scandinavia, Germany and Russia.

Racial variation Differences are of size and colour: Races *exsul* of Cape Verde Islands and *bannermani* of Canary Islands are both small and sandy. Nominate *cursor* from Morocco to south-west Iran is larger and pale creamy. Race *bogolubovi* from northern Iran to India is larger and pinker.

Similar species Somali Courser *Cursorius somalensis* (Ethiopia, Somalia and Kenya; not known in northern hemisphere but might occur) is marginally smaller (L 21cm, 8.5"; WS 42cm, 16.5"), darker above (brownish-buff rather than sandy-cream), has only outer underwing dark (complete underwing dark in Cream-coloured Courser), proportionally shorter wings, but slightly longer bill and legs, tail with narrow dark terminal band. Juvenile Somali Courser is more strongly marked than juvenile Cream-coloured, with more extensive dark markings on upperparts and strongly barred tail tip. Temminck's Courser (23; Africa south of Sahara) is slightly smaller, darker above and below, with entirely rusty crown and a black area on belly.

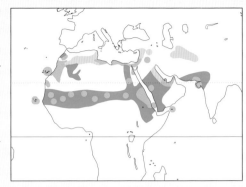

Indian Courser (24; overlaps with Cream-coloured in Pakistan and north-west India) is very similar to Temminck's, darker above than Cream-coloured, with bright rufous neck, breast and belly and a small black area on lower belly; has black lores and white uppertail-coverts which show in flight.

References Gutiérrez (2001), Hazevoet (1988), Pearson & Ash (1996), Koch & Hazevoet (2000), Schekkerman & van Wetten (1988), Thorup (2006).

▲ **22b. Juvenile/first non-breeding wing-stretching**. Southwest England, late September. Same individual as 22a; the black primaries and coverts contrast with the buff inner-wing. Bryan Thomas.

▲ **22c. First non-breeding/first-breeding**. Egypt, early April. Some presumably late-hatched individuals retain much of their juvenile plumage, as here; others moult completely to adult plumage. Ingo Weiss.

▲ **22d. Adult *cursor***. Cyprus, early May. Adults of the nominate race are relatively large, and a paler, creamy buff. RJC.

▲ **22e. Adult *exsul***. Cape Verde Islands, March. This race is relatively small compared to the nominate, and is a slightly darker, more rusty buff. René Pop.

▲ **22f. Juvenile Somali Courser *C. somalensis*.** Kenya, mid September. Formerly lumped with Cream-coloured Courser, with which it may occasionally occur in northern Kenya. This and 22g are included for comparative purposes. Juvenile is darker and more heavily marked than juvenile Cream-coloured Courser; this individual has acquired a few darkish-buff first non-breeding upperpart feathers and coverts. RJC.

▲ **22g. Adult Somali Courser *C. somalensis*.** Kenya, mid September. Longer bill and legs than Cream-coloured Courser; see main text for further distinctions. RJC.

23. TEMMINCK'S COURSER
Cursorius temminckii

A small, dark African courser with a black belly that is regular in Mauritania.

Identification L 20cm (8"); **WS** 38cm (15"). Small, with an upright stance, longish neck; short, black, slightly decurved bill with pale base to lower mandible, and long off-white legs. Forehead (often flecked black) and crown rufous brown, narrowly bordered black at rear; narrow white supercilia join at lower nape. Lores pale, black mask around eye extends narrowly to nape. Upperparts, wing-coverts and tertials pale greyish brown; throat white, neck pale brown, breast chestnut with large black central patch extending to belly. Flanks and remainder of underparts white. *In flight* from above uniformly brown, white-sided tail darker at tip; primaries, primary coverts and secondaries black, narrow white trailing edge to secondaries. Underwing black; feet extend beyond tail. Sexes similar year-round, juvenile separable; lacks hind toe. *Feeds* on insects and other invertebrates, often making short runs to catch prey.

Juvenile Duller version of adult. Forehead and crown brown, streaked blackish; buffy supercilium. Upperparts and wing-coverts have dark subterminal bar and pale buff fringes; inconspicuous dark belly patch.

First non-breeding As adult; moult sequence probably much as Cream-coloured Courser. Until moult is complete may perhaps be aged by retained juvenile feathers, particularly wing-coverts.

Adult Described under Identification.

Call Metallic, grating *keer keer keer* given in flight.

Status, habitat and distribution Occurs regularly in Mauritania, just within the Western Palearctic. Widespread in Africa south of the Sahara, making both local and longer distance movements. Usual habitat is short-grassland and savanna, particularly freshly burned areas; avoids more arid habitat.

Racial variation Only the nominate race occurs in the northern hemisphere.

Similar species Cream-coloured Courser (22), for which see.

References Snow & Perrins (1998), Hockey & Douie (1995).

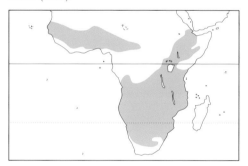

▼ **23a. Adult Temminck's Courser**. Kenya, late May. Appears very similar to Cream-coloured (and Somali) Courser until dark breast/belly patch is seen. Note all-rufous crown, lacking grey to rear shown by the other two species. RJC.

▼ **23b. Adult Temminck's Courser**. Kenya, late March. Dark belly is diagnostic amongst coursers in Africa, and in flight both the primaries and secondaries are black, unlike both Cream-coloured and Somali Coursers, where only primaries are black. RJC.

24. INDIAN COURSER
Cursorius coromandelicus

A striking rufous-brown courser of the Indian subcontinent.

Identification L 23cm (9"); **WS** 48cm (19"). Medium-sized with upright stance, longish neck; short, black, slightly decurved bill with pale base to lower mandible; very long off-white legs. Forehead and crown dark rufous brown, narrowly bordered black at rear; narrow white supercilia join at lower nape, and dark lores extend as black line through eye to nape. Upperparts, wing-coverts and tertials pale dull brown; throat white, neck rusty brown, darkening on flanks, breast and belly. Large black diffusely-bordered central breast patch reaches to legs; remainder of underparts white. *In flight* from above uniformly brown with white rump and white tip to tail; primaries, primary coverts and outer secondaries black, narrow white trailing edge to secondaries. Underwing dark grey-brown, with black primaries and primary coverts; feet extend beyond tail. Sexes similar year-round, juvenile separable; lacks hind toe. *Feeds* mainly on insects and other invertebrates, often making short runs to catch prey.

Juvenile Forehead and crown dark brown, with pale streaks; buffy supercilium running from forehead to join at nape; pale lores. Upperparts and wing-coverts strikingly chequered brown and buff, with extensive pale buff bars and notches; primaries narrowly fringed buff. Underparts buff, lacking adult's dark belly patch.

First non-breeding As adult. Moult sequence probably similar to Cream-coloured Courser. Until moult is complete may be aged by retained juvenile feathers, particularly wing-coverts.

Adult Described under Identification.

Call A low clucking *gwut* or *gut* on taking flight.

Status, habitat and distribution Widely distributed but local through Pakistan, southern Nepal, India, and northern Sri Lanka. May make local movements outside breeding season.

Racial variation None.

Similar species Cream-coloured Courser (22), which see. Jerdon's Courser (25) has a very restricted distribution and is slightly larger with a distinctive head pattern and breast-band.

Reference Ali & Ripley (1983).

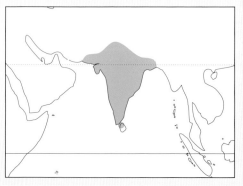

▲ **24a. Juvenile Indian Courser.** Western India, early August. Juveniles are distinctive; ageing is straightforward so long as some juvenile feathers remain. Raja Purohit.

◄ **24b. Adult Indian Courser**. Western India, February. Similar to Temminck's Courser in Africa but with stronger head pattern. Amano Samarpan.

▼ **24c. Adults**. Western India, early November. The most striking of the northern hemisphere coursers; as with all coursers, they have the same appearance year-round. Arpit Deomurari.

◄ **24d. Adult Indian Courser in flight**. Western India, late January. Distinguished from Cream-coloured Courser (the only possible confusion species, apart from the very restricted Jerdon's Courser) by stronger head markings, darker, rufous upperparts and dark secondaries. Sumit Sen.

25. JERDON'S COURSER
Rhinoptilus bitorquatus

A courser of very restricted distribution, and the only *Rhinoptilus* courser in the northern hemisphere. Critically Endangered.

Identification L 27cm (10.5"); **WS** 52cm (20.5"). Medium-sized, with a more horizontal stance than *Cursorius* coursers; longish neck; very short, slightly decurved bill with black tip and straw-yellow base, and very long yellowish-flesh legs. Forehead and crown to nape black with narrow white centre stripe, broad creamy supercilia meet at lower nape; lores, area below eye and ear-coverts brown with white speckling. Upperparts, wing-coverts and tertials dull greyish-brown, with small area of white at bend of folded wing; chin to foreneck cream, with rusty brown central throat patch; neck and breast dull greyish-brown, darker marginally around cream foreneck, narrow creamy breast-band. Remainder of underparts off-white. *In flight* from above brown; black primaries and secondaries, with white at tips of outer primaries and at bend of wing. Uppertail-coverts white, tail black with white margins; feet extend beyond tail. Sexes similar year-round, juvenile probably separable; lacks hind toe. *Feeds* probably as other coursers on insects and other invertebrates.

Juvenile Unknown, but likely to be similar to juvenile of the closely related Bronze-winged Courser *Rhinoptilus chalcopterus* of Africa, which has pale-fringed upperpart feathers, a less well defined breast-band than the adult and distinctive narrow, pointed juvenile primaries.

Adult Described under identification.

Call Poorly described: 'occasional plaintive cry'.

Status, habitat and distribution Very rare; in hilly, rocky areas with patchy open scrub. Restricted to a small area in Andhra Pradesh, India. Sedentary.

Racial variation None.

Similar species Indian Courser (24), which see.

References Ali & Ripley (1983).

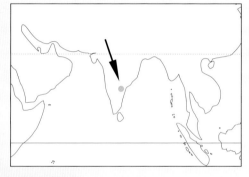

◄ **25a. Adult**. East-central India, date not known. The only possible confusion species is Indian Courser, which has a different head pattern and lacks Jerdon's Courser's breast-band. As is so often the case with coursers, this bird is facing away from the photographer! Simon Cook.

PRATINCOLES

Four species (in one genus) occur in the northern hemisphere: Collared Pratincole *Glareola pratincola*, Oriental Pratincole *G. maldivarum*, Black-winged Pratincole *G. nordmanni* and Little Pratincole *G. lactia*. There are eight species (in two genera) worldwide. Pratincoles are placed with the coursers in the family Glareolidae.

Pratincoles are medium-sized birds with short bills and short legs. They are particularly aerobatic shorebirds, resembling large swallows or terns (particularly marsh terns) in their swooping flight and their forked tails. No doubt this similarity is the result of convergent evolution resulting from their aerial foraging for insects. Pratincoles are insectivorous, usually catching their prey on the wing, though they also feed on the ground. Gregarious birds, they generally breed in loose colonies and also occur in flocks outside the breeding season. The larger species of pratincole covered here (and possibly also Little Pratincole) are unusual among the shorebirds in having a complete post-juvenile moult. Once this moult is complete it is impossible in the field to separate first non-breeding and adult non-breeding birds.

Table 5. Distinguishing features of the larger northern hemisphere pratincoles. There is much variation and some overlap in all characters.

	26. Collared Pratincole	27. Oriental Pratincole	28. Black-winged Pratincole
Trailing edge to wing	Broad, white, but may be narrow by June owing to wear	Virtually absent in fresh plumage, narrow buffish in juvenile	Absent
Tail projection beyond folded wings	Usually beyond wing tip; >10 mm longer diagnostic of Collared	Usually short of wing tip; >15 mm shorter diagnostic of Oriental	Usually short of wing tip, but longer than Oriental
Tail fork	Deeper fork; longer dark tip to outer feather	Shallower fork; shorter dark tip to outer feather	Intermediate between 26 and 27
Outer secondaries paler than inner	Diagnostic of Collared if shown	No	No
Nostril shape	Slit, length 2.5 to 4x width	Oval, length <2x width	Elongate to oval, 3.5 to 2x width
Outer primary shaft	Usually white/creamy	Usually deep buff brown	Usually off-white
Legs	Shorter	Longer	Longer, as Oriental

26. COLLARED PRATINCOLE
Glareola pratincola

The common pratincole throughout much of Europe, western Asia and Africa.

Identification L 25cm (10"); **WS** 58cm (23"). Medium-sized, with short neck; very short, broad, decurved dark bill and short grey legs. Elongated shape owing to long, pointed wings and long tail, which reaches to or just beyond folded wings. Upperparts and wing-coverts uniform dusty brown (adult), primaries and secondaries much darker; throat and upper foreneck creamy, bordered with streaks (adult non-breeding) or thin black line (adult breeding); breast brownish, rest of underparts white. *In flight* from above has white shaft to outer primary, shows contrast between brown wing-coverts and darker flight feathers, has white trailing edge to secondaries, white rump and uppertail-coverts and deeply-forked tail. From below, white body contrasts with dark underwings which have red-brown and black coverts; white trailing edge to secondaries can be difficult to see, feet do not extend to base of forked tail. Plumage varies seasonally, sexes similar, juvenile separable; toes partially webbed. *Feeds* on insects in erratic but graceful, tern-like flight; gregarious.

Juvenile Mantle and scapulars are broadly tipped pale buff with dark subterminal bands; wing-coverts are less patterned, with pale fringes; dark primaries have neat pale fringes. Throat pale buff, unmarked; breast coarsely streaked. Has narrow off-white trailing edge to secondaries in flight. Bill is all-dark.

First non-breeding/adult non-breeding Both ages are as adult breeding but head is streaked and pale throat has streaked border. Post-juvenile moult is complete but moult of flight feathers suspended prior to migration, with some bleached and worn juvenile primaries often retained until February, occasionally longer. White trailing edge to secondaries may wear away completely by about June. Orange-red at base of bill.

First-breeding/adult breeding Moult completed about February, sometimes earlier; many first-breeding indistinguishable from full adult. Unstreaked head; creamy throat neatly outlined by black line. Lores black in male, browner in female, though the difference is difficult to see in the field. Red at base of bill typically reaches beyond feathering below bill, this greater extent of red being useful in separation from Oriental Pratincole.

Calls A sharp nasal *kik* or *kittik*, or combination of the two.

Status, habitat and distribution Locally common in suitable habitat; frequents open areas, usually near water, but is not necessarily coastal. A southern species, breeding in loose colonies around the Mediterranean (particularly Spain), in Kazakhstan and east to Pakistan. Migrates to sub-Saharan Africa, south at least to Kenya; the Pakistan population moves south to India. Vagrant in north-west Europe, Norway, Madeira, Cape Verde Islands, Barbados.

Racial variation and hybridisation The nominate *pratincola* occurs north of the Sahara. Other races (*erlangeri* and *fuelleborni*) occur in Africa south of Sahara. North African breeding birds seem to be intergrades between *erlangeri/fuelleborni* and the nominate. Races *erlangeri* and *fuelleborni* differ from the nominate in being richer, darker brown above, particularly on crown, having less contrast between the dark lores and chestnut-brown crown, a brighter buffy-yellow throat patch, and generally darker underparts. Hybrids of Collared x Black-winged Pratincole, and possibly Collared x Oriental Pratincole (intermediate in some characters, but could have been an aberrant individual of either species) have been reported.

Similar species Oriental Pratincole (27). Black-winged Pratincole (28) is darker, has entirely black underwing-coverts and lacks white trailing edge to secondaries. See Table 5 (p. 105) for distinctions between these three species.

References Driessens (2005), Driessens & Svensson (2005), Ebels (2002), Thorup (2006), Votier (1997).

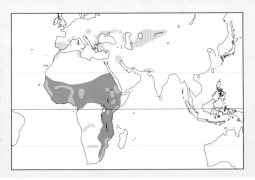

▲ **26a. Juvenile**. Oman, early September. Shows some feather wear, particularly the tertial tips, but is in complete juvenile plumage. Note the pale primary fringes of this plumage, particularly the just-visible inners. Hanne and Jens Eriksen.

▲ **26b. Juvenile**. Oman, early November. An older, very worn individual, but still in virtually complete juvenile plumage. Ray Tipper.

▲ **26c. First non-breeding**. Eastern England, mid October. Most body feathers are adult-type, and the inner primaries have been replaced, though a few juvenile coverts are retained that show either pale or dark tips depending on the degree of wear. This individual is much more advanced than 26b. Robin Chittenden.

▲ **26d. Adult breeding**. Cyprus, early April. Note the tail-streamers reaching to the tips of the folded wings, which in adults separates from both Oriental and Black-winged Pratincoles. RJC.

▲ **26e. Adult breeding in flight**. Kuwait, early March. Note the contrast between the dark primaries and the rest of the upperwing, and the narrow white trailing-edge; these are both features that assist specific identification. The white trailing-edge may wear away by June. Pekka Fagel.

▲ **26f. Adult breeding in flight**. Cyprus, late April. The contrast between the black wing-coverts and the brown flight feathers on the underwing, together with the white trailing edge, confirm that this bird is a Collared Pratincole. RJC.

27. ORIENTAL PRATINCOLE
Glareola maldivarum

A shorter-tailed Asian breeding species that migrates to southeast Asia and Australia.

Identification L 24cm (9"); **WS** 57cm (22.5"). In many respects intermediate between Collared and Black-winged Pratincoles. Medium-sized, with short neck; very short, broad, dark bill and medium-length grey legs. In all plumages closely resembles Collared and Black-winged Pratincoles, being particularly similar to Collared. Tail typically falls short of folded wings, sometimes very much so, particularly in juveniles. Upperparts and wing-coverts uniform dull brown (adult), primaries and secondaries darker; throat and upper foreneck creamy, bordered with streaks (adult non-breeding) or thin black line (adult breeding). Breast brownish, lower breast warm peach, showing more colour than Collared Pratincole, rest of underparts white. *In flight* from above has brown shaft to outer primary, shows limited contrast between wing-coverts and darker primaries, lacks Collared Pratincole's white trailing edge to secondaries, has white rump and uppertail-coverts, and shallowly forked tail; below white body contrasts with dark underwings which have red-brown coverts. Feet do not extend to base of forked tail. Plumage varies seasonally, sexes similar; juvenile separable. Toes partially webbed. *Feeds* on insects in tern-like flight.

Juvenile Very similar to juvenile Collared Pratincole. Mantle and scapulars are broadly tipped whitish with dark subterminal bands, more contrasting than Collared; wing-coverts less strongly patterned, with pale fringes, dark primaries have neat pale fringes. Throat pale buff, unmarked; breast coarsely streaked. Has very narrow buffish trailing edge to secondaries in flight; tail shorter, less forked than adult. Bill all-dark

First non-breeding/adult non-breeding Similar to adult breeding but head is streaked and pale throat has streaked border. As in Collared Pratincole post-juvenile moult is complete and primary moult is suspended, so birds with retained juvenile primaries are first non-breeding. Bill may be all dark.

First-breeding/adult breeding Unstreaked head; creamy throat neatly outlined by black line. Upper breast usually rich buff to yellow-orange, typically brighter than Collared Pratincole. Red at base of bill typically just reaches end of feathering below bill, but variable.

Calls A sharp nasal *kik* or *kittik*, as Collared Pratincole.

Status, habitat and distribution Breeds colonially and opportunistically in open, freshwater areas (including recently reclaimed land) from north-east Mongolia and China to Taiwan and Japan, perhaps Korea, northern India, Sri Lanka and south-east Asia. The northern population spends the non-breeding season from Indonesia to northern Australia (where more than 2.5 million were reported in February 2004); populations from the Indian subcontinent

▼ **27a. Juvenile**. Taiwan, mid August. Similar to other juvenile pratincoles in having upperparts and wing-coverts with dark subterminal marks and pale tips, and having pale tips to the primaries. Ming-Li Pan.

are largely sedentary. Vagrant to North America, Europe, western Indian Ocean and New Zealand.

Racial variation and hybridisation No races recognised; for possible Collared x Oriental Pratincole, see Collared Pratincole.

Similar species Collared Pratincole (26) and Black-winged Pratincole (28); see Table 5 (p. 105) for distinctions between these species.

References Barter (2002), Burns (1993), Driessens & Svensson (2005), Driessens & Zekhuis (2007), Higgins & Davies (1996), Rosair & Gilbert (1988), Sitters *et al.* (2004).

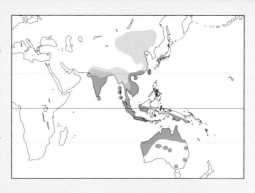

◄ **27b. Juvenile in flight.** Malaysia, early May. Still largely in juvenile plumage, though has just started replacing the inner primaries. The tail streamers are short at this age, but the basic wing pattern, with dark trailing edge to wing, are as adult. Hatching occurs at least as early as March in Malaysia. David Bakewell.

◄ **27c. Adult breeding.** Western Australia, late March. The oval nostril typical of Oriental Pratincole is clear on this individual. RJC.

▲ **27d. Adult non-breeding**. Hong Kong, late November. Much as adult breeding, but more diffuse head pattern with much less black; has outer primaries still to replace. Tail streamers fall short of primary tips. Ray Tipper.

▲ **27e. Adult breeding in flight**. Hong Kong, mid April. Similar to Collared Pratincole in flight, but lacks pale trailing edge to wing and has much shorter tail streamers. Martin Hale.

▲ **27f. Adult in flight**. Western Australia, late October. This bird has suspended its primary moult with the outer two still to be replaced; note the brownish outer primary shaft of this species. Rohan Clarke.

28. BLACK-WINGED PRATINCOLE
Glareola nordmanni

A scarce species of western Asia that migrates
to southern Africa. Near Threatened.

Identification L 24cm (9.5"); **WS** 61cm (24").
Medium-sized with a short neck; very short, broad
bill with tiny area of red at base in breeding adult,
and medium-length grey legs. In all plumages very
similar to Collared Pratincole (26), though slightly
darker. The major distinguishing features from
Collared Pratincole are as follows. Wings slightly
longer and tail marginally shorter (with shallow fork),
so wing-tips generally fall beyond tail (but note that
juvenile Collared also has shorter, less forked tail).
Legs slightly longer. *In flight* from above darker, with
less contrast between wing-coverts and flight feathers than Collared Pratincole and, particularly, lacks
white trailing edge to secondaries; below, uniformly
blackish underwings contrast with white body. Male
is marginally larger than female; plumage varies seasonally and with age. Toes partly webbed. *Feeds* in
flocks, on insects in tern-like flight.

Juvenile Very similar to juvenile Collared Pratincole.
Mantle and scapulars are broadly tipped whitish

with dark subterminal bands, more contrasting
than Collared; wing-coverts less strongly patterned,
with pale fringes, dark primaries have neat pale
fringes. Throat pale buff, unmarked; breast coarsely
streaked. Bill all-dark.

First non-breeding/adult non-breeding As with
Collared, Black-winged Pratincole has complete
post-juvenile moult, so first non-breeding and adult
non-breeding are indistinguishable once juvenile
feathers have been shed, though adult has brighter
colour at bill base. Bill may be all-dark.

First-breeding/adult breeding As Collared
Pratincole, with unstreaked head, creamy throat
neatly outlined by black line, but darker above, with
dark or black lores (male blacker than female). Red
at base of bill usually restricted, not extending to
end of feathering below bill.

Call Similar to Collared Pratincole, but lower in
pitch.

Status, habitat and distribution Rare. Breeds
in open steppes north and east of Black Sea, to
about 80°E; migrates to western and southern
Africa. Vagrant to Europe, north to Iceland and
Scandinavia, mostly August–September.

Racial variation and hybridisation No races
recognised. For Collared x Black-winged Pratincole
see Collared Pratincole.

Similar species Collared Pratincole (26) and
Oriental Pratincole (27). See Table 5 (p. 105) for
distinctions from these species.

References Dring (1988), Thorup (2006), van den
Berg (1985).

▼ **28a. Juvenile.** Kazakhstan, late June. Rather darker
overall than either juvenile Collared or Oriental Pratincole.
The bird behind is an adult. Askar Isabekov.

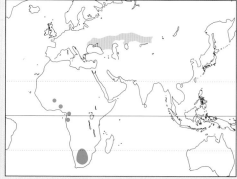

▲ **28b. Adult breeding**. Western Kazakhstan, late April. Note the short tail of this species. Igor Karyakin.

▲ **28c. Adult breeding**. Western Kazakhstan, late April. Darker above than Collared or Oriental Pratincole, lacking both contrast on upperwing and a white trailing edge to the wing; has short tail-streamers. Igor Karyakin.

29. LITTLE PRATINCOLE
Glareola lactia

A small, pale Asian pratincole that is particularly associated with rivers and estuaries.

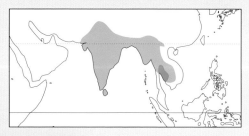

Identification L 18cm (7"); **WS** 48cm (19"). Small, with a short neck; very short, broad, dark bill and short brown to black legs. Elongated shape similar to other pratincoles owing to long pointed wings; in flight short, shallowly forked tail may appear square-ended when spread. In breeding plumage lacks black line surrounding pale area on throat shown by the larger northern hemisphere pratincoles. *In flight* from above generally light grey-brown, with dark grey or black primaries and primary coverts (often showing skua-like white flash), white secondary bar with black trailing edge, rump white, forked tail black, with white sides and narrow white tip. Below, underwing black with white secondary bar just reaching to primaries; feet do not reach tail. Sexes similar, plumage varies seasonally, juvenile separable; toes partly webbed. *Feeds* on insects in large flocks in bat-like flight, and also feeds on ground like small plover. Crepuscular and gregarious.

Juvenile Upperparts buff-grey with wing-coverts, mantle and scapulars having indistinct brownish sub-terminal bar. Crown greyish with paler brown spots; chin white, streaked throat. Bill all-black.

Non-breeding Much as adult breeding but lacks black lores and has pale throat with brownish streaking.

Adult breeding Upperparts and wing-coverts pale greyish brown, darker forehead and crown, black lores, prominent white eye-ring, particularly below eye. Throat has pinkish wash, breast very pale grey, remainder of underparts white. Red base to black bill.

Call A high pitched *prririp*, given in flight.

Status, habitat and distribution Breeds in extensive colonies in coastal marshes and estuaries, and on sandbanks in rivers and streams, from Afghanistan through Nepal, India, Sri Lanka, east to Thailand and Cambodia. Largely sedentary but moves locally when non-breeding, becoming more coastal, in response to changing water levels. Vagrant west to Arabian Gulf.

Racial variation No races recognised.

Similar species All other northern hemisphere pratincoles are considerably larger, darker and lack the white in the wing shown by Little Pratincole.

◄ **29a. Adult non-breeding.** Oman, early November. For a short period when moulting there is less black on the lores than when in the breeding plumage. Hanne and Jens Eriksen.

▲ **29b. Adult breeding**. Sri Lanka, mid January. Ray Tipper.

▼ **29c. Adult in flight**. Eastern India, early February. White in the flight pattern is unique among northern hemisphere pratincoles. Sujan Chaterjee.

▼ **29d. Adult in flight.** Eastern India, early April. The underwing pattern is just as striking as that of the upperwing. Nikhil Devasar.

RINGED PLOVERS

There are 13 species (in one genus) of ringed plovers in the northern hemisphere: Little Ringed Plover *Charadrius dubius*, Common Ringed Plover *C. hiaticula*, Semipalmated Plover *C. semipalmatus*, Long-billed Plover *C. placidus*, Wilson's Plover *C. wilsonia*, Killdeer *C. vociferus*, Piping Plover *C. melodus*, Kittlitz's Plover *C. pecuarius*, Three-banded Plover *C. tricollaris*, Kentish Plover *C. alexandrinus*, Malaysian Plover *C. peronii*, 'White-faced' Plover *Charadrius* sp. and Collared Plover *C. collari*. There are about 23 *Charadrius* species worldwide, part of the family Charadriidae, to which all plovers and lapwings belong.

The ringed plovers are a group of very similar species. Four are essentially North American (Semipalmated, Wilson's and Piping Plovers, and Killdeer), two are primarily European and Asian (Little Ringed and Common Ringed Plovers), while the Kentish Plover complex occurs throughout the northern hemisphere. Long-billed Plover and Malaysian Plover are east or south-east Asian, and the other three species discussed here (Kittlitz's, Three-banded and Collared Plovers) are vagrants to the northern hemisphere, the former two from Africa, the latter from South America. Ringed plovers are rather dumpy, short-necked, shortish-legged and short-billed shorebirds, all of which, at least when breeding, have complete or near-complete breast-bands or, in the case of Kentish, Malaysian and 'White-faced' Plovers, small breast-side patches. The Killdeer, the largest of the group, and Three-banded Plover have two black breast-bands. Ringed plovers are generally birds of the water's edge, particularly outside the breeding season when most species are to be found on open sandy or muddy shores. The Killdeer also occurs on open grassland. They all share the 'stop-start' or 'run-stop-peck' feeding technique, which is also characteristic of the larger plovers and lapwings. This involves alternately standing motionless and then abruptly walking – or running – in an apparently random direction, then bending forward, picking from the ground or water surface. On wet ground all plovers often use the 'foot-trembling' method of feeding (p. 25).

The seasonal sequence of plumages is similar in all species. The juvenile upperparts are characteristically lightly scalloped, the feathers having thin dark submarginal lines and pale fringes. In first non-breeding plumage the juvenile mantle and scapulars are replaced with much plainer adult feathers, as are many of the more visible coverts on the folded wing. Thus it is often not possible to distinguish first non-breeding from adult non-breeding birds from about December onwards.

There is an interesting correlation between migratory status and adult breeding plumage. Adults of several ringed plover species (or races) have breeding-type plumage all the year round: Little Ringed *C. d. dubius* and *C. d. jerdoni* (but not *C. d. curonicus*), Common Ringed *C. h. hiaticula* (but not other races), Long-billed Plover, Killdeer, Kittlitz's, Three-banded, Malaysian, 'White-faced' and Collared Plovers. Most of these are sedentary, or nearly so, but the correlation is not absolute; the northern populations of Killdeer move south when non-breeding, and 'White-faced' Plover is probably a short-distance migrant. The corollary also holds – those species that are migratory generally have distinct non-breeding and breeding plumages, for example Piping and Kentish Plovers, though again there is at least one exception: Wilson's Plover, the males of which have a distinct breeding plumage, is either sedentary or only a short-distance migrant.

30. LITTLE RINGED PLOVER
Charadrius dubius

A small, slender ringed plover with a distinctive yellow orbital ring.

Identification L 15cm (6"); **WS** 35cm (14"). Small, short-necked, with short, black, rather pointed bill and medium-length pinkish-yellow to bright yellow legs (see below). Prominent yellow orbital ring at all ages. Upperparts pale brownish-grey (adults) or buff (juveniles); white hindneck collar has a black lower border which continues as band across breast. Remainder of underparts white. *In flight* from above uniform (no wing-bar, though a thin line is formed by white tips of greater coverts), with white sides to dark-centred rump and tail; underwing white, feet do not extend beyond tail. Sexes of similar size, separable when breeding; some populations vary seasonally; juvenile distinct. Webbing between outer and middle toe, slight between middle and inner; no hind toe. *Feeds* in typical 'run-stop-peck' manner of small plovers.

Juvenile Head brown-buff, with small off-white area on forehead; dull-yellow orbital ring. Upperparts and wing-coverts have darkish subterminal lines and pale fringes; breast-band brown, often broken at centre. Bill all black, legs pale flesh becoming yellowish-flesh.

First non-breeding/adult non-breeding Head pattern much as juvenile but darker across lores and ear-coverts. Upperparts uniform pale brownish-grey, breast-band brownish-black. Bill all black, or with small pink spot at base of lower mandible; legs typically yellow (but see Racial variation below). First non-breeding retains some worn juvenile wing-coverts.

First-breeding/adult breeding Many, though not all, breed in second calendar year. Head pattern has white forehead, black band over forecrown (main feature of breeding plumage) and from bill across lores onto ear-coverts, enclosing eye; white above forecrown bar extends behind eye as short supercilium. Upperparts uniform pale brownish-grey. Bill is black but has a small pink (sometimes pink-orange) area at the base of the bill, which can be difficult to see; yellow legs tinged brownish or pink, so duller than when non-breeding. Male has black lores and breast-band, the black forcrown bar is wider on average, and has slightly more prominent bright yellow orbital ring; on female, black areas have brown admixed. First-breeding male often retains some brown feathering in the breast-band.

Call A disyllabic *pee-oo*, with downward inflection; also see race *dubius* below.

Status, habitat and distribution Uncommon to common, generally breeding inland near fresh water in open gravelly areas, dried river beds, etc. Race *curonicus* occurs from western Europe (not Iceland or Ireland) and North Africa east through Asia to Korea and Japan. European population winters in Africa south of Sahara. Central Asian and Siberian populations migrate to region from East Africa to southern Asia, Philippines, Borneo, and Australia where rare but regular. Vagrant to Ireland, southern Africa, West Indies, Aleutians and Alaska. Two other races are largely sedentary: *jerdoni* from India to south-east Asia, and extralimital nominate *dubius* in Philippines and New Guinea.

Racial variation Race *curonicus*, described above, has a pre-breeding moult on non-breeding grounds (like migratory races of Common Ringed Plover), when blacker head markings and breast-band are acquired; this may be by late November. Eastern populations of *curonicus* may have brighter yellow legs, particularly when non-breeding. Adults of races *dubius* and *jerdoni* seem to be very similar; both have breeding-type plumage year-round. Characters of these latter races are not well known owing to confusion with *curonicus*, with which they occur when non-breeding. In breeding-type plumage, *dubius* and *jerdoni* differ from *curonicus* in usually having a broad white forehead, wider black forecrown band, bill with pink (not yellow) on basal third/half of lower mandible that spreads to upper mandible, broader orbital ring when breeding, legs pale flesh to horn darkening when breeding (sometimes with yellow tinge),

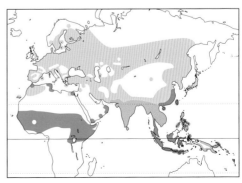

and (*dubius*) apparently a different call (a harsh *chit chit chit*). However, there is probably overlap in characters; young birds perhaps cannot be separated. The three races usually have a mark on the otherwise white outer tail feather: in *curonicus* a large black patch or bar, in *dubius* it is entirely white or with a smaller dark spot, and *jerdoni* usually has a smaller dark spot.

Similar species Slighter build, prominent yellow orbital ring, absence of obvious wing-bar in flight, and largely black bill (*curonicus*) are sufficient to eliminate other ringed plovers.

References Archer & Iles (1998), Barter (2002), Carter & Rogers (1998), Dowsett *et al.* (1999), Thorup (2006).

▲ **30a. Juvenile *curonicus*.** Eastern England, mid August. Upperparts and wing-coverts have narrow pale fringes with thin dark submarginal line; dull, thin orbital ring. RJC.

▼ **30b. First non-breeding *curonicus*.** Thailand, early December. Has replaced upperparts, but retains juvenile wing-coverts; when non-breeding lacks black forecrown bar. Individuals of eastern populations of *curonicus* often have particularly bright yellow legs when non-breeding. RJC.

▲ **30c. Adult non-breeding** *curonicus*. Greece, mid September. Less bright legs than 30b. RJC.

▲ **30d. Adult male breeding** *curonicus*. Cyprus, mid April. Males have more black in the facial pattern than females, especially on the ear-coverts. Individuals from western populations of *curonicus* tend to have less white on forehead and forecrown, and a narrower black forehead bar than both eastern *curonicus* and race *jerdoni*. RJC.

▲ **30e. Adult female breeding** *curonicus*. Cyprus, early May. Ear-coverts less black than male; this individual has more white and black on forehead than 30d. RJC.

◄ **30f. Adult female breeding** *jerdoni*. Nepal, early December. This sedentary race has breeding plumage year-round when adult, and typically has more extensive white on forehead and black on forecrown, a wider orbital ring (particularly in males), and duller legs than breeding individuals from western populations of *curonicus*. Amano Samarpan.

◀ **30g. Adult male breeding**
curonicus. South Korea, early May.
Eastern populations of *curonicus*
have a head pattern similar to
jerdoni, but the orbital ring is
perhaps less wide and the legs
are rather brighter than male
jerdoni. RJC.

◀ **30h. Adult female breeding**
curonicus. South Korea, early May.
The mate of 30g. RJC.

▼ **30i. Adult male breeding**
***curonicus* in flight**. Finland, late
May. Virtual lack of wing-bar
separates from Common Ringed
Plover; the black mark on the outer
tail is larger than in other races of
Little Ringed Plover. Antti Below.

31. COMMON RINGED PLOVER
Charadrius hiaticula

The common small ringed plover over much of Europe and Asia, with both migrant and resident races.

Identification L 19cm (7.5"); **WS** 36.5cm (14.5"). Forms a species pair with Semipalmated Plover (32) Small, short-necked, with a very short blunt-tipped bill and medium-length yellow-orange legs. Brownish-grey upperparts, shade varying with race; white hind-collar has black lower border which continues as black or brown band around breast. Remainder of underparts white. *In flight* from above shows a white central wing-bar, white patches on sides of rump, longish dark tail with white on sides and tip; underwing white, feet do not extend beyond tail. Sexes of similar size, separable when breeding; plumage varies seasonally and with age; slight racial variation. Restricted webbing between outer and middle toes only, which separates from Semipalmated Plover; no hind toe. *Feeds* in typical 'run-stop-peck' small-plover manner.

Juvenile Upperparts and wing-coverts dull greyish-brown, with dark subterminal lines and pale fringes; ear-coverts and fairly broad breast-band dark-brown, latter sometimes broken at centre. The time of retention of this plumage depends on race (see below). Bill black; legs dull yellow-orange.

First non-breeding Uniform greyish-brown upperparts; breast-band and ear-coverts dark-brown, black from end of year. Retains some worn juvenile wing-coverts; small yellow area at base of black bill, legs orange. This plumage is acquired by all populations, though the non-migratory nominate race *hiaticula* quite quickly assumes adult breeding.

Adult non-breeding Acquired by the migratory populations, which moult to adult breeding prior to northward migration. Adult nominate *hiaticula* has no (or restricted) moult to adult breeding, and has breeding-type plumage for much of year.

Adult breeding Most birds probably breed in second calendar year, and gain adult breeding plumage. As non-breeding but head pattern has black from base of bill, around eye to ear-coverts, and over forecrown, white supercilium behind eye only; bill orange, with black tip. Male has thin yellow-orange orbital ring when breeding, black ear-coverts and breast-band, first-breeding perhaps showing limited brown in these areas; female has narrower black forecrown bar, brown in black of head and breast-band (particularly at sides of latter), and much browner ear-coverts.

Call Fluty *too-i*, with upward inflection.

Status, habitat and distribution Common, breeding in stony coastal areas, inland along rivers, and on tundra in north of range. Race *psammodroma* breeds from north-east Canada (overlapping with Semipalmated Plover on Baffin Island), Greenland and Iceland; *tundrae* occurs in northern Scandinavia and northern Russia to far-eastern Siberia (a few on St Lawrence Island); nominate *hiaticula* in western Europe as far north as southern Scandinavia. Generally coastal when non-breeding; nominate is sedentary, while the other two races migrate south to Mediterranean, widely around Africa, the Arabian peninsula, and in small numbers to Japan and south-east Asia. Vagrant on Aleutians. Rare vagrant to eastern north America as far south as Caribbean.

Racial variation Differences are clinal and there is much overlap. Races *psammodroma* and, particularly, *tundrae* are smaller and darker (both as adults and juveniles), and have subtly rounder head and slighter bill. Race *tundrae* (and also probably *psammodroma*) has a different moult schedule from *hiaticula*, so when breeding may appear slightly darker above than *hiaticula* (which lacks significant moult to adult breeding and is thus more faded when breeding). The similarities between *psammodroma* and *tundrae* are such that they are perhaps best regarded together as '*tundrae*-types'. Juvenile *tundrae*-types, which often have quite dark submarginal fringes and also lack an orange tone in legs, appear not to undergo body moult until they reach non-breeding areas, and can often be in juvenile plumage well into October. From October to at least late February the

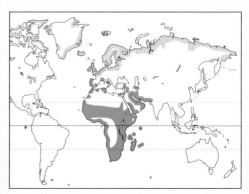

nominate has breeding-type head pattern and bill colour, which is generally not the case with *tundrae*-types. Breeding *psammodroma* has a narrower white forehead on average (particularly females), but there is much overlap with the other races.

Similar species Semipalmated Plover (32) has different call, has webbing between all three toes (not just between outer two as in Common Ringed Plover) and is marginally smaller with less bulky build, more rounded head and slightly more stubby bill. Common Ringed (except breeding male) lacks a yellow orbital ring (present at all ages

in Semipalmated Plover) and has wider breast-band. Black/dark area on lores of Semipalmated generally meets bill base above gape, so narrower than on Common Ringed where it meets at gape. Adult Semipalmated typically has smaller white supercilium behind eye than Common Ringed. Upperparts and coverts of juvenile Semipalmated often lack dark submarginal lines shown by juvenile Common Ringed Plover, particularly *tundrae*-types.

References Chandler (1987b), Ebels (2002), Engelmoer & Roselaar (1998), Thorup (2006), Wallace *et al.* (2001).

▲ **31a. Juvenile *hiaticula*.** Wales, early September. Breast-band often incomplete in juveniles of all races. Race *hiaticula* is typically paler above than other races, with less obvious dark submarginal lines around upperpart and covert feathers, as here, but with wide pale supercilium behind eye. This individual has already replaced a few upperpart feathers, which does not occur until much later in the other (migratory) races. RJC.

▼ **31b. Juvenile, probably *tundrae*.** Eastern England, mid October. Still in complete juvenile plumage despite the relatively late date. Compared to juveniles of race *hiaticula*, is darker above, with extensive dark submarginal lines around upperpart and covert feathers, narrow white supercilium behind eye, and dull brownish-yellow legs. RJC.

▼ **31c. Juvenile, possibly *psammodroma*.** Southwest England, mid September. Very similar to *tundrae*. Note the presence of a partial web between the outer toes only; this aids specific separation from Semipalmated Plover, which has webbing between all three toes (see 32c). RJC.

▲ **31d. Adult non-breeding, possibly *psammodroma* from location**. Canary Islands, mid September. The migratory races of Ringed Plover have a non-breeding plumage; this individual has a non-breeding head pattern and is acquiring new upperparts and wing-coverts. RJC.

▲ **31e. Adult non-breeding, probably *tundrae* from location**. Israel, early March. This bird is starting to acquire its breeding facial pattern. RJC.

▼ **31f. Adult male breeding *hiaticula***. Eastern England, late February. The colour rings showed that this bird was more than three years old. Adults of race *hiaticula* have breeding-plumage head pattern virtually year-round; males have a largely black head pattern and breast-band, and a narrow orange orbital ring. RJC.

◄ **31g. Adult female breeding** *hiaticula*. Eastern England, early May. Much grey in the black of the head pattern and breast-band, and lacks the male's orange orbital ring. RJC.

◄ **31h. Adult male breeding** *tundrae*. Finland, June. Compared to nominate *hiaticula* has restricted supercilium and slightly finer bill. Gordon Langsbury.

▼ **31i. Adult male breeding** *psammodroma*. Iceland, late June. Small supercilium, with white band above bill often narrow; bill shape perhaps nearer *hiaticula* than *tundrae*. RJC.

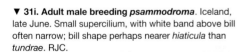

▼ **31j. Adult male** *tundrae* **in flight**. Finland, early July. Antti Below.

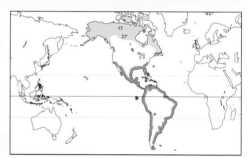

32. SEMIPALMATED PLOVER
Charadrius semipalmatus

The common small ringed plover of North America, wintering south to southern South America.

Identification L 18cm (7"); **WS** 37cm (14.5"). Forms a species pair with Common Ringed Plover (31). Small, short-necked and round-headed with a very short bill and medium-length yellow-orange legs. Narrow yellow orbital ring in all plumages. Dull medium-brown upperparts; white hindneck collar has black lower border which continues as black or brown breast-band, remainder of underparts white. *In flight* from above shows a white wing-bar, white patches at sides of rump, and dark tail outlined white on sides and tip; underwing white; legs do not extend beyond tail. Sexes of similar size, separable when breeding; plumage varies seasonally and with age. Has some webbing between all three toes; no hind toe. *Feeds* in typical 'run-stop-peck' small plover manner.

Juvenile Upperparts and wing-coverts dull are medium-brown with dark subterminal lines (often much fainter than in Common Ringed) and pale fringes when fresh; ear-coverts and breast-band dark-brown. Bill probably all black initially, but quickly developing small yellow patch at base; legs dull yellow.

First non-breeding/adult non-breeding Uniform medium grey-brown upperparts. The breast-band, ear-coverts and forecrown are dark-brown in first non-breeding, black in adults; first non-breeding retains some worn juvenile wing-coverts. Small yellow-orange area at base of bill; legs orange.

Adult breeding Most probably breed in their second calendar year. As adult non-breeding, but head pattern has black from base of bill, around eye to ear-coverts, and over fore-crown. White supercilium behind eye only, smaller and often lacking on male. Breast-band black in male, brownish in female. Bill orange with black tip.

Call A distinctive *chee-wit*, with rising inflection.

Status, habitat and distribution Common, breeding on gravelly or sandy areas both on the coast and inland in the tundra, from Alaska through northern Canada (overlapping with Common Ringed Plover on Baffin Island) to Newfoundland and Nova Scotia. Non-breeding in coastal North America from the Carolinas to the Gulf coast, and from Washington (in small numbers) south to California and northern South America. Vagrant to Azores and Europe.

◄ **32a. Juvenile.** Florida, mid September. Still in complete juvenile plumage, but the all-black bill of younger birds is already acquiring an orange base. Dark area on lores narrows towards upper mandible, unlike juvenile Common Ringed Plover, where it is wider. RJC.

Racial variation No races recognised.

Similar species Common Ringed Plover (31), which see. Piping (36) and Snowy Plovers (39) often occur with Semipalmated Plovers, particularly outside the breeding season, but these two species are considerably paler and, with the exception of Piping Plover in breeding plumage, lack a breast-band.

References Chandler (1987a), Dennis (1986), Dukes (1980), Engelmoer & Roselaar (1998), Mullarney (1991), Nol & Blanken (1999), Walsh (1985), Yésou (1982).

▲ **32b. First non-breeding**. Florida, late December. Note contrast between darker non-breeding upperparts and slightly faded (and worn) wing-coverts. Bulbous bill-tip of Semipalmated Plover is shown well by this individual. RJC.

▼ **32c. Adult breeding male**. Florida, late May. Note the web between the inner toes, diagnostic in separating Semipalmated from Common Ringed Plover. RJC.

▲ **32d. Adult breeding**. Florida, mid June. Although most Semipalmated Plovers probably breed in their second calendar year, not all do so. This individual is well south of the breeding grounds at the height of the breeding season, and seems most unlikely to breed. On the breeding grounds this would be confidently identified as a female because of the restricted black in the head pattern and breast-band, but non-breeders usually fail to develop full breeding plumage. Semipalmated Plovers of all ages and plumages show a narrow orbital ring (unlike in Common Ringed Plover, where this is a feature of the breeding male only). RJC.

▲ **32e. First non-breeding in flight**. Florida, late December. RJC.

33. LONG-BILLED PLOVER
Charadrius placidus

An elegant east Asian ringed plover associated with stony river beds.

Identification L 20cm (8"); **WS** 45cm (17.5"). In many respects a large version of Little Ringed Plover, with which it shares its breeding habitat. Elegant, long-tailed and less compact than other ringed plovers. Small, with short, slender, pointed black bill with a small yellow patch at base of lower mandible (which despite the species' name is not particularly long for a plover), and medium-length flesh/straw legs. White forehead, black band across forecrown to eye, white supercilium behind eye; narrow, pale-yellow orbital ring; crown, lores and ear-coverts grey-brown. White hindneck collar has black lower border which continues as black breast-band; upperparts, wing-coverts plain grey-brown; underparts entirely white. **In flight** from above largely uniform grey-brown, with an indistinct wing-bar formed by the white in the primaries and the white tips of the greater secondary coverts; has long tail. Sexes are of similar size, adults have similar plumages year-round, juvenile is separable. Very restricted web between outer toes only; no hind toe. **Feeds** in typical 'run-stop-peck' small plover manner.

Juvenile As adult, but has narrow pale-buff fringes on upperparts and wing-coverts, lacks black forehead bar and breast-band brown, not black. Bill all-dark, orbital ring indistinct, legs yellow.

First non-breeding/adult non-breeding Aged by worn wing-coverts contrasting with fresh mantle feathers; gradually acquires black in forecrown and breast-band; legs become pale straw. Adults perhaps briefly lose black on head and breast-band.

Adult breeding Age at first breeding not known. Adult breeding described under Identification; males typically have a broader black forecrown bar, more black on front of breast-band and on nape, and perhaps yellower legs than females, but there is much overlap and even breeding pairs may not be easily separable. This plumage is lost only briefly during August–October when moulting.

Call *Piwee*, with rising inflection.

Status, habitat and distribution Fairly common in sandy and gravel-strewn river beds and dry ricefields from south-east Siberia, east and north-east China to Korea and Japan; an uncommon non-breeding visitor to north-east India, Nepal and Bangladesh; vagrant to Bali and Singapore. Siberian and more northerly Chinese and Japanese populations move south when non-breeding.

Racial variation None.

Similar species With its complete breast-band, Long-billed Plover is only likely to be confused with Little Ringed (30) and Common Ringed Plovers (31) in its range. Both are smaller, more dumpy, with shorter yellow or orange legs and shorter tails; Little Ringed Plover has a more obvious bright yellow orbital ring, and an even more obscure wing-bar in flight; Common Ringed Plover a has broader black breast-band, orange legs and a conspicuous white wing-bar.

References Meeth & Meeth (1989), Schols (2005), Konrad (2005).

▼ **33a. Juvenile**. Japan, late September. Buff fringes to upperparts, tertials and wing-coverts, with a rather diffuse head pattern. Orbital ring is indistinct. Nobuhiro Hashimoto.

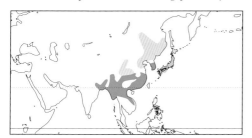

▲ **33b. Adult male**. South Korea, early May. The wide black forecrown bar suggests this is a male; breeding pairs can often be difficult to sex as the amount of black is quite variable. RJC.

▶ **33c. Adult female**. South Korea, late May. The more poorly marked of a breeding pair. RJC.

▼ **33d. Adult female**. South Korea, late May. The flight pattern is similar to that of Little Ringed Plover, though the latter is considerably smaller. RJC.

34. WILSON'S PLOVER
Charadrius wilsonia

A relatively large New World ringed plover with a heavy bill.

Identification L 18.5cm (7.5"); **WS** 36cm (14"). Small, short-necked, rather square-headed, with short heavy black bill, large dark eye and medium-length flesh-coloured legs. White forehead joins supercilium; white eye-ring is particularly obvious below eye, where it contrasts with dark ear-coverts. Upperparts uniform sandy-brown; white hindneck collar not quite continuous across nape. Breast-band sandy brown, sometimes incomplete (black in breeding males), remainder of underparts white. *In flight* from above has white wing-bar, white patches at side of darkish-centred rump, and dark tail fringed white; underwing white, feet extend just to tail tip. Sexes of similar size, separable when breeding; plumage varies with age and season. Webbing between outer and middle toes; no hind toe. *Feeds* largely on small crabs, which it catches by running them down when they emerge from their burrows, sometimes chasing for surprising distances.

Juvenile Upperparts and wing-coverts appear scalloped, having pale-buff fringes; pale-brown breast-band is sometimes incomplete. Legs dull flesh.

First non-breeding/adult non-breeding Uniform upperparts; breast-band pale to medium brown, sometimes broken at centre. First non-breeding retains at least some worn juvenile wing-coverts. Legs dull flesh.

Adult breeding Age of first breeding not known but some perhaps breed in second calendar year. Upperparts replaced pre-breeding, which at least initially are slightly greyer than retained wing-coverts. Male gains black bar on forecrown, black line from bill across lores to lower margin of ear-coverts, and (complete) black breast-band; area of black variable, perhaps more extensive on older males. Female as adult non-breeding. Legs pinkish-flesh.

Call High pitched, whistled, *whit*.

Status, habitat and distribution A fairly common, entirely coastal species, occurring on open beaches, muddy areas and saltmarsh. The nominate race breeds south from New Jersey to Florida, around the Gulf coast and the Caribbean; occurs non-breeding from southern Florida south to Brazil, occasionally wandering coastally north of breeding range. Race *beldingi* breeds along Pacific coasts of Mexico, and central America to Peru; *cinnamominus* along coastal northern South America and nearby islands, south to Brazil. These two latter races are largely sedentary.

Racial variation Nominate described above. Race *beldingi* averages darker above, with more extensive dark areas on head (forehead less extensively white); *cinnamominus* differs only marginally from the nominate, but both these sedentary races probably acquire breeding plumage sooner than the more migratory nominate race.

Similar species Head pattern and, particularly, the heavy bill and flesh-coloured legs separate Wilson's Plover from other small North American ringed plovers.

Reference Corbat & Bergstrom (2000).

▼ **34a. Juvenile**. Florida, early September. Heavy black bill, rather plain plumage, and pale legs are common to all ages and plumages, except breeding male. Small, neat upperparts and wing-coverts with diffuse pale fringes indicate a juvenile. RJC.

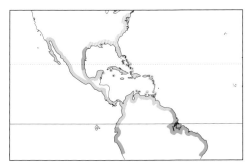

▶ **34b. First non-breeding**.
Florida, late October. Aged by the
contrast between fresh upperparts
and paler, slightly faded juvenile
wing-coverts. RJC.

◀ **34c. Adult non-breeding**.
Florida, mid October. The
upperparts and wing-coverts are
both newly grown. RJC.

▶ **34d. First-breeding male**.
Florida, early April. Breeding males
acquire a black forecrown bar,
black lores, and a black breast-
band, but the extent of black is
variable with age. The restricted
black and worn wing-coverts
suggest that this individual is a
first-breeding male. RJC.

▲ **34e. Adult breeding male**. Texas, mid April. The extensive black areas and relatively unworn wing-coverts show this to be an adult. RJC.

▲ **34f. Adult breeding female**. Florida, mid June. Females have similar plumage year-round, though they acquire fresh upperparts before breeding. RJC.

▼ **34g. Juvenile/first non-breeding in flight**. Florida, mid September. RJC.

35. KILLDEER
Charadrius vociferus

The largest of the ringed plovers, long-tailed and with a distinctive double breast-band.

Identification L 25cm (10"); **WS** 48cm (19"). Medium-sized, with short black bill, rounded head, longer neck than other ringed plovers, and medium-length pale yellowish or pinkish legs (but see juvenile below). Head pattern with black from bill below eye to ear-coverts, and black fore-crown bar; short white supercilium behind eye. Crown and upper-nape brown. Tail is long and projects well beyond folded wings, giving an elongate appearance. Upperparts dull-brown; two black breast-bands, the upper of which continues as lower border to white hindneck collar. Remainder of underparts white. Orange-red orbital ring, yellow in juvenile. *In flight* from above shows white wing-bar and longish rusty-orange, white-edged but dark-ended tail; underwing white, feet do not reach tip of tail. Sexes of similar size, adult plumages similar, juvenile separable. Restricted webbing between outer toes; no hind toe. *Feeds* in typical plover manner, sometimes in small feeding flocks outside the breeding season.

Juvenile As described above, with upperparts and wing-coverts neatly fringed buff, acquiring adult-type plumage quite quickly; often retains tail-down streamers for a week or more after fledging. Legs off-white.

First non-breeding Replaces upperparts, which have rufous fringes, but retains many juvenile wing-coverts, often until the end of the year.

Adult non-breeding As adult breeding, but all plumage fresh with rufous fringes.

Adult breeding Will breed in second calendar year. Replaces some or all upperparts, which appear fresher than the retained wing-coverts. Black of head and breast-bands perhaps more intense than non-breeding, particularly on male, but even in breeding pairs sexes often not obviously different.

Calls A plaintive, far-carrying *kill-dee*, and a rapid *sissi-si-sip*.

Status, habitat and distribution Common in open lowland grassy areas inland and on the coast. The nominate race breeds throughout North America, south from Alaska, Ontario, Quebec and Newfoundland; non-breeding occurs south from Washington east to Massachusetts. Wanders north of breeding areas; vagrant to Hawaii, the Aleutians, east to Europe, most frequently between November and March when displaced from usual non-breeding areas by bad weather. Race *ternominatus* breeds in the West Indies, with *peruvianus* in South America; these races are both sedentary.

▼ **35a. Juvenile**. Texas, early August. Neat, fringed upperparts and wing-coverts, yellow (not red) orbital ring, and tail-down, but otherwise remarkably similar to the adult. Only as a chick do they lack a second breast-band. RJC.

Racial variation Races *ternominatus* and *peruvianus* differ only marginally from the nominate race.

Similar species The double breast-band separates Killdeer from all other northern hemisphere plovers except the much smaller Three-banded Plover (38) of Africa, which also has a different head pattern. The downy young of Killdeer, however, has only one breast-band.

References Evans (1994), Jackson & Jackson (2000).

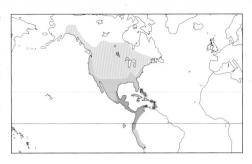

◄ **35b. First non-breeding**. California, early November. Aged by contrast between newly acquired upperparts and faded juvenile wing-coverts. Already has red orbital ring of adult. RJC.

▼ **35c. Adult non-breeding**. California, early November. The fresh buff-fringed upperparts and coverts were grown during the moult to non-breeding. RJC.

▲ **35d. Adult breeding**. Texas, mid April. Upperparts freshly moulted, but wing-coverts remain from non-breeding; otherwise no significant difference from non-breeding. Sexing may be possible in breeding pairs if male has more black in head pattern. RJC.

▼ **35e. Adult non-breeding in flight**. Florida, late December. The longish rufous tail is immediately obvious. RJC.

36. PIPING PLOVER
Charadrius melodus

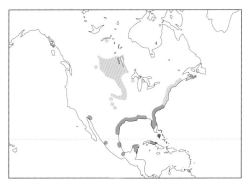

A small, pale grey-brown, stubby-billed plover. Near Threatened.

Identification L 18.5cm (7.5"); **WS** 36cm (14"). Small, with a round head, very short blunt bill and short yellow-orange legs. Upperparts uniform pale sandy-brown or sandy-grey with white hind-neck collar. Underparts white except for complete or incomplete black or grey breast-band. *In flight* from above shows a prominent white wing-bar; off-white uppertail-coverts form band across full width of dark-ended tail (diagnostic among the ringed plovers). Underwing white, feet do not extend beyond tail. Sexes of similar size, separable when breeding; plumages vary seasonally and with age. Lacks hind toe. *Feeds* generally on sandy beaches, in similar manner to other ringed plovers.

Juvenile Head pattern rather plain, coloured as upperparts, with white forehead and supercilium, and dark eye. Indistinctly scaly above, with pale fringes to upperparts and wing-coverts. Greyish breast-band is incomplete. Bill black.

First non-breeding/adult non-breeding Uniformly pale above. Greyish breast-band sometimes broken; otherwise much as juvenile. First non-breeding retains some worn juvenile wing-coverts, but soon becomes difficult to age in field. Bill black.

Adult breeding Some breed in second, most by third calendar year. As non-breeding, but acquires black forecrown bar and black breast-band (which extends around neck to form lower border to white hindneck collar), which is sometimes incomplete, especially in females. Bill black with orange-yellow base; both bill and legs brighter orange in male.

Female has narrower forecrown bar, and brown in black of breast-band.

Calls A plaintive *peep* or *peep-lo*.

Status, habitat and distribution Scarce and decreasing. Breeds on sandy shorelines, both inland by lakes in the prairies, around the eastern Great Lakes, and along the Atlantic coast from Quebec and Newfoundland south to Virginia. Atlantic coast breeders are sometimes separated as *C. m. melodus*, inland breeders as *C. m. circumcinctus*. When non-breeding the Atlantic population moves south down Atlantic coast, prairie birds to Gulf coast, and Great Lakes birds to south Atlantic and Florida Gulf coasts, but the populations overlap.

Racial variation Atlantic breeders ('*melodus*-type') often have a completely white lores and narrow or incomplete breast-band; inland breeding birds ('*circumcinctus*-type') often have a fairly broad

◄ **36a. Juvenile.** Florida, early September. Very pale in all plumages. Juvenile with small, neat upperparts and wing-coverts. RJC.

complete breast-band and some black markings on lores, but there is much overlap. No differences when non-breeding.

Similar species Juvenile might be confused with juvenile Semipalmated Plover (32), but the latter is marginally smaller and darker, and more obviously scaly above, with a more conspicuous breast-band and a more tapered bill. Snowy Plover (39) is smaller and usually has greyish (not yellow) legs. In flight, white uppertail-coverts are diagnostic of Piping Plover.

Reference Haig & Elliott-Smith (2004).

▲ **36b. First non-breeding**. Florida, late December. Ringed as a chick in Saskatchewan, Canada, six months previously; otherwise would be difficult to age with any certainty. RJC.

▼ **36c. Adult non-breeding**. Florida, late December. Ringed as a breeding adult female six months previously. RJC.

◄ **36d. Adult breeding male,
melodus-type**. Florida, late March.
This form has a completely white
lores and a broken breast-band;
the all-black breast-band suggests
a male. The very worn, 'shredded'
wing-coverts can be shown by
birds of all ages and have no
particular significance. RJC.

▲ **36e. Adult breeding male,
circumcinctus-type (right), and
possible *melodus*-type breeding
female (left)**. Florida, late March.
Birds of the *circumcinctus*-type
are particularly pale, and often
have a complete breast-band and
small black flecks at the base of
the bill. The left-hand bird is darker
above, has a poorly developed
black forecrown bar and has brown
in its incomplete breast-band,
suggesting a female. It could,
however, be an immature male, or
even an adult male still to moult
completely. RJC.

◄ **36f. Non-breeding Piping Plover
in flight**. Florida, late December.
The off-white uppertail-coverts are
diagnostic of the species. RJC.

37. KITTLITZ'S PLOVER
Charadrius pecuarius

A small African plover with a prominent white supercilium.

Identification L 13cm (5"); **WS** 33cm (13"). Very small with a medium-length all-black bill and very long greyish legs. Breeding adult has white forehead, brown-grey crown, black line from bill to eye, black forehead bar through eye to nape, and broad white supercilium that extends to form hind-neck collar; chin, throat, sides of face white, breast usually pale peach, remainder of underparts white. Upperparts dark grey-brown, with broad diffuse greyish fringes, giving slightly mottled appearance. *In flight* from above rather dark, with narrow white wing-bar formed by the white on the inner primaries and white tips to greater secondary coverts, pale secondary panel formed by median and greater coverts, and extensive white on sides of tail; feet project beyond tail. Plumage varies with season, sexes similar (but see below); juvenile is separable. Restricted web between middle and outer toe; no hind toe. *Feeds* by picking in typical small plover manner.

The timing of the various plumages is apparently variable, and is associated with opportunistic, near year-round (ten-month) breeding period.

Juvenile Rather plain, lacking breeding adult's distinctive black head pattern and breast colour. Pale forehead, broad off-white supercilium runs to white hindneck collar; crown brown-grey, and diffuse greyish area below eye. Upperparts are grey-brown with broad sharply defined buff fringes;

▼ **37a. Juvenile**. Kenya, mid September. A particularly long-legged small plover. Broad buff fringes to dark-centred upperparts; wide pale buff supercilium. RJC.

upper breast washed very pale peach, remainder of underparts white.

First non-breeding/adult non-breeding Similar to juvenile, including head pattern, but upperparts like adult breeding; supercilium and hind-collar variably pale buff to off-white. Has grey-brown breast-side patch that is usually absent from both juvenile and adult breeding. First non-breeding retains some juvenile wing-coverts.

Adult breeding Some breed within 12 months of hatching. Plumage described under Identification; male of breeding pair perhaps distinguished by blacker facial markings and deeper breast colour, but much variation.

Call An abrupt, trilled *drrr* or *kittip*, in flight.

Status, habitat and distribution Locally common, often in small groups, throughout sub-Saharan Africa and Madagascar, in flat open areas by or near water; small breeding population in Egypt. Apparently makes seasonal movements within Africa, which probably explains occasional vagrancy to Mediterranean, Norway and Saudi Arabia; most November–March.

Racial variation Variation is slight; probably best regarded as monotypic.

Similar species Head pattern of Kittlitz's, with broad white supercilium continuing to hindneck

▼ **37b. First non-breeding**. Kenya, early March. Rather faded juvenile wing-coverts; upperparts being replaced with more diffusely fringed feathers than juvenile. Vagrants to the Mediterranean region have mostly been in this or adult non-breeding plumage, but there has been at least one individual in breeding plumage. RJC.

collar, is distinctive, and among northern hemisphere small plovers sufficient to separate from all but Three-banded (38), which has pinkish legs. Non-breeding birds separated from Little Ringed Plover (30) and Common Ringed Plover (31) by lack of dark breast-band, longer bill and long grey legs, and from non-breeding or female Kentish Plover (39) by latter's more dumpy shape, more horizontal stance, shorter and usually darker grey legs and completely white underparts. Non-breeding Lesser Sand Plover (43) has no white hindneck collar.

References Ludvigsen (1991), Shirihai (1996), Shirihai & van den Berg (1987), Snow & Perrins (1998), Vlachos (2006).

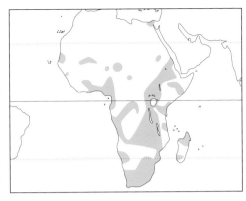

▲ **37c. Adult non-breeding**. Kenya, mid September. Rather dark above, with broad, pale, rather indistinct supercilium, and off-white hind-collar. It is not known how long this plumage is retained, as breeding in Africa occurs almost year-round. RJC.

▲ **37d. Adult (probably male) breeding**. Kenya, mid September. White forecrown bar joins broad supercilium, which meets to form hind-collar. Sexes are not always separable, but typically male has more extensive black in head pattern; underparts are pale to mid terracotta, off-white on undertail. RJC.

▼ **37e. Adult (probably female) breeding**. Kenya, mid September. Compared to male, shows less black in head pattern, with peach breast and belly fading to white on undertail. RJC.

38. THREE-BANDED PLOVER
Charadrius tricollaris

An unmistakable small African plover with two black breast-bands.

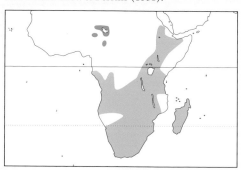

Identification L 18cm (7"); **WS** 34cm (13.5"). Small, long-tailed, with short red-based black-tipped bill and medium-length dull pink legs. Crown dark brown, flecked buff; forehead white, white supercilium from above eye encircles rear of crown; face grey, merging to white on chin and throat. Upperparts and wing-coverts uniform dark brown; two black breast-bands, remainder of underparts white. Conspicuous red orbital ring. *In flight* from above dark, with narrow white inner-wing bar formed by tips of secondary coverts, tail outlined white; feet do not reach end of tail. Sexes similar, with similar plumage year-round, juvenile separable. Restricted webbing between outer toes only; no hind toe. *Feeds* by picking in typical small plover manner.

Juvenile Very similar to adult, but crown paler brown, face paler grey, with more extensive white on chin and throat; upperparts and wing-coverts fringed pale buff (fringe composed of fine spots). Often retains tail-down. Legs duller pale pink than adult. Bill all black. Narrow red orbital ring, iris dark brown.

First non-breeding Mantle feathers are quickly replaced but many juvenile wing-coverts retained, and often thin strings of tail-down. Trace of red at bill-base.

Adult Opportunistic breeder, generally in second half of year; age at first-breeding not known, but seems likely to be within 12 months of hatching. Plumage described above. Iris pale brown.

Calls A variety of fairly loud *wick-wick* and *tsip-tsip* calls.

Status, habitat and distribution Fairly common at and near the water's edge, including coasts, throughout east and southern Africa and Madagascar. Vagrant to Egypt.

Racial variation Only the nominate race is likely to occur in the northern hemisphere.

Similar species Forbes's Plover *Charadrius forbesi* of central and west Africa is very similar to Three-banded Plover but it is unlikely to occur in the northern hemisphere. A double black breast-band is also shown by Killdeer (35), which see.

Reference Snow & Perrins (1998).

▼ **38a. Juvenile/first non-breeding**. Kenya, mid September. Still has the paler crown of the juvenile, has some tail-down and the red orbital ring is still narrow, but has perhaps replaced some upperpart feathers. RJC.

▼ **38b. Adult**. Kenya, mid September. Adults have the same plumage year-round; the sexes are not usually separable. RJC.

39. KENTISH PLOVER
Charadrius alexandrinus

A small, widely distributed ringed plover.
Alternative name: Snowy Plover.

Identification L 16.5cm (6.5"); **WS** 34cm (13"). North American races are referred to as 'Snowy', Palearctic ones as 'Kentish'. A small plover with a breast-band restricted to patches on the sides of the breast. Round-headed, short-necked, with short slender black bill; medium-length rather variable greyish to blackish legs (Kentish very rarely may have pale yellowish-brown legs), averaging paler grey in Snowy. Upperparts medium grey-brown (Kentish), paler and greyer in Snowy, with white hindneck collar; black or brown side-patches form a vestigial breast-band. *In flight* from above shows white wing-bar and extensive white on sides of tail, below entirely white, including underwing; feet do not extend beyond tail. Sexes are of similar size, separable when breeding; plumages vary seasonally and with age. Lacks hind toe. *Feeds* in typical small plover manner.

Juvenile *Kentish:* head brown-grey with white forehead running into white supercilium. Upperparts and wing-coverts fringed pale-buff, quite scaly when fresh. Breast-side patches relatively inconspicuous. *Snowy*: as Kentish, but averages paler, usually with pale grey legs.

First non-breeding/adult non-breeding *Kentish:* as juvenile, but upperparts uniform, with narrow slightly paler fringes when fresh; breast-side patches mid-brown. Sexes generally similar, but some birds, presumably males, have dark grey-brown facial markings and breast-side patches. First non-breeding retains some juvenile buff-fringed wing-coverts. *Snowy*: similar to Kentish but averages paler.

Adult breeding First breeds in second calendar year. *Kentish*: male has black line of variable extent across lores through eye extending onto ear-coverts, and black forecrown bar. White forehead runs into white supercilium and usually has rusty crown, brightest at rear; black breast-side patches. Female head pattern and breast-side patches brown where male is black, lacking rufous crown, only marginally more contrasted than juvenile or non-breeding. A very small number of either sex have complete breast-bands, a trait more frequent in Asian populations. *Snowy*: male has similar black head markings to Kentish, but with extent of black on lores reduced (completely white on some); rusty-buff tinge to crown, but usually less bright than Kentish. Female head pattern and breast-side patches brown as Kentish, but a few (perhaps older birds) have some black in these areas; crown drabber than male, same colour as mantle. There are no records of Snowy Plover having a complete breast-band.

Calls *Kentish*: a sharp but quiet *wit* and a harsher *prrr*; *Snowy*: a quiet *ku-wee* and a low *knut*.

Status, habitat and distribution In northern hemisphere fairly common on coastal sandy beaches and inland on saline wetlands. *Kentish*: Nominate race *alexandrinus* breeds in Azores, Canary and Cape Verde Islands, Europe from extreme south Sweden and Denmark, south to Mediterranean and North Africa, Black Sea, east to south-east China; vagrant to Britain and Ireland. Non-breeding spent on coasts of Mediterranean, Africa to just south of Sahara, Arabia east to India and south-east Asia. Race *seebohmi* in Sri Lanka and south-east India, where largely sedentary; *dealbatus* breeds in north-east China and Japan, non-breeding in Borneo and Philippines. *Snowy*: Race *nivosus* breeds in the west from south Washington to Mexico (vagrant north to British Columbia and Alaska), inland in Great Plains, and coastally in Gulf of Mexico, where the population east from Louisiana to Florida and from Yucatán peninsula to Greater and Lesser Antilles is sometimes separated as *C. a. tenuirostris*. Both Nearctic Snowy taxa are only occasional or vagrant on eastern coast of North America north to Ontario; non-breeding period spent in southern part of range, south to Panama. There is another race in South America.

Racial variation *Kentish*: *seebohmi* similar to nominate *alexandrinus*, but breeding male generally lacks or has restricted black forehead bar, and often has little or no black on lores. Race *dealbatus* is very similar to *alexandrinus*, though some have more rufous

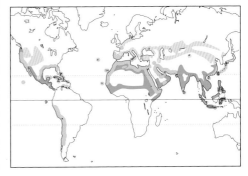

crowns, and others (often females) pink-flesh legs. Complete breast-bands when breeding have been reported for *alexandrinus* in middle East, and slightly more frequently for *dealbatus* in Asia. **Snowy**: race *nivosus* averages paler above than nominate *alexandrinus*, with slightly shorter legs; *tenuirostris* is even paler than *nivosus*, and has a largely white lores in all plumages.

Similar species Piping (36), Kittlitz's (37) and 'White-faced' Plovers (41), all of which see. Malaysian Plover (40) has mottled sandy brown upperparts and sandy-brown legs, a different head pattern and a complete black or brown breast-band.

References Chandler & Shirihai (1995), Page *et al.* (1995), Sharrock (1980), Thorup (2006).

▲ **39a. Juvenile *alexandrinus*.** Greece, mid September. Juvenile upperparts and wing-coverts are small, pale-fringed and neatly arranged. Leg colour is quite variable in this species, but mid to darkish grey is quite typical. RJC.

▲ **39b. First non-breeding *alexandrinus*.** Cyprus, late September. Similar to 39a, but has replaced most of the upperparts, retaining juvenile wing-coverts. RJC.

▼ **39c. Adult non-breeding *alexandrinus*.** Canary Islands, mid September. Upperparts and wing-coverts larger and rather 'untidy'. RJC.

◄ 39d. Adult first-breeding male
alexandrinus. Cyprus, early May.
First-breeding males often acquire
only restricted black on head;
compare with full adult 39e. RJC.

► 39e. Adult breeding male
alexandrinus. Cyprus, late April.
Extensive black on forecrown, lores
and ear-coverts, and some rufous
on rear-crown. RJC.

◄ 39f. Adult breeding female
alexandrinus. Cyprus, late April.
Brown on head and breast-side
patches where male is black. RJC.

▲ **39g. Adult breeding male *dealbatus*.** South Korea, early May. Very similar to *alexandrinus*, but with wider black forecrown bar; some, but by no means all, have more rufous on rear crown. RJC.

▲ **39h. Adult breeding male *dealbatus*.** South Korea, early May. In a few individuals, particularly of race *dealbatus*, the breast-side patches extend to a complete (or near-complete) breast-band. RJC.

▲ **39i. Adult breeding female *dealbatus*.** South Korea, late April. Some females of this race have pale, fleshy legs, as here. RJC.

▲ **39j. Adult breeding male *dealbatus* in flight.** South Korea, late April. RJC.

▲ **39k. Adult breeding male seebohmi**. Sri Lanka, mid May. Male has rather narrow forehead bar, and the extent of black on lores is variable, sometimes absent. No rufous on crown. Ray Tipper.

◄ **39l. Adult non-breeding male nivosus**. California, early November. This race averages paler than those of the Palearctic, but the difference is not great. From the colour bands this is a three-year-old male, though sex cannot be determined in this plumage. RJC.

▼ **39m. Adult male breeding nivosus**. California, mid April. Males usually show only a trace of rufous on the hindcrown; worn coverts suggest this might be a first-breeding individual, but the extent of black is typical of adult males. RJC.

▲ **39n. Adult non-breeding**
tenuirostris. Florida, mid October.
This race is particularly pale above,
with pale grey legs. RJC.

◄ **39o. Adult breeding male**
tenuirostris. Florida, late March.
As in 39m, worn coverts suggest
first breeding, but again the extent
of black is typical of adults. Some
breeding *tenuirostris* males show
a completely white lores. RJC.

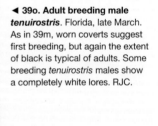

◄ **39p. Non-breeding *tenuirostris***
in flight. Florida, mid September.
RJC.

40. MALAYSIAN PLOVER
Charadrius peronii

A small plover of coastal beaches, similar to Kentish but with mottled upperparts. Near Threatened.

Identification L 15cm (6"); **WS** 31cm (12"). Very small (slightly smaller than Kentish Plover), round-headed, short-necked, with a short black, blunt-tipped bill and long pale yellowish-brown or grey legs. Sandy-brown upperparts appear distinctively mottled owing to dark-centred buff-fringed feathers; white (adult male) or pale rusty-brown (female and juvenile) hindneck collar; black or brown breast-band, narrow (often broken) at centre. *In flight* from above shows white wing-bar and extensive white on sides of tail, as Kentish Plover; entirely white below, feet do not extend beyond tail. Adult plumages same year-round; sexes of same size but separable, juvenile identifiable. Webbing probably small or lacking; no hind toe. *Feeds* in typical small plover manner.

Juvenile Very similar to the adult female but with duller, less contrasted head markings and incomplete breast-band, and smaller, neater, more obviously pale-fringed upperpart feathers.

First non-breeding Soon resembles adult female, from which it is probably indistinguishable in the field.

Adult Age of first breeding is not known but probably in the second calendar year. Male has a white forehead; the forecrown, loral stripe, ear-coverts and breast-band that runs around the hindneck are black. The female is rusty-brown where male is black. Both sexes have similar plumages year-round.

Call A quiet *wit.*

Status, habitat and distribution Uncommon, usually in pairs or small parties, on more remote sandy beaches, sometimes mudflats, of coastal Thailand, Cambodia, Malaysia, Philippines, Borneo and Indonesia. Sedentary.

Racial variation None.

Similar species Kentish (39) and 'White-faced' Plovers (41), which see.

▼ **40a. Adult female Malaysian Plover**. Thailand, late November. Slightly smaller than Kentish Plover, with paler legs, and 'mottled' upperparts owing to the wide pale fringes and darker feather centres. RJC.

▲ **40b. Adult male (left) and female (right)**. Thailand, late November. Once adult plumage is attained, it remains the same year-round. RJC.

▼ **40c. Adult male Malaysian Plovers in flight**. Thailand, early February. Hanne and Jens Eriksen.

41. 'WHITE-FACED' PLOVER
Charadrius sp.

A recently recognised taxon with a distinctive white face; known only from coastal south-east Asia, it seems likely to represent a valid species.

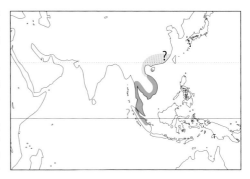

Identification Estimated dimensions **L** 17cm (6.5"); **WS** 35cm (14"). Small, with an extensive white forehead, short black bill, prominent dark eye, and black or brown breast-side patches similar to but more restricted than Kentish Plover; white hindneck collar, sandy-brown upperparts and white underparts, and medium to long dull greyish-pink or flesh legs. *In flight* from above uniform sandy-brown, with narrow white wing-bar, broader on primaries, outer-tail white; toes just reach tail-tip. Sexes similar size, but separable; plumages similar year-round once adult, juvenile probably separable but plumage unknown. Presumably lacks hind toe. *Feeds* in typical small plover manner.

Juvenile Undescribed but likely to have rather plain face and be generally similar to non-breeding adult, probably with sandy-brown pale-fringed upperparts and wing-coverts.

First non-breeding Lacks any black on head or breast-side patches; crown sandy-brown, lores white in male, those of female perhaps smudged pale buff. Apparently retains juvenile wing-coverts.

First-breeding/adult Age at first-breeding not known, but seems likely to be in second calendar year. Male has a broad black forecrown bar, bright orange crown and ear-coverts, and small black breast-side patches; lores white. Female lacks any black, but has warm orange-buff crown and ear-coverts, slightly less bright than male, and has diffuse pale orange-buff lores. Adult wing-coverts have white fringes when fresh; first-breeding appears to retain worn juvenile coverts.

Calls Undescribed.

Status, habitat and distribution Rare; probably breeds on coastal sandy areas in south China. Prefers dry sandy coastal areas when non-breeding, particularly reclaimed ground; reported from Sumatra, Singapore, Malaysia, Thailand and Vietnam.

Racial variation None known.

▼ **41a. Juvenile/first non-breeding** (left) with Kentish Plover. Malaysia, late November. Paler, more sandy, and with slightly longer, paler legs than Kentish Plover. Has probably replaced its juvenile upperparts, but retains its juvenile pale-fringed wing-coverts. David Bakewell.

Similar species Both Kentish (39) and Malaysian Plovers (40) are smaller and more delicate, with proportionally shorter legs and finer bills, and lacking the 'white face', having dark on lores; above both are darker, while Malaysian is distinctively mottled. 'White-faced' is slightly larger, and larger-headed, than Kentish, which usually has grey rather than dull greyish-pink or flesh legs.

References Bakewell & Kennerley (2008); Kennerley *et al.* (2008).

▲ **41b. Adult female**. Malaysia, early February. Shows the large head and blunt-tipped bill of this taxon. Adults of both sexes have very restricted breast-side patches, brown in the female and black in the male; female has no black on head. David Bakewell.

▲ **41c. Adult male**. Malaysia, early February. Bright rufous, almost orange on crown, brighter than female. Once adult plumage is attained it apparently remains the same year-round. David Bakewell.

▲ **41d. Adult male in flight**. Malaysia, early February. Flight pattern very similar to Kentish Plover, but with the wing-bar slightly broader on inner primaries and with more white on trailing edge of secondaries. David Bakewell.

▲ **41e. Adult male**. South China, early July. Losing black of forehead, which it will probably replace quite quickly. John and Jemi Holmes.

42. COLLARED PLOVER
Charadrius collaris

A small ringed plover that lacks the white hindneck collar of most other species.

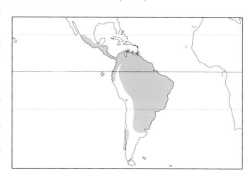

Identification L 15cm (6"); **WS** 32cm (12.5"). Very small, with white forehead, a rather slender, short black bill, narrow black breast-band in adults, brown upperparts and white underparts, longish tail and medium-length flesh or pale-yellow legs. Adults have white forehead with black forecrown, black loral stripe from bill to eye; black breast-band does not continue on hindneck. *In flight* from above uniform grey-brown, with narrow white wing-bar, broader on primaries, outer tail white; feet do not reach tail-tip. Sexes are of similar size with similar plumage year-round, may perhaps be separated when breeding; juvenile distinctive. Lacks hind toe. *Feeds* in typical plover manner, on small insects and other invertebrates.

Juvenile Rather plain; face and throat white, remainder of head buff, lacking black markings of adult. Small brown breast-side patches recall Kentish Plover.

First non-breeding Gradually acquires adult plumage, breast-side patches becoming black and joining to form complete breast-band.

Adult Age of first breeding apparently not known but probably in second calendar year. Adult plumage described in Identification. Males typically have a broader black forecrown bar, more black on ear-coverts, and more rufous on hind-crown and nape, but differences are slight and sexes can perhaps only be separated in well-marked breeding pairs.

Calls Include a high pitched *chip ... chip*, and a dry *prrrip*.

Status, habitat and distribution Common in sandy areas, both coastally and inland (when generally close to water) from western central Mexico, south through Central and South America (largely east of the Andes) to northern Argentina; also central Chile. Largely sedentary; a very rare vagrant to southern Texas.

Racial variation None.

Similar species Separated from all other ringed plovers in the Americas except Wilson's Plover (34) by lack of white hindneck collar. Wilson's is larger, with a heavier black bill, and the adult has much less extensive black head markings than adult Collared.

Reference Yovanovich (1995).

▼ **42a. Adult.** Venezuela, mid November. Adults have the same plumage year-round; differences between male and female are slight, and the sexes are not safely separable except perhaps in breeding pairs. RJC.

▼ **42b. Adult in flight.** Venezuela, mid November. RJC.

OTHER SMALL PLOVERS

These include plovers in the genus *Charadrius* other than the ringed plovers. There are six such species in the northern hemisphere: Lesser Sand Plover *Charadrius mongolus*, Greater Sand Plover *C. leschenaultii*, Caspian Plover *C. asiaticus*, Oriental Plover *C. veredus*, Eurasian Dotterel *C. morinellus* and Mountain Plover *C. montanus*.

These six species are larger than most of the ringed plovers. All are migratory and have distinctive juvenile, non-breeding and breeding plumages. Mountain Plover differs from the others in having similar coloration to the ringed plovers, but the remaining five species have a colourful breeding plumage, with much terracotta red on their underparts. In the case of Greater Sand Plover, the race *columbinus* (which breeds relatively early) appears to have a less 'complete' breeding plumage, paler than other races; this is a characteristic of several other relatively early-breeding species, such as European Golden Plover.

Eurasian Dotterel attains a breeding plumage with a dark belly; the female has a somewhat brighter plumage, reflecting the reversed breeding roles of this species. Mountain Plover is also unusual since it apparently retains its juvenile plumage and does not acquire a first non-breeding plumage, similar to most Grey Plovers, but this is poorly documented.

▲ **Adult breeding male Oriental Plover in flight**. Mongolia, early June. Jari Peltomäki.

43. LESSER SAND PLOVER
Charadrius mongolus

A widespread Asian plover with two distinctive subspecies groups. Alternative name: Mongolian Plover.

Identification L 20cm (8"); **WS** 41cm (16"). A rather plain plover, except when breeding. Small, short necked, with rounded head, short blunt-tipped black bill and medium-length usually greyish-black legs. Upperparts generally grey-brown; lacks hind-neck collar. *In flight* from above shows white wing-bar of uniform width on primaries, narrower on secondaries; tail pattern varies with subspecies group (see below), toes just reach end of tail. Sexes of similar size but can usually be separated; plumage varies seasonally and with age. Lacks hind toe. *Feeds* in typical small plover manner, though tending to stop to pick more frequently than Greater Sand Plover.

Juvenile Crown, upperparts and wing-coverts grey-brown, with buff fringes, wearing and fading with age. Forehead and supercilium white, the latter usually prominent behind eye. Underparts are white with grey-brown breast-band of variable size and extent, sometimes incomplete at centre. Leg colour more variable and paler than older birds, dull yellowish or greyish.

First non-breeding/adult non-breeding Moults to non-breeding after migration, from late August. Much as juvenile, but lacks latter's pale fringed upperparts. First non-breeding retains some pale-fringed juvenile wing-coverts. Legs medium to dark grey.

First-breeding Many remain in non-breeding areas, with some attaining partial breeding plumage.

Adult breeding Most probably do not breed until their third calendar year. Male has extensive black on forehead, around eye to ear-coverts. Forecrown and upper breast, breast sides and/or flanks orange or bright terracotta, depending on race, as does extent of white above bill. Orange-red of underparts may or may not be bounded on upper breast by narrow black line, depending on race. Female has most or all of the black areas on the head replaced by brown, and is generally duller. Legs are usually dark grey.

Calls Variable, probably of no identification value.

Status, habitat and distribution Fairly common, breeding above the tree-line in sparsely vegetated montane areas, often near water; when non-breeding, coastal, generally on muddy and silty substrates, sometimes on fresh water. Five subspecies usually recognised, separable for field identification into the '*atrifrons* group' (including races *pamirensis*, *atrifrons* and *schaeferi*), and the '*mongolus* group' (including *mongolus* and *stegmanni*). The '*atrifrons* group' breeds in central Asia and Himalayas, and when non-breeding moves to coasts of East Africa, the Middle East, India, Bangladesh, Thailand, Malaysia and Borneo; the '*mongolus* group' breeds from eastern Siberia to Kamchatka, with non-breeding birds on the coasts of the Arabian Gulf, India, China, Japan, eastern Indonesia, New Guinea and Australia. Scarce vagrant to North America and Europe.

Racial variation When breeding the two groups can be reliably separated on plumage, but not individual races, which differ mainly in morphology. The '*atrifrons* group' has the forehead all-black or with two tiny white spots, and orange on the breast (lacking a black upper margin) extending a short distance down flanks; rear flanks white. Breeding '*mongolus* group' has extensive white on the forehead, which may be divided by a black central line; bright terracotta on breast extends well down flanks, terminating with scattered grey-brown feathers, and often with a narrow black upper-breast margin. When non-breeding, flanks of '*atrifrons* group' are clear white, but many '*mongolus* group' have scattered grey/brown on flanks. In flight, '*atrifrons* group' has paler tail-band and extensive white at sides of rump and tail, *mongolus* group has broad dark tail-band with narrow white at sides of rump and tail.

Similar species Greater Sand Plover (44) is larger, though very similar in all plumages, particularly juvenile and non-breeding. Differs in structure, having larger and squarer head, larger eye; longer, more slender bill (usually lacking the prominent bulge at tip of Lesser's upper mandible), and longer, generally paler legs than Lesser Sand Plover.

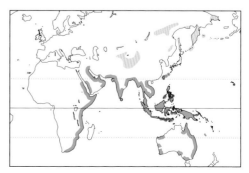

Tarsus: bill length ratio is 1.55–1.65 for Greater, 1.85–2.00 for Lesser Sand Plover. When breeding (except for some male *columbinus*), rufous-orange breast of Greater does not extend so far down flanks as on Lesser. See Greater Sand Plover for distinctions from Caspian (45) and Oriental Plovers (46).

References Balachandran & Hussain (1998), Garner *et al.* (2003), Hirshfeld (1994), Hirshfeld *et al.* (2000), Hirshfeld & Stawarczyk (1993), Hockey (1993).

▲ **43a. Juvenile, *mongolus* group**. Japan, late August. This group often shows some markings on flanks. Proportionally shorter bill, with more bulbous tip, than Greater Sand Plover. Leg colour in Lesser Sand Plover averages darker than Greater Sand Plover, but is often paler in juveniles, as here. Nobuhiro Hashimoto.

▼ **43b. Juvenile/first non-breeding, *atrifrons* group**. UAE, early October. Moulting to first non-breeding, as shown by a few slightly greyer scapulars; birds of the *atrifrons* group always have white flanks. Hanne and Jens Eriksen.

◄ 43c. Adult non-breeding, *atrifrons* group. Thailand, mid November. RJC.

◄ 43d. Probable first-breeding, *mongolus* group. Western Australia, mid March. These birds do not breed until their third calendar year; some first-breeding birds acquire a reduced amount of breeding plumage, but this individual could also be a poorly marked breeding female. RJC.

◄ 43e. Adult breeding male, *mongolus* group. South Korea, late April. When breeding, the *mongolus* group have variable amounts of white on the forehead, and extensive dark markings on the flanks. RJC.

▲ **43f. Adult breeding female, *mongolus* group**. Japan, late April. Females have restricted black in the face pattern. Nobuhiro Hashimoto.

▼ **43g. Adult breeding male, *atrifrons* group**. Oman, early May. Usually lacks white on forehead, has no black upper border to the breast, underpart colour is less saturated, and has no or very restricted markings on flanks. Hanne and Jens Eriksen.

▼ **43h. Adult breeding male, *mongolus* group, wing-stretching**. Japan, early May. The extent of white on the primaries is less than that present in Greater Sand Plover (44j). RJC.

44. GREATER SAND PLOVER
Charadrius leschenaultii

A widespread Asian plover with striking
orange-buff and black breeding plumage.

Identification L 24cm (9"); **WS** 44cm (17.5").
Medium-sized, with squarish, flat-crowned head;
large eye; short, black, pointed bill and medium-
length, generally greyish or dull olive legs.
Upperparts generally pale grey-brown; lacks hindneck collar. *In flight* from above shows white wing-
bar, widening towards trailing edge on primaries,
narrower on secondaries; white at sides of rump
and tail, white-tipped tail with dark subterminal
band; toes reach beyond tail. Sexes of similar size,
separable when breeding; plumage varies seasonally and with age. Small web between outer and
middle toes; no hind toe. *Feeds* in typical plover
manner, tending to stop to pick less frequently
than Lesser Sand Plover.

Juvenile Crown, upperparts and wing-coverts
pale grey-brown, with buff fringes, wearing and
fading with age. Forehead and supercilium white.
Underparts white with pale grey-brown, buff-washed
breast-band of variable size and extent, typically
incomplete at centre. Some may retain juvenile
plumage well beyond year's end. Leg colour can be
slightly darker than older birds.

First non-breeding/adult non-breeding Much as
juvenile, but lacks latter's pale fringed upperparts.
Breast-band of variable extent. First non-breeding
retains some pale-fringed juvenile wing-coverts.

First-breeding Much as adult non-breeding, but
upperparts fresh, contrasting with worn wing-coverts.
Many probably remain in non-breeding area.

Adult breeding The majority probably do not breed
until their third calendar year, and even then may
have poorly developed breeding plumage. Males
have black from forehead to eye and on the cheeks,
two small white areas enclosed by black above bill,
and an indistinct supercilium; rufous on the nape.

Broad rufous-orange breast-band, wider at sides,
sometimes with narrow black upper border. Female
has much less, if any, black on face, and a rather
poorly developed rufous breast-band.

Calls Variable, probably not of use for identification
purposes.

Status, habitat and distribution Breeds in lowland deserts and dry sparsely vegetated areas;
races *columbinus* in Middle East, *crassirostris* from
Armenia to Kazakhstan and nominate *leschenaultii* from eastern Kazakhstan through north-west
China to Mongolia. Non-breeding birds found on
coasts, in eastern Mediterranean (*columbinus*), Red
Sea and Arabian Gulf (*columbinus* and *crassirostris*),
East Africa, and south-east Asia to Australia (*leschenaultii*). Rare vagrant in western Europe, north-west
Africa and California.

Racial variation Race *columbinus* moults (and
breeds) about two months earlier than the other
races; juvenile paler above owing to wider pale
fringes of upperparts. Male breeding has black on
face less well developed than other races, white above
bill more diffuse; broad rufous fringes on upperparts. Races *crassirostris* and *leschenaultii* are more
similar to one another, with the black on face more
sharply defined, and the breast-band darker and

▼ **44a. Juvenile *leschenaultii*.** Japan, early August.
A young individual. Nobuhiro Hashimoto.

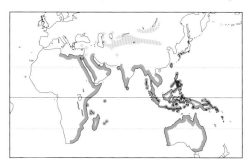

narrower than *columbinus*. Race *leschenaultii* has the darkest and narrowest breast-band, and largest white areas above bill.

Similar species Lesser Sand Plover (043), which see. Both Caspian (45) and Oriental Plovers (46) are of similar body size, and juvenile and non-breeding might be confused with Greater Sand, but both

differ in structure with proportionally smaller, more rounded head, longer neck and legs, and extensive brown on breast. Considerably larger than Common Ringed (31) or Semipalmated (32) Plovers.

References Hirshfeld (1994), Hirshfeld *et al.* (2000), Hirshfeld & Stawarczyk (1993), Hockey (1993).

▲ **44b. Juvenile *crassirostris* (from location)**. Oman, late September. An older individual than 44a. Hanne and Jens Eriksen.

▲ **44c. Adult non-breeding *columbinus* (from location)**. Cyprus, late September. When non-breeding all races have a grey breast-band, sometimes incomplete. RJC.

▼ **44d. First-breeding *leschenaultii***. Western Australia, mid March. Does not breed in second calendar year, and remains in non-breeding type plumage; some of those that breed in their third calendar year may have a poorly developed breeding plumage. RJC.

▼ **44e. Adult male *columbinus***. Cyprus, mid April. With a poorly developed black area on the face, perhaps a third calendar year individual. RJC.

▲ **44f. Adult male *columbinus*.** Cyprus, late April. Even adult males have incompletely developed plumage as here, though the orange-brown underparts are much more extensive than in the other two races. The broad orange-buff fringes to the upperparts are also characteristic of *columbinus*. RJC.

▲ **44g. Adult female *columbinus*.** Cyprus, mid April. Females lack black on the face. RJC.

▲ **44h. Adult breeding male *crassirostris*.** Eastern Azerbaijan, mid April. This race resembles *leschenaultii*, but has less white on the lores. Kai Gauger.

▲ **44i. Adult breeding male *leschenaultii*.** Western Australia, mid March. Red on the underparts in this race is confined to a wide breast-band, but the black on the head is intense. RJC.

▼ **44j. Adult *leschenaultii*, wing-stretching**. Western Australia, late March. White on the primaries extends closer to the trailing edge of the wing than in Lesser Sand Plover (see 43h). The new, fresh primaries show this to be an adult, either a female, or a male with a late pre-breeding body-feather moult. RJC.

45. CASPIAN PLOVER
Charadrius asiaticus

A strongly migratory Central Asian plover of grasslands and open ground.

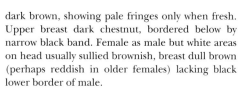

Identification L 19cm (7.5"); **WS** 46cm (18"). Very similar to Oriental Plover (46) in all plumages. Small, round-headed, with large eye, broad white or whitish forehead and supercilium contrast with dark crown; short, pointed, dark-grey bill and long (usually) greenish legs. Pale grey-brown breast (except in breeding male), remainder of underparts white. *In flight* from above brown with narrow wing-bar formed by white on inner primaries (enhanced by prominent white shafts of outer primaries), and white covert tips; below, dark primaries contrast with whitish underwing-coverts and axillaries, toes just project beyond tail. Sexes same size but differ in breeding plumage; plumage varies seasonally; juvenile separable. Lacks hind toe. *Feeds* in dry, short grassland in typical plover manner.

Juvenile Plain brown-grey upperparts with reddish-buff fringes that quickly wear; wing-coverts are fringed pale buff. Crown finely scalloped with whitish fringes. Breast variable, buff to medium brown.

First non-breeding/adult non-breeding Juvenile upperparts are quickly replaced with slightly larger and darker buffish-fringed feathers. Adult has reddish upperpart fringes and buffish wing-covert fringes, so difficult to distinguish from juvenile/first non-breeding except by larger and more irregularly placed feathers. Legs may be yellowish.

First-breeding Similar to adult non-breeding.

Adult breeding Probably breeds in third calendar year, although there is no evidence that first-breeding birds remain in non-breeding areas. Male has forehead, supercilium and throat white, dark behind eye; upperparts are uniform medium to dark brown, showing pale fringes only when fresh. Upper breast dark chestnut, bordered below by narrow black band. Female as male but white areas on head usually sullied brownish, breast dull brown (perhaps reddish in older females) lacking black lower border of male.

Calls A sharp *tee-up*, and shorter *tup*, or *trit*.

Status, habitat and distribution Common but unevenly distributed in poorly vegetated areas in Russia north-west of Caspian Sea, Kazakhstan, Uzbekistan and Turkmenistan. Non-breeding birds occur in dry grasslands in east and southern Africa. Vagrant in Europe west to Britain and Norway, east to Maldives, India, Sri Lanka and northern Australia.

Racial variation None.

Similar species Greater (44) and Lesser Sand Plovers (43), which see. Oriental Plover (46) is very similar, but is slightly larger, lacks a significant wing-bar, has diagnostic brownish, not whitish, underwing and usually paler, yellowish legs; breeding male Oriental has either largely white or rather pale head, very different to male (and many female) Caspian Plover, which has prominent dark crown and patch around eye.

References Nielsen & Colston (1984), Taylor (1983), Thorup (2006).

▼ **45a. Juvenile/first non-breeding**. Kenya, early September. Has gained a few (darker) non-breeding upperpart feathers. RJC.

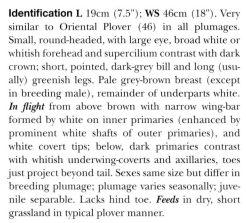

◄ **45b. Adult non-breeding**. Kenya, early September. Moulting to non-breeding. RJC.

◄ **45c. Adult breeding male**. UAE, May. A rather dull individual; compare with the breeding male in 45d. Ian Boustead.

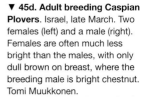

▼ **45d. Adult breeding Caspian Plovers**. Israel, late March. Two females (left) and a male (right). Females are often much less bright than the males, with only dull brown on breast, where the breeding male is bright chestnut. Tomi Muukkonen.

46. ORIENTAL PLOVER
Charadrius veredus

An inland plover of open ground with long, yellowish legs.

Identification L 24cm (9.5"); **WS** 52cm (20.5"). Medium-sized, usually with a darkish crown (but see breeding male), broad supercilium, large eye, rather plain pale face; short, pointed, grey-black bill and long, buff-yellow/orange legs. Pale grey-brown breast (except for breeding male); remainder of underparts white. *In flight* from above uniformly brown, but (particularly juveniles) can show narrow white line formed by tips of primary and secondary coverts; below, wing uniformly darkish, with primaries, underwing-coverts and axillaries being similarly toned, toes project beyond tail. Sexes same size but differ in breeding plumage; plumage varies seasonally; juvenile separable. Lacks hind toe. *Feeds* in typical plover manner, with quick runs, largely on insects in dry, short grassland.

Juvenile Supercilium sullied buff; plain brown-grey upperparts with reddish-buff fringes, wing-coverts fringed pale buff. Crown finely scalloped with whitish fringes. Breast variable, buff to medium brown.

First non-breeding/adult non-breeding Juvenile upperparts are quickly replaced with slightly larger, darker buffish-fringed feathers. Adult has reddish upperpart fringes and buffish wing-covert fringes, so difficult to distinguish from juvenile/first non-breeding except by larger and more irregularly placed feathers. Darker breast shows a clear demarcation from white belly.

First-breeding/adult breeding Probably breeds in second calendar year, since very few first years remain in non-breeding area and both age groups attain full adult-type breeding plumage by late March. Male can be striking, with head, neck, nape and upper breast white, sometimes with a darker (rear) crown; upperparts show pale fringes only when fresh. Dark chestnut band across breast has broad black lower boundary with white belly. However variable, with some males much duller, head pattern much as non-breeding and pale chestnut breast. Female with paler chestnut breast-band, lacking black lower border. Legs brighter, more orange than non-breeding. The majority of first breeding do not replace any primaries.

Call Bubbly *chip-ip-chip*.

Status, habitat and distribution Breeds in dry steppe and desert in southern Siberia, northern Mongolia, east as far as north-east China. Most spend non-breeding season in northern and central Australia; scarce in remainder of Australia and New Zealand, also in Indonesia and Papua New Guinea. Vagrant to Greenland, Finland, Andaman Islands, Thailand, Japan and South Korea.

Racial variation None.

Similar species Greater Sand Plover (44) and Caspian Plover (45), which see.

References Arkhipov (2005), Branson & Minton (2007), Barter (2002), Cox (1988a), Dunn (2003), Marchant & Higgins (1993), Ogle (1992), Ozerskaya & Zabelin (2006).

▼ **46a. Juvenile**. China, early April. The early date is surprising, but the neat buff-fringed upperparts and wing-coverts show this to be an undoubted juvenile. Yuan Xiao.

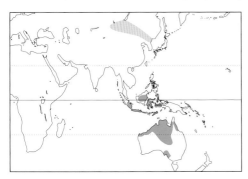

▲ **46b. Non-breeding**. Western Australia, late October. Has two generations of wing-coverts, which could be either juvenile/first non-breeding or adult breeding/non-breeding. The earlier generation of feathers are those with off-white fringes. Rohan Clarke.

▲ **46c. Adult non-breeding**. East Timor, mid November. Upperparts as first non-breeding (46b), but wing-coverts larger and more irregular. Colin Trainor.

▲ **46d. Adult male breeding**. Mongolia, mid June. Males are very variable, this individual having a particularly white head; some have a dark rear-crown. Jari Peltomaki.

◀ **46e. Adult male breeding.** China, early April. A much duller individual; it is possible that individuals like this are first-breeding birds. Yuan Xiao.

◀ **46f. Adult breeding female in flight.** Mongolia, early June. Note reduced contrast of the breast-band compared to breeding male (46d). This bird has already started wing moult, in spite of the early date. Jari Peltomäki.

◀ **46g. Oriental Plover in flight.** Mongolia, early June. Uniformly brown above apart from the white outer primary shaft. Jari Peltomäki.

47. EURASIAN DOTTEREL
Charadrius morinellus

An often confiding Eurasian plover
of open, sparsely vegetated areas.

Identification L 21cm (8.5"); **WS** 46cm (18").
Small, with rounded head, prominent dark eye
and short black bill, medium-length neck and
medium-length yellowish legs. Rather scaly brown
upperparts, dark crown, broad pale supercilium
meeting in V at hindneck. The underparts are
yellowish-buff with a pale crescentic breast-band;
dark belly in breeding adult. *In flight* from above
uniformly brown; pale below, with dark belly of
breeding adult clearly visible; feet do not extend
beyond tail. Sexes same size but can differ slightly
in plumage when breeding; plumage varies seasonally and with age. Narrow web along toe between
outer and middle toes only; no hind toe. *Feeds* by
picking from ground or vegetation, in rather less of
a 'stop-start' manner than other plovers.

Juvenile Crown dark, with buff supercilium.
Mantle and scapulars dark brown, with strongly
contrasting fairly broad whitish fringes, incomplete at feather tips; wing-coverts similar, but less
contrasted. Indistinct pale breast-band; underparts
buff with strongly mottled foreneck; lower belly and
undertail-coverts white.

First non-breeding/adult non-breeding Mantle,
scapulars and tertials more uniform than juvenile,
with grey-brown centres and orange-buff fringes
giving overall paler appearance. First non-breeding
retains some juvenile wing-coverts and tertials,
which are more contrasting than the remainder of
the upperparts. Adult commences moult to non-
breeding from August, completing in non-breeding
area; first and adult non-breeding is rarely seen away
from non-breeding grounds.

First-breeding/adult breeding Some may breed
in second calendar year, remainder in third.
Crown dark, contrasting with white supercilium;
throat white, neck and upper breast greyish, white
breast-band having dark outline. Lower belly dark
chestnut-brown. Sexes are basically similar and
often difficult to separate but female is generally
brighter and more contrasted than male. First-
breeding acquires variable amounts of dark belly
feathers.

Call A quiet trill, given particularly at take-off.

Status, habitat and distribution Uncommon,
breeding on Arctic or montane tundra in Scotland,
Scandinavia, Finland, and from northern Russia
to far north-eastern Siberia. Often occurs in small
flocks (or 'trips') on migration, when it uses tradi-
tional staging locations. Whole population migrates
to semi-desert in Mediterranean and North Africa
east to Iran. Scarce or vagrant in Europe outside its
breeding range; also vagrant to the Aleutians and
along the Pacific coast of North America, south to
California and Mexico.

◄ **47a. Juvenile**. Southwest
England, mid September. In
complete juvenile plumage apart
from one newly acquired, buff-
fringed non-breeding scapular.
Juvenile feathers have more
contrast than non-breeding, and
have irregular pale fringes that do
not extend to feather tip. RJC.

Racial variation No races recognised.

Similar species Could be confused with any of the three slightly larger golden plover species (49-51), but Dotterel's prominent supercilium, whitish breast-band and yellowish legs should make identification straightforward. Sociable Lapwing (59) shares prominent supercilium meeting at nape, but is larger and has dark legs and a completely different flight pattern.

Reference Thorup (2006).

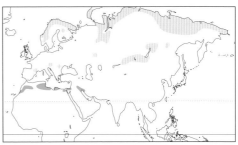

▲ **47b. Juvenile**. Southwest England, mid September. Wing-stretching, showing feather detail. A different individual from 47a. RJC.

▲ **47c. First non-breeding**. Oman, early January. The upper tertial is non-breeding, with a buff fringe, but underlying it there are at least two white-fringed juvenile tertials. Hanne and Jens Eriksen.

▼ **47d. Adult breeding**. Scotland, mid June. Note the dark chestnut-brown lower belly. RJC.

▼ **47e. Adult breeding in flight**. Finland, mid June. Markus Varesvuo.

48. MOUNTAIN PLOVER
Charadrius montanus

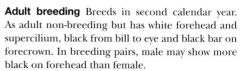

A plover of sparsely vegetated upland areas in North America. Vulnerable.

Identification L 22cm (9"); **WS** 47cm (18.5"). Medium-sized, longish necked with round head, grey-brown crown; short, black, fairly pointed bill and long pale buff or grey legs. Upperparts buff-brown, sides of neck and breast buff, remainder of underparts white. *In flight* from above has white wing-bar on inner primaries and broad black band on white-tipped tail; underwing white, toes just reach tail. Sexes similar; plumage varies with age and season. Lacks hind toe. *Feeds* in typical plover manner with short runs, sometimes probing, typically on ground that is either bare or with very short vegetation; prey includes small insects and other invertebrates. Usually in loose flocks when non-breeding.

Juvenile Crown lightly flecked white, supercilium poorly developed; upperparts and wing-coverts brown with medium-buff fringes; sometimes with obscure, dark, subterminal bars. Brown-black primaries have buff fringes at tips when fresh. This plumage may be retained for a significant period but detailed data are lacking.

First non-breeding/adult non-breeding First non-breeding retains much of juvenile plumage. Adults have plain sandy brown upperparts and coverts with dull rufous fringes, so that when characteristic juvenile feathers wear or are lost the distinction between the two age classes is difficult.

First-breeding As adult breeding, but retains some juvenile wing-coverts and may perhaps still show buff-fringed primary tips.

Adult breeding Breeds in second calendar year. As adult non-breeding but has white forehead and supercilium, black from bill to eye and black bar on forecrown. In breeding pairs, male may show more black on forehead than female.

Call Often silent on ground away from breeding areas but shrill *kip* may be given in flight.

Status, habitat and distribution Scarce, breeding locally in sparsely vegetated uplands in the United States, in Montana, Wyoming and Colorado south to New Mexico and Texas. Non-breeding birds occur from central California and southern Texas to north and central Mexico, usually on open fields. Vagrant within the United States to Florida and Oregon.

Racial variation None.

Similar species None. Lacks breast-band or breast-side patches of most other similar-sized plovers.

Reference Knopf & Wunder (2006).

◀ **48a. Juvenile**. California, early November. Aged by neat, small feathers; this plumage is retained for some time. RJC.

◀ **48b. Adult non-breeding**.
California, early November. Similar
to juvenile, but feathers less even in
size, and fresher, the darker centres
showing more contrast with the
buff fringes. RJC.

◀ **48c. Adult breeding**. Colorado,
mid April. Breeding adult gains
black on the lores and fore-crown.
RJC.

▼ **48d. Mountain Plovers in
flight**. Colorado, late July. Most
of the birds are adults in wing
moult; the others are juveniles.
Bill Schmoker.

PLUVIALIS PLOVERS

All four *Pluvialis* (or 'tundra') plovers occur in the northern hemisphere. These are American Golden Plover *Pluvialis dominica*, Pacific Golden Plover *P. fulva*, European Golden Plover *P. apricaria* and Grey or Black-bellied Plover *Pluvialis squatarola*. They are all migratory Arctic or subarctic breeding species.

These larger plovers share many of the characters of the smaller plovers, though they tend to be less compact, with longer necks, and are more attenuated at the rear. All have short bills and rounded heads, and share the plover 'start-stop-tilt' feeding style.

The moult sequences are largely similar to other shorebirds, though unusually American Golden Plover replaces its primaries in its first non-breeding season, and many first non-breeding Grey Plovers retain much worn juvenile plumage. They all have finely spotted or 'spangled' upperparts as juveniles and, in the case of the three golden plovers, have very similar first non-breeding and adult non-breeding plumages. Ageing of these three species is difficult but the juvenile of each has a characteristic mottled (in fact spotted or barred) breast and belly. Grey Plovers are more easily aged, particularly when non-breeding, by their many retained juvenile feathers.

In breeding plumage the *Pluvialis* plovers all have a striking black face, throat, breast and belly, males with more black than females. Both American and Pacific Golden Plovers are monotypic, but European Golden Plover has two races: nominate *apricaria*, which breeds in Britain, Ireland and southern Scandinavia, and *altifrons*, which breeds in Iceland, northern Scandinavia, Finland and Russia. These two taxa are perhaps best regarded as forms, rather than races, differing mainly in the extent of black on the face. Both taxa commence body moult to breeding plumage at the same time, but *apricaria*, which breeds earlier and thus has less time to acquire its breeding plumage, has a reduced amount of black on its face.

Some European Golden Plovers start their post-breeding moult (both wing and body) on the breeding grounds, when a few black breast feathers are replaced by feathers with yellow and brown stripes, and male *altifrons* start to lose their intensely black faces. These yellow breast feathers, rather than the off-white feathers of the non-breeding plumage, seem to be a consequence of the hormonal state of the breeding birds, resulting in brighter feathers than are produced later in the moult to non-breeding.

Table 6. Structural differences are useful for identifying the golden plovers. Details apply to both juveniles and adults. Note that in Pacific Golden Plover the tips of the longest primaries (9 and 10) lie close together, but they are slightly more spaced in European Golden Plover and more so again in American Golden Plover. The primary extension beyond the tail is < ½ bill length in Pacific and European Golden Plovers, > ½ bill length in American Golden Plover. Adults of all three species, and also first non-breeding American Golden Plovers, moult their outer primaries (and probably tertials) on the non-breeding grounds.

Points of comparison	American Golden Plover *P. dominica*	Pacific Golden Plover *P. fulva*	European Golden Plover *P. apricaria*
Number of primary tips beyond tertials	4–5	2–3	usually 4
Number of primary tips beyond tail (distance beyond tail)	2–3 (12–22mm)	0–2 (0–9mm)	1–3 (<10mm)
Tertial length relative to tail	extends variably from middle of tail to near tail-tip	usually reaches almost to tail-tip	extends to about mid-tail
Toes relative to tail in flight	reach end of tail, sometimes marginally beyond	reach beyond tail	do not reach end of tail

49. AMERICAN GOLDEN PLOVER
Pluvialis dominica

A strongly migratory North American golden plover.

Identification L 26cm (10.5"); **WS** 54cm (21"). Needs care to separate from Pacific Golden Plover (50), with which it forms a species pair; see Table 6 (p. 172). All individuals may not be specifically identifiable. Medium-sized, round-headed with a prominent dark eye, short black bill and medium-length black legs. Underparts, including flanks and undertail, largely black in adult breeding, and folded wings which usually extend noticeably beyond tail, aid separation from Pacific Golden. *In flight* from above is uniform apart from narrow white bar on outer wing; below underwing (including axillaries) is buff-grey; toes do not usually reach beyond tail. Sexes of similar size; plumages vary with age, sex and season. Small web between outer and middle toes, lacks hind toe. *Feeds* in typical plover manner. Often in flocks when non-breeding.

Juvenile Head and neck greyish, white supercilium; upperparts, especially scapulars, greater coverts and tertials, spotted or spangled dull gold; remainder spangled white. Pale brown barring on belly.

First non-breeding/adult non-breeding As juvenile but with limited spangling on upperparts; foreneck and breast grey; belly white in adults. First non-breeding replaces primaries but retains some barring on underparts, particularly flanks.

First-breeding/adult breeding Many gain largely complete breeding plumage (perhaps later than adults) and apparently breed in their second calendar year. Upperparts strongly spangled gold and white; a continuous white area, composed of forehead, supercilium, broadening on sides of neck (but not extending onto flanks as in Pacific Golden Plover) separates the upperparts from the black underparts, which extend from face and foreneck over whole of breast, flanks and belly to include the undertail-coverts. Female usually has some white feathers on underparts but can be nearly as dark as male. The species-diagnostic pattern of white and black on neck and flanks can often still be seen late in the season when in advanced body-moult to non-breeding.

Calls Disyllabic, fairly high pitched *too-ee*, and a plaintive *too-oo-eet*, often with the emphasis on the first syllable, similar to but perhaps higher-pitched than Pacific Golden Plover.

Status, habitat and distribution A fairly common inland species, breeding in North American Arctic and subarctic tundra from west and north-west Alaska (where sympatric with Pacific Golden Plover) east to Baffin Island. Migrates south through interior North America or via Hudson Bay to reach the coast around New England, then over the Atlantic to South America; non-breeding season is spent in inland grassland and wetlands in northern Argentina and Uruguay, with the birds returning north via the Mississippi and Missouri valleys. Uncommon away from breeding and non-breeding areas. Vagrant to Europe, most frequently in September and October; also to northern and western Africa.

Racial variation and hybridisation No races recognised. May possibly hybridise with Pacific Golden Plover in Alaska but good evidence is lacking. An American Golden Plover was recorded nesting with a Grey Plover on the Chukotka Peninsula in 2005, though the clutch was taken by a predator. This was also the first breeding record for American Golden Plover in the Palearctic.

Similar species Eurasian Dotterel (47), which see. From Pacific Golden Plover (50) by structure (summarised in Table 6, p. 172). Juvenile American is usually greyer, duller, much less yellow in appearance and has a more barred belly; when breeding has black on flanks and undertail-coverts (Pacific Golden has white flanks and undertail-coverts). Breeding Pacific moults earlier to non-breeding than American Golden Plover, so individuals still largely in breeding plumage in late September are likely to be American Golden Plover. From European Golden Plover (51) by structure (see Table 6),

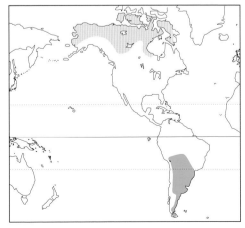

smaller size, more prominent pale supercilium, proportionally longer legs, and buff-grey (not white) underwing-coverts and axillaries. From Grey Plover (52) by smaller size, less bulky head and bill, less grey appearance in all plumages, and different flight pattern lacking the diagnostic black axillaries of Grey Plover.

References Byrkjedal & Thompson (1998), Alexander-Marrack (1992), Alström (1990), Connors (1983), Connors *et al.* (1993), Dunn *et al.* (1987), Fishpool & Demey (1991), Golley & Stoddart (1991), Jaramillo (2004), Johnson & Connors (1996), Johnson & Johnson (2004), McGeehan & Meininger (2000), Taldenkov (2006).

▲ **49a. Juvenile**. Southwest England, mid October. Aged by neatly edge-spotted upperparts, and uniform greyish spotting on belly. Juvenile American Golden Plovers are generally much greyer than juvenile Pacific, which are extensively spotted golden-yellow above. Gordon Langsbury.

▼ **49b. Adult non-breeding**. Florida, mid March. Note long primary extension beyond tertials, which aids separation from Pacific Golden Plover. Has just started to acquire a few dark upperpart breeding feathers. RJC.

▼ **49c. Adult male breeding**. Alaska, early June. Lacks the white line down the flanks shown by adult breeding Pacific Golden Plover. René Pop.

▲ **49d. Adult female breeding**. Alaska, June. Female American Golden Plovers are often very similar to males (more so than in Pacific), though less intensely black, paler in the face and with a few white underpart feathers. Peter LaTourrette.

▼ **49e. Adult breeding male in flight**. Alberta, Canada, mid April. The upperwing flight pattern with a narrow white wing-bar is common to all three golden plovers. Gerald Romanchuk.

▼ **49f. Juvenile showing underwing**. Alberta, Canada, mid September. The buff-grey axillaries that help separate American and Pacific Golden Plovers from European Golden Plovers (which have white axillaries; see 51e) are well shown here, as is the neat but faint barring on breast and flanks that characterises juveniles of all three species. Gerald Romanchuk.

50. PACIFIC GOLDEN PLOVER
Pluvialis fulva

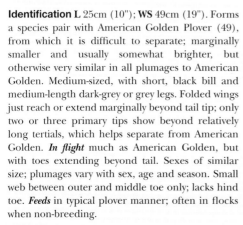

A largely Siberian equivalent of American Golden Plover which is also strongly migratory.

Identification L 25cm (10"); **WS** 49cm (19"). Forms a species pair with American Golden Plover (49), from which it is difficult to separate; marginally smaller and usually somewhat brighter, but otherwise very similar in all plumages to American Golden. Medium-sized, with short, black bill and medium-length dark-grey or grey legs. Folded wings just reach or extend marginally beyond tail tip; only two or three primary tips show beyond relatively long tertials, which helps separate from American Golden. *In flight* much as American Golden, but with toes extending beyond tail. Sexes of similar size; plumages vary with sex, age and season. Small web between outer and middle toe only; lacks hind toe. *Feeds* in typical plover manner; often in flocks when non-breeding.

Juvenile Head and neck strongly washed yellow, with yellowish supercilium; upperparts extensively spangled gold. Limited barring on sides of belly.

First non-breeding/adult non-breeding Duller than juvenile, with less yellow especially on fore-neck, and less obvious spangling. First non-breeding retains some barring on belly; adult has a more uniformly pale greyish-brown belly.

▼ **50a. Juvenile.** Thailand, late November. Still largely in juvenile plumage in spite of the late date. Pacific Golden Plovers often look larger-headed and thicker-billed than American, and have a generally 'yellow' appearance, juveniles particularly so. RJC.

First-breeding Some do not breed, and remain in the non-breeding area, largely or completely in non-breeding plumage; others acquire full breeding plumage and migrate north. All retain worn juvenile primaries, replaced from June by non-migrants, later (on same schedule as adults) by migrants, so birds with very worn primaries moulting to non-breeding are second calendar year.

Adult breeding As American Golden Plover, but supercilium is broader and the white area separating upperparts from the black underparts extends continuously from side of neck, along flanks (with a few black bars) to undertail-coverts, which are mainly white. Younger birds may be less well marked. Females have more white on underparts than males, some extensively, and are typically less black below than female American Golden Plover. The species-diagnostic pattern of white and black on the neck and flanks can often still be seen late in the season when in advanced body-moult to non-breeding.

Calls Much as American Golden, *chu-it*, recalling Spotted Redshank, and an extended *chu-ee-uh*, with emphasis on second syllable.

Status, habitat and distribution Breeds in Arctic tundra in Siberia east from the Yamal peninsula to western Alaska. Found non-breeding coastally from East Africa, Arabia, to south and south-east Asia, the Pacific islands and Australasia; small numbers in southern California. Vagrant in Western Europe, West Africa.

Racial variation and hybridisation No races recognised; see hybridisation under American Golden Plover. Forms (possibly hybrids) intermediate between Pacific and European Golden Plover occur; these are smaller than European, with the latter's

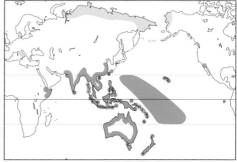

white axillaries, but with other characters of Pacific.

Similar species Eurasian Dotterel (47) and American Golden Plover (49), which see; see also Table 6, p. 172.

References Alström (1990), Byrkjedal & Thompson (1998), Connors (1983), Connors *et al.* (1993), Dunn *et al.* (1987), Fishpool & Demey (1991), Golley & Stoddart (1991), Jaramillo (2004), Johnson & Connors (1996), Johnson & Johnson (1983, 2004), Meadows (2003), Pym (1982), Savage & Johnson (2005), Vinicombe (1988), Wallace (1983).

◄ **50b. Non-breeding**. South Korea, May. Those that do not breed in their second calendar year remain south of the breeding grounds in non-breeding plumage. Aurélien Audevard.

◄ **50c. Adult male breeding**. China, late April. Note the white line down the flanks (not shown by adult breeding American Golden Plover); the black on the face will be even darker by June. Short primary extension beyond tertials, and restricted projection beyond tail, separate from American. Yuan Xiao.

▼ **50d. Adult female breeding**. Alaska, early June. Has much less black on the face and belly than the male. René Pop.

◀ **50e. Adult moulting to non-breeding**. Queensland, Australia, mid September. Although there is now much white in the belly, the impression of the white line down the flanks of breeding plumage still remains. RJC.

◀ **50f. Adult.** Hong Kong, mid April. Wing-stretching, showing flight pattern. Probably still moulting to breeding plumage, making it difficult to sex at this stage. Martin Hale.

▼ **50g. Non-breeding Pacific Golden Plovers in flight**. Eastern India, late March. Narrow white wing-bar and buff-grey axillaries are shared by American Golden Plover, but toes extend beyond tail (not usually the case in American). European Golden Plover has white axillaries. Sumit Sen.

51. EUROPEAN GOLDEN PLOVER
Pluvialis apricaria

Largely European in distribution year-round, this is the largest of the three golden plovers.

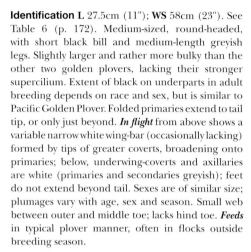

Identification L 27.5cm (11"); **WS** 58cm (23"). See Table 6 (p. 172). Medium-sized, round-headed, with short black bill and medium-length greyish legs. Slightly larger and rather more bulky than the other two golden plovers, lacking their stronger supercilium. Extent of black on underparts in adult breeding depends on race and sex, but is similar to Pacific Golden Plover. Folded primaries extend to tail tip, or only just beyond. *In flight* from above shows a variable narrow white wing-bar (occasionally lacking) formed by tips of greater coverts, broadening onto primaries; below, underwing-coverts and axillaries are white (primaries and secondaries greyish); feet do not extend beyond tail. Sexes are of similar size; plumages vary with age, sex and season. Small web between outer and middle toe; lacks hind toe. *Feeds* in typical plover manner, often in flocks outside breeding season.

Juvenile The whole of the upperparts are brown, spangled gold, with some white spotting intermixed, particularly on the mantle. Scapulars and tertials have a broad dark shaft-streak reaching feather tip. Breast is closely spotted, with brown barring on paler belly.

First non-breeding/adult non-breeding Upperparts much as juvenile, though less bright; breast streaked, rather than spotted. Scapulars and tertials have thin or absent shaft-streak at tip. First non-breeding retains some juvenile wing-coverts and perhaps some barring on flanks; adults have white belly and lower flanks. Difficult to age in the field in the absence of flank barring.

Adult breeding Most probably breed in second calendar year; plumage as adult breeding. Face, throat, foreneck, breast and belly black or brown, depending on race, dark areas having white border; mantle and scapulars spangled bright gold and

(to a lesser extent) white. In breeding pairs, face pattern of female is generally paler, and belly less extensively black with scattered white feathers. Some individuals start their post-breeding moult (of both wings and body) on the breeding grounds, gaining a few yellow-and-brown striped breast feathers; at the same time males of the northern form *P. a. altifrons* start to show reduced black in the face.

Call A clear, yodelled *loo-ee*.

Status, habitat and distribution A fairly common species, breeding on Arctic and subarctic tundra in Iceland, northern Scandinavia, Finland and northern Russia (*altifrons*), and Britain, Ireland and southern Scandinavia (nominate *apricaria*). Non-breeding birds occur on grassland, often inland, in Britain and Ireland, Netherlands, and south to North Africa. Regular during migration in southern Greenland, vagrant in eastern Canada, rare vagrant in Alaska.

Racial variation The two races, which differ only when breeding, are probably best regarded as forms. Individuals vary but typical *altifrons* male has prominent white supercilium, black face and

▼ **51a. Juvenile**. Wales, late October. Has the mottled breast and upper belly of the other juvenile golden plovers, but is larger, bulkier, and lacks their obvious pale supercilia. RJC.

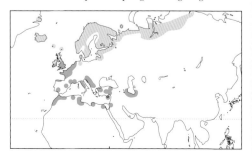

foreneck, while *apricaria* is duller, with a less obvious supercilium and brown face and foreneck. Males of form *apricaria* can be less well marked than some *altifrons* females.

Similar species Eurasian Dotterel (47); American (49) and Pacific Golden Plovers (50), all of which see; see also Table 6 (p. 172).

References Byrkjedal & Thompson (1998), Engelmoer & Roselaar (1998), Jukema & Piersma (1987), Piston & Heinl (2001).

▲ **51b. Adult moulting to non-breeding.** Southwest England, early September. Has lost its dark face and much of its black underparts. Primary moult is almost complete; note new dark primary adjacent to a remaining faded one. RJC.

▼ **51c. Adult male breeding** *altifrons*. Iceland, late June. Males of this form have completely black faces, which start to fade from mid-June. RJC.

▲ **51e. Adult male breeding *altifrons***. Iceland, late June. The white axillaries help to distinguish European from the other two golden plovers, which have buff-grey axillaries (49f). European Golden Plovers not only commence body moult on the breeding grounds, but also primary moult; this individual is in the process of replacing its inner primaries, an earlier stage of the moult shown by 51b. RJC.

▲ **51d. Adult female breeding *altifrons***. Iceland, late June. The facial pattern of the female in this form is similar to males of the southern form *apricaria*. The yellow feathers on the belly are newly grown on the breeding grounds. RJC.

▼ **51f. Adult breeding in flight**. Eastern England, mid April. Showing the narrow wing-bar of this species; the flock probably includes both *apricaria* and *altifrons*. RJC.

52. GREY PLOVER
Pluvialis squatarola

Near-circumpolar breeding species, and more coastal than the other *Pluvialis* plovers on breeding grounds. Alternative name: Black-bellied Plover.

Identification L 28.5cm (11.5"); **WS** 59cm (23"). Medium-sized and round-headed, with a prominent dark eye; short, slightly bulbous, black bill and medium-length grey legs, darker when breeding. Larger and more bulky than the superficially similar golden plovers; basically grey in all plumages and, as with the golden plovers, has black underparts when breeding. *In flight* from above has outer wing blackish, with broad white wing-bar, inner wing uniformly grey; rump white, tail barred grey. Below whitish but shows striking diagnostic black axillaries, which in adult breeding are contiguous with the black underparts; toes reach just beyond tail-tip. Sexes of similar size, plumages vary with age and sex. The only *Pluvialis* plover with a (rudimentary) hind toe; some webbing between all toes. *Feeds* in typical, deliberate, stop-watch-pick plover manner.

Juvenile Darkish grey-brown crown, white supercilium; mantle, scapulars and tertials dark grey-brown with extensive, very pale yellowish, edge-spotting when fresh, giving whitish spangled appearance at a distance; wing-coverts similar, but paler and duller; foreneck and breast finely but uniformly streaked. Some individuals retain the majority of this plumage at least until the end of the year.

First non-breeding/adult non-breeding Head pattern less contrasting than juvenile; upperparts pale to mid-grey, irregularly notched and fringed whitish. Neck and upper breast mottled or streaked grey, remainder of underparts white. First non-breeding usually retain a variable proportion (sometimes nearly all) of darker, worn, juvenile upperparts, wing-coverts and tertials.

First-breeding Most probably do not breed until their third calendar year, and remain south of their breeding grounds. Often much as adult non-breeding but the extent of this plumage is apparently variable; some gain a few (rarely many) breeding feathers on both upperparts and underparts, others appear to moult directly to adult non-breeding.

Adult breeding Crown white, variably streaked greyish, nape and sides of neck white, upperparts with many black-and-white barred feathers; face, foreneck, breast and belly black, undertail-coverts white. Underparts black, with a few white-fringed feathers when fresh. Female acquires fewer breeding feathers on upperparts, underparts brownish-black with many whitish feathers, face is often much less black than male. Since breeding females can be similar to incompletely moulted males, plumage differences can only be used with confidence for sexing females on the breeding grounds. Legs dark grey or black.

Call Plaintive *pee-oo-eee*, with downward inflection on middle syllable.

Status, habitat and distribution A fairly common species, with a near-circumpolar Arctic breeding distribution (though not Greenland or northern Scandinavia), breeding on the tundra; non-breeding found coastally on mudflats and beaches from mid-latitudes in northern hemisphere south to South America, South Africa, India, south-east Asia and Australia; rare New Zealand. Females on average migrate earlier, and spend non-breeding season further south than males. Vagrant to Greenland, Iceland and Spitsbergen.

Racial variation and hybridisation Three races recognised but the differences are small; nominate *squatarola* from Russia east to Alaska, *tomkovichi* on Wrangel Island and *cynosurae* in northern Canada. All are very similar but *cynosurae* is marginally smaller in most measurements, the breeding male is often more contrastingly black-and-white, and juveniles typically have larger pale spots on upperparts. For hybrid American Golden Plover x Grey Plover, see American Golden Plover (49).

Similar species Separated from the golden plovers (49–51) by larger size, heavier bill and greyer appearance in all plumages and, particularly, by flight pattern with white rump and diagnostic

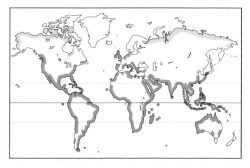

black axillaries. When breeding, extent of black most resembles American Golden Plover (49), but undertail-coverts are white.

References Byrkjedal & Thompson (1998), Engelmoer & Roselaar (1998), Howell & Pyle (2002), Loftin (1962), Paulson (1995).

◄ **52a. Juvenile**. Florida, mid October. The yellowish spots on the upperparts gradually fade to white. RJC.

◄ **52b. Juvenile**. Florida, mid December. Juvenile plumage is typically retained until the year-end, by which time it is faded and worn, so is less contrasting and somewhat untidy. The retained (though worn) juvenile pale-notched tertials are a useful feature for ageing at this stage. RJC.

◄ **52c. First-breeding**. Florida, late June. Birds of this age remain south of the breeding grounds. Most seem to slowly replace their juvenile feathers with either a few black breeding-type feathers and/or grey adult non-breeding type feathers. RJC.

▲ **52d. Adult non-breeding**. Eastern England, mid December. The black feather protruding from beneath the wing is a misplaced axillary; see 52g. RJC.

▼ **52e. Adult breeding male**. Florida, early May. Note the near-black legs, which are grey when non-breeding. RJC.

◄ **52f. Adult breeding female**.
South Korea, early May. Females
can show very variable amounts
of breeding plumage, though this
individual is fairly typical; legs are
rarely quite so dark as those of the
male. Caution is needed in sexing
females away from the breeding
grounds, as individuals like this
might be incompletely moulted
males, or birds that will not breed.
RJC.

◄ **52g. Non-breeding Grey Plover
in flight**. Florida, mid December.
Showing the diagnostic black
axillaries. RJC.

▼ **52h. Adult breeding in flight.**
South Korea, early May. Males
front and rear, possible female
in the middle, and the bird in the
background is in non-breeding
type plumage. RJC.

LAPWINGS

Ten species of lapwings occur in the northern hemisphere: Spur-winged Lapwing *Vanellus spinosus*, River Lapwing *V. duvaucelii*, Black-headed Lapwing *V. tectus*, Yellow-wattled Lapwing *V. malabaricus*, Grey-headed Lapwing *V. cinereus*, Red-wattled Lapwing *V. indicus*, Sociable Lapwing *V. gregarius*, White-tailed Lapwing *V. leucurus*, Southern Lapwing *V. chilensis* and Northern Lapwing *V. vanellus*. There are 23 lapwing species worldwide, in one genus.

The lapwings are a distinctive group of large plovers. As with all plovers they have relatively short bills and, apart from the crested Black-headed, Southern and Northern Lapwings, all have rounded heads. They have broad wings and a 'flappy' flight, particularly so in the case of Northern Lapwing, from which the name 'lapwing' is derived. Several lapwings have carpal spurs, the function of which is unknown. Like all plovers they have the 'start-stop-tilt' feeding style and they also regularly use foot-trembling to obtain prey.

The majority of the northern hemisphere lapwings have, as adults, either similar plumage year-round or only very minor differences in breeding plumage. The exceptions are Grey-headed, Sociable and Northern Lapwings, which have distinctive breeding plumages, and in the latter two species the sexes can be separated when breeding. The juveniles of all species have distinctive plumages, but in most it is difficult to separate first non-breeding from adult non-breeding, an exception being Grey-headed Lapwing, where the first non-breeding bird lacks the dark lower boundary to the grey breast and has a dark, not red, iris.

The migratory status of the northern hemisphere lapwings is variable. Some (Spur-winged, Grey-headed, Sociable, White-tailed and Northern Lapwings) are migratory, others (including River, Yellow-wattled and Red-wattled Lapwings) are largely sedentary. Of the remainder, Southern Lapwing is a South American species that occurs north to Panama, and Black-headed Lapwing has been recorded as a vagrant (Israel in the 19th century) from sub-Saharan Africa.

53. SPUR-WINGED LAPWING
Vanellus spinosus

A striking African lapwing with black carpal spurs, reaching the Nile Valley, eastern Mediterranean and Middle East.

Identification L 27cm (10.5"); **WS** 64cm (25"). Medium-sized, with black forehead, crown, and nape; face, sides and back of neck bright white, black of chin and throat joins black belly. Short black bill. Upperparts and folded wings grey-buff; the breast, upper belly and flanks black, rest of underparts white; very long, dark grey-black (rarely mid-grey) legs. *In flight* from above primaries and secondaries and broad band at end of tail black, white greater coverts form band separating grey-buff back and inner/median coverts from black primaries and secondaries; rump and tail-coverts white, feet extend beyond tail. Sexes similar in size with similar plumage year-round; juvenile separable. Has black carpal spurs, usually only seen when in flight; often lacks hind toe. *Feeds* in typical lapwing manner, sometimes in small flocks when non-breeding.

Juvenile A duller version of the adult, with black areas tinged brownish and flecked white; upperparts and wing-coverts with wide pale fringes; lacks adult's prominent longer scapulars.

First non-breeding Quickly becomes indistinguishable from adult in the field.

Adult Age at first breeding not known. Adult described in Identification; grey-brown scapulars are elongate, overlapping folded wing. Carpal spurs longer in male than female.

Call Noisy when breeding; alarm call *trek, trek, trek.*

Status, habitat and distribution Relatively common on near-coastal drier areas and (more usually) on inland, sometimes brackish, wetlands and irrigated areas. Visitor to Greece, Turkey, Cyprus and Syria in breeding season; year-round in Egypt and eastern Mediterranean, Nile valley, Arabia and sub-Saharan Africa north of equator from Senegal to Ethiopia and Kenya; occasional or vagrant to Tanzania, Iran and western Europe, where birds generally thought to be escapes occasionally occur.

Racial variation and hybridisation None. Apparently hybridises with Blacksmith Plover *Vanellus armatus*, a southern hemisphere African species, in central Kenya, where the ranges of the two species just overlap.

Similar species White-tailed Lapwing (60) has pale, plain head, very long bright yellow legs, lacks strongly pied appearance of Spur-winged, and shows a short white tail in flight. Red-wattled Lapwing (58) has a somewhat similar pied appearance and flight pattern but has narrow red wattle from eye to forehead, a largely red bill, and very long yellow legs. Spur-winged Lapwing is unlikely to occur with closely related River Lapwing (54) of north-east India and south-east Asia, which has a similar flight pattern, but would be quickly separated by its pale (not black) breast and belly.

References Gantlett (1997), Pearson (1983), Thorup (2006).

▼ **53a. Immature**. Kenya, mid March. RJC.

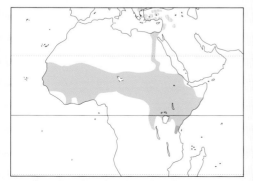

▲ **53b. Adult**. Cyprus, early April. Males and females are indistinguishable on the ground when the spurs are not visible; both have similar plumages year-round. RJC.

▼ **53c. Adult male in flight**. The Gambia, mid January. Note the spurs at the bend of the wing, which are not usually visible unless bird is in flight; their length shows this to be a male. RJC.

▼ **53d. Adult female in flight**. Israel, early April. The female has very short or (as here) vestigial spurs. RJC.

54. RIVER LAPWING
Vanellus duvaucelii

A distinctive Asian lapwing strongly associated
with rivers and lakes.

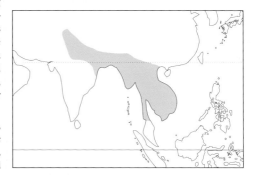

Identification L 31cm (12"); **WS** 62cm (24.5").
Medium-sized, with black face and black throat
extending to upper breast; extensive black crest,
usually held folded down nape giving a large-headed
appearance. Sides of head pale grey, separated from
black areas by diffuse white line, short black bill,
upperparts pale buff-brown. Underparts with broad,
buff, breast-band with darker lower edge; small black
patch between legs, remainder of underparts white;
long black legs. When standing usually has hunched,
horizontal stance; raises crest only rarely. *In flight*
from above primaries and secondaries and band at
end of tail are black, greater coverts white, separating
buff back and inner/median coverts from black primaries and secondaries; rump and uppertail-coverts
white, feet extend beyond tail. Sexes of similar size
and plumage year-round; juvenile separable. Black
carpal spurs usually only seen in flight. Lacks hind
toe. *Feeds* in typical lapwing manner on snails,
worms, insects, small fish, etc. Usually seen singly or
in pairs.

Juvenile A duller version of the adult plumage,
with duller bare-part colours. Black areas of the
head tinged brownish; crest shorter than adult,
extending less far down the nape; upperparts and

wing-coverts with pale fringes, breast-band lacks
darker lower margin.

First non-breeding Quickly becomes indistinguishable from adult in the field.

Adult Age at first breeding not known. Adult
plumage described in Identification. Carpal spurs
longer in male than female.

Calls A high pitched *kek, kek, kek, ..., keer, keer, keer,
...,* and *keer-ker rit.*

Status, habitat and distribution Bare areas (such
as sandbanks) adjacent to larger rivers and lakes

▼ **54a. Adult**. India, late December. The Asian equivalent of Spur-winged Lapwing, with similar upperparts. RJC.

to 1,000m altitude, in north-central India, Nepal, south-east Asia to northern Thailand and south-east China. Largely sedentary, but may make seasonal, weather-induced movements.

Racial variation None.

Similar species Spur-winged Lapwing, which see. Red-wattled Lapwing (58) has a similar flight pattern, but has a largely red bill, a small red wattle, and bright yellow legs.

Reference Ali & Ripley (1983).

▲ **54b. Adult**. India, late December. Has a large crest which is raised when alarmed. RJC.

▼ **54c. Adult in flight**. North India, late April. The upperwing pattern is similar to that of Spur-winged Plover (53c). Sumit Sen.

55. BLACK-HEADED LAPWING
Vanellus tectus

An elegant, crested African lapwing, vagrant to the region.

Identification L 25cm (10"); **WS** 59cm (23"). Medium-sized; crown black, with slender black crest at rear, white forehead (but see below), tiny red wattle at base of short, black-tipped, red bill; iris yellow. White from throat to nape forms hind-neck collar; black around neck extends narrowly down middle of breast. Upperparts plain brown; sides of breast variably pale grey, buff or off-white; underparts white; very long red legs. Narrow vertical black breast-stripe diagnostic amongst lapwings. *In flight* broad white area formed by primary coverts and greater secondary coverts separates brown mantle and secondary coverts from black flight feathers; rump white, tail largely black with narrow white terminal band, toes just extend beyond tail. Sexes similar in size, plumage same year-round; juvenile separable. Lacks hind toe. *Feeds* in usual plover manner, on insects, snails and other invertebrates.

Juvenile Similar to adult but with duller bare-part colours, shorter crest, smaller wattles, and buff fringes to upperparts and black areas of head and neck.

First non-breeding Quickly becomes indistinguishable from adult in the field.

Adult Age at first breeding not known. Adult described above.

Calls A harsh, shrill *kiarr*, and *kir-kir-kir*.

Status, habitat and distribution A fairly common dry-country African species, occurring in bare or grassy (sometimes scrubby) areas, including playing fields, in sub-Saharan Africa, almost entirely north of the equator; from Senegal east to Ethiopia and north-west Kenya (nominate race *tectus*), and Kenya east of the Rift Valley and southern Somalia (race *latifrons*). Largely sedentary, though occasionally moves in response to local rains; in small parties when not breeding.

Racial variation Race *latifrons* is a vagrant north to Israel, with a single 19th century record. The two races differ only in the amount of white on the forehead, which extends to above the eye in *latifrons* but barely reaches above the wattle in nominate.

Similar species White-tailed Lapwing (60) has a similar flight pattern (though has yellow legs, and feet extend well beyond tail in flight), but Black-headed Lapwing is easily identified by combination of black-and-white head with crest and black line down upper breast.

▼ **55a. Adult *latifrons*.** This is the race that has occurred in the northern hemisphere. Differs from nominate *tectus* in the amount of white on forehead, which extends to above the eye in *latifrons*, but is much less in *tectus*. RJC.

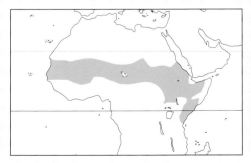

▲ **55b. Adult *tectus*.** Kenya, mid March. Compare extent of white on forehead with 55a. RJC.

▼ **55c. Adult *tectus* in flight.** The Gambia, mid January. Both races have the same flight pattern. RJC.

56. YELLOW-WATTLED LAPWING
Vanellus malabaricus

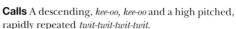

A dry-country lapwing with conspicuous yellow wattles, confined to the Indian subcontinent.

Identification L 27cm (10.5"); **WS** 63cm (25"). Medium-sized, with black or brownish-black crown narrowly bordered white; upperparts, neck and upper breast buff-brown; remainder of the underparts white. Short black bill with yellow base and prominent yellow wattle; iris pale brown; very long yellow legs. *In flight* from above brown, with black flight feathers, white innerwing-bar formed by white on secondaries and tips to greater coverts; rump white, tail white with subterminal black band and white outer feathers; feet extend well beyond tail. Sexes similar in size, adult plumages have minor seasonal differences; juvenile is separable. Vestigial yellow carpal spur difficult to see, even in flight; lacks hind toe. *Feeds* in typical lapwing manner, mainly on insects. Can raise crown feathers when excited.

Juvenile A dull version of adult, generally brown where adult is black. Upperparts with dark subterminal marks, coverts and tertials fringed buff; white chin, small, dull yellow wattles.

First non-breeding Top of crown brown, remainder black; chin and throat brown, concolorous with uniformly brown breast; wattle smaller than adult. May retain some juvenile feathers.

Adult non-breeding As adult breeding but has much brown in crown and no dark at lower margin of brown breast.

Adult breeding Age at first breeding not known. Crown black or largely so, chin and throat black, lower margin of brown breast narrowly bordered black. Male may have blacker crown; wattles variable in size, some hanging well below bill and extending quite high up forehead, but it is not clear if this varies with age, season, sex, or simply individually.

Calls A descending, *kee-oo, kee-oo* and a high pitched, rapidly repeated *twit-twit-twit-twit*.

Status, habitat and distribution Widely distributed and fairly common in arid grassland, though sometimes near wetter areas, in southern Pakistan, India, Bangladesh and Sri Lanka. Largely sedentary but a short-range migrant in north-west of range.

Racial variation None.

Similar species Grey-headed Lapwing (56) is larger, has only tiny yellow wattle and has brown-grey or grey crown, lacking Yellow-wattled Lapwing's prominent black crown. Red-wattled Lapwing (58) has a similar flight pattern, but is marginally larger, and has black face and upper breast. Sociable Lapwing (59) has somewhat similar head pattern but lacks yellow wattles, has dark legs, and has a different flight pattern.

Reference Ali & Ripley (1983).

▼ **56a. Juvenile**. Sri Lanka, early November. Note the tiny yellow wattle, largely brown crown and extensive pale fringes on the upperparts and wing-coverts. Gehan de Silva Wijeyeratne.

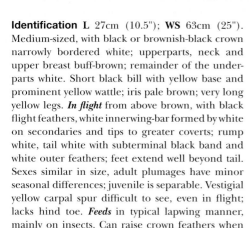

▲ **56b. Adult non-breeding**. India, late December. Extensive brown on crown, and no black at margin of the brown/white on upper breast; small wattle (compare with 56c) suggests this individual may be first non-breeding. RJC.

▼ **56c. Adult breeding**. India, late December. All-black crown, larger wattle, and some black between brown and white on upper breast. RJC.

57. GREY-HEADED LAPWING
Vanellus cinereus

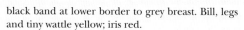

A large migratory Asian lapwing with a distinctive black breast-band.

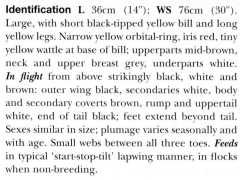

Identification L 36cm (14"); **WS** 76cm (30"). Large, with short black-tipped yellow bill and long yellow legs. Narrow yellow orbital-ring, iris red, tiny yellow wattle at base of bill; upperparts mid-brown, neck and upper breast grey, underparts white. *In flight* from above strikingly black, white and brown: outer wing black, secondaries white, body and secondary coverts brown, rump and uppertail white, end of tail black; feet extend beyond tail. Sexes similar in size; plumage varies seasonally and with age. Small webs between all three toes. *Feeds* in typical 'start-stop-tilt' lapwing manner, in flocks when non-breeding.

Juvenile Head and breast dull grey-brown, forehead and chin off-white; lower margin of breast with brown or blackish band. Grey-brown upperparts and wing-coverts have narrow buff fringes when fresh. Bill and legs dull yellow, wattle absent or rudimentary; iris dark.

First non-breeding Acquires adult-type upperparts but may retain some worn juvenile coverts; lacks dark lower boundary to grey breast. Bill and legs yellow, wattle rudimentary; iris dark.

Adult non-breeding Much as adult breeding but with more diffuse black lower margin to grey-brown upper breast. Bill and legs yellow, tiny yellow wattle (but larger than on juvenile and first non-breeding); iris red.

Adult breeding Age at first breeding not known. Head and breast grey, upperparts mid-brown; broad black band at lower border to grey breast. Bill, legs and tiny wattle yellow; iris red.

Calls A plaintive *chee-it* and a harsher *cha ha eet* ('did all eat?'); generally silent when non-breeding.

Status, habitat and distribution Breeds north-eastern China and Japan, non-breeding Nepal, north-east India and Bangladesh, south and east to Burma and Thailand, and southern Japan; vagrant to Andaman Islands, Korea, Philippines, Indonesia and Australia.

Racial variation None.

Similar species Yellow-wattled Lapwing (56), which see. White-tailed Lapwing (60) is smaller, has all-black bill, proportionally longer brighter yellow legs and short, entirely white tail.

References Ali & Ripley (1983), Barter (2002), Duckworth (2006), Herbert (2006).

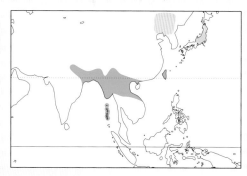

◄ **57a. Juvenile**. Japan, late July. Dark-fringed upperparts and coverts. Nobuhiro Hashimoto.

◀ **57b. First non-breeding.**
Thailand, late November. This bird
is acquiring an adult-type mantle,
but retains worn juvenile wing-
coverts. Note lack of black at lower
margin to grey-brown breast. RJC.

◀ **57c. First non-breeding.**
Japan, mid December. An older
individual than 57b; this bird has
largely replaced its wing-coverts
and is just acquiring its tiny yellow
wattle on the lores, but still has the
dark iris of the immature. Nobuhiro
Hashimoto.

▼ **57d. Adult non-breeding.**
Japan, mid December. Some
brown in breast-band, but
otherwise as adult breeding.
Nobuhiro Hashimoto.

◄ **57e. Adult breeding**. Japan, late April. Tiny yellow wattle and black breast-band; more worn than non-breeding. RJC.

◄ **57f. Juvenile wing-stretching**. Japan, late May. Nobuhiro Hashimoto.

▼ **57g. Adults in flight**. Hong Kong, mid November. Lack of dark breast-band suggests that the two birds on the left are first non-breeding. Martin Hale.

58. RED-WATTLED LAPWING
Vanellus indicus

A noisy and colourful lapwing with long yellow legs and a distinctive red wattle.

Identification L 34cm (13"); **WS** 69cm (27"). Large, with mainly black-and-white head and a small, narrow red wattle from eye to bill-base; iris red, bill is short, red with black tip. Upperparts pale olive-brown, black from throat to breast; remainder of underparts white; white on side of head and neck varies with race, very long yellow legs. *In flight* from above dark, with diagonal white wing-bar, white rump and white uppertail-coverts; tail white with broad black subterminal band; feet extend beyond tail. Sexes similar in size; plumage varies slightly seasonally, and with age. Small black spur at bend of wing difficult to see, even in flight. *Feeds*, mainly at night, in typical lapwing manner, by picking following a short run; food consists of invertebrates, usually insects. When non-breeding usually in ones and twos or in small groups.

Juvenile Like adult but duller; crown is a dull black, chin and throat off-white, upper breast is

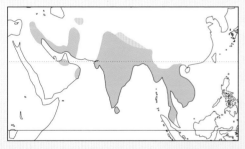

dull black with white flecks. Upperparts and wing-coverts lack adult's gloss and have narrow buff fringes; duller bare-part colours. Side-neck pattern as adult, depending on race.

First non-breeding Quickly becomes like adult non-breeding, with greyish forehead, brown-tinged crown and whitish flecks on sides of head; can only be aged with certainty if some pale-fringed juvenile scapulars, coverts or tertials are retained.

Adult non-breeding May have brown-tinged crown and/or white flecking on throat, but many apparently as adult breeding.

Adult breeding Age of first breeding not known. Plumage described under Identification; purplish gloss on wing-coverts perhaps most obvious when breeding, but probably shown year-round.

Call Noisy; rapid, repeated *did-did-do-weet*, the first syllable often repeated several times ('did, did, did, he do it?').

Status, habitat and distribution Common in wetlands and wet grassland from eastern Turkey, eastern Iraq, southern Iran, northern Oman to west Pakistan (race *aigneri*); east Pakistan, India and Bangladesh (nominate *indicus*); Sri Lanka (*lankae*); north-east India, Nepal and east through Burma and Thailand to peninsular Malaysia and Cambodia (*atronuchalis*). Altitudinal and short-distance migrant in the northern part of the range, otherwise sedentary. Vagrant west to Israel.

◄ **58a. Juvenile *indicus*.** Oman, late August. Commencing moult to first non-breeding, having replaced a few upperpart feathers. Hanne and Jens Eriksen.

Racial variation The races *aigneri*, *indicus* and *lankae* form the '*indicus* group'. All appear similar in the field, differing only slightly in size (*lankae* is the smallest, with darkest and most glossy upperparts), with a white side-neck stripe that extends from rear of eye to upper flanks. Race *atronuchalis* has white restricted to a patch behind and below eye, with the rest of the neck black, and has a narrow, white, hind-neck collar not shown by the other races.

Similar species Spur-winged Lapwing (53) and Yellow-wattled Lapwing (56), which see.

References Ali & Ripley (1983), Roos & Schrijvershof (1993).

▲ **58b. Adult non-breeding *indicus*.** India, late December. Non-breeding may have some brown in black of crown, and white flecks on sides of head as here. RJC.

▲ **58c. Adult breeding *indicus*.** India, early January. Many (including birds of both races) appear to have breeding-type plumage year-round. RJC.

▼ **58d. Adult *atronuchalis*.** Thailand, late November. This race has a small white area behind eye and narrow white hindneck collar, and often has a more obvious purple gloss on the wing-coverts. As with *indicus*, many individuals have similar plumage year-round. RJC.

▼ **58e. An *atronuchalis* Red-wattled Lapwing in flight.** Southern Thailand, mid January. Andrew W. Clarke.

59. SOCIABLE LAPWING
Vanellus gregarius

An attractive lapwing of the steppes of western Asia, wintering in north-east Africa, the Middle East and north-west India. Critically Endangered.

Identification L 29cm (11"); **WS** 64cm (25"). Medium-sized; pale brown above, with dark crown, white or off-white forehead, white supercilium to nape; narrow, dark, eye-stripe behind eye; upperparts buff-brown; foreneck and upper breast pale brown, lower breast and remainder of underparts white (black and chestnut lower belly when breeding), short black bill, iris brown; long, dark-grey or black legs. *In flight* from above body and inner wing-coverts brown, primaries and primary coverts black, secondaries white, rump and tail white with black subterminal band in centre of tail only; toes just reach beyond tail. Sexes differ slightly when breeding; plumage varies seasonally and with age. *Feeds* in typical lapwing manner, making short runs then picking from ground, largely on insects.

Juvenile Crown brown, streaked white, supercilium off-white; lores pale, narrow black eye-stripe behind eye only. Upperparts and wing-coverts dull brown with pale buff fringes, some feathers with dark subterminal lines; neck and upper breast pale brown, streaked dark brown.

▼ **59a. Juvenile**. Oman, early November. Similar to adult non-breeding, but aged by very neat feathering, particularly on throat and upper breast. Hanne and Jens Eriksen.

First non-breeding/adult non-breeding Crown grey-brown, forehead and supercilium tinged buff; upperparts plain grey-brown with narrow white fringes when fresh. Throat and upper breast heavily streaked brown, belly white. First non-breeding quickly becomes like adult non-breeding but may retain pale-fringed juvenile coverts.

First-breeding May not breed in second calendar year. Much as adult non-breeding but some acquire partial breeding plumage with scattering of dark feathers on white belly.

Adult breeding Crown black, forehead and supercilium bright white, black from bill through lores and behind eye; upperparts pinkish brown, throat white, neck and breast pinkish buff, large black belly patch; vent and undertail white. Males have darker crown and belly.

Calls Various harsh, rasping calls when breeding, such as *krech, krech, krech*. Generally silent away from breeding grounds but may occasionally give harsh calls, particularly when taking flight.

Status, habitat and distribution Scarce migrant, breeding from Kazakhstan east to central Russia, in grassland or scrubby areas; non-breeding birds occur in Arabia, Turkey, Syria, Israel, Eritrea and northern Ethiopia, and north India, in semi-desert and on cultivated fields with little vegetation. Vagrant, often with flocks of Northern Lapwings, to most of western Europe as far as Ireland, and to Morocco, southern India and Sri Lanka.

Racial variation None.

Similar species Eurasian Dotterel (47) and Yellow-wattled Lapwing (56), which see.

References Ali & Ripley (1983), Hofland (2007), Thorup (2006).

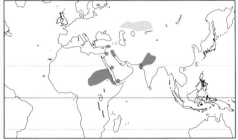

▲ **59b. Adult non-breeding**. Oman, early November. Feathering far less neat than 59a, and throat and upper breast are coarsely and irregularly streaked. Markus Varesvuo.

▲ **59c. Adult male breeding**. Dubai, late February. Breeding plumage is striking, with black crown, pale unstreaked throat and upper breast, and black belly. The intensity of black on the crown and belly suggests this is a male. Clive Temple.

▼ **59d. Sociable Plovers in flight**. Israel, late December. Markus Varesvuo.

60. WHITE-TAILED LAPWING
Vanellus leucurus

A long-legged Asian lapwing with a diagnostic short white tail.

Identification L 28cm (11"); **WS** 55cm (22"). Medium-sized, with plain buff-brown head and body, paler on sides of head and at base of bill; neck buff-brown, breast grey-brown; remainder of underparts white or off-white. Short all-black bill, very long yellow legs. Typically exhibits upright stance. *In flight* from above body brown, extensive white from primary coverts across secondaries; outer wing black, rather short all-white tail, legs and feet extend well beyond tail. Sexes similar, with similar plumage year-round; juvenile separable. *Feeds* in typical lapwing manner, by picking, on insects and aquatic invertebrates, sometimes wading quite deeply and completely submerging head.

Juvenile Uniform buff-brown above, with plain face; crown streaked darker brown; mantle, scapulars and wing-coverts have dark-brown, broad, subterminal marks and narrow buff fringe at tips. Foreneck and breast washed buff; remainder of underparts white.

First non-breeding Quickly becomes indistinguishable from adult; though retains juvenile wing-coverts.

Adult Probably breeds in second or third calendar year. Breast grey-brown, fringed white when fresh; belly has peach wash, fading to white, as do paler areas of face.

Calls Recalls Northern Lapwing in tone and pitch: *kee-vic*, and *kee-vee-ik*; more noisy when breeding.

Status, habitat and distribution Fairly common in lowland wetlands, breeding from Romania, Turkey, Syria and Iraq eastwards somewhat discontinuously to central Kazakhstan; the breeding range appears to be expanding westward. Northern birds are migrants; non-breeding season spent in north-east Africa, the Middle East and northern India. Vagrant to western and northern Europe.

Racial variation None.

Similar species Short white tail eliminates any other lapwing.

References Ali & Ripley (1983), Helbig (1985), Kiss & Szabó (2000), Pettet (1982), Thorup (2006).

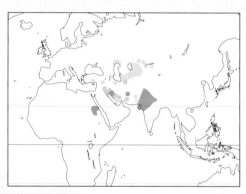

◄ **60a. Juvenile**. Dubai, July. Aged by broad dark upperpart subterminal marks and pale fringes. Ian Boustead.

◀ **60b. First non-breeding/first-breeding.** Israel, mid March. Aged by the few juvenile wing-coverts that still remain. RJC.

◀ **60c. Adult**. India, late December. Once adult has similar plumage year-round; the short white tail is diagnostic. RJC.

▼ **60d. Adult in flight**. Oman, early November. Markus Varesvuo.

61. SOUTHERN LAPWING
Vanellus chilensis

A large, noisy, crested South American lapwing that just reaches the southern limits of our region.

Identification L 35cm (13.5"); **WS** 70cm (27.5"). Large, with grey head and neck, wispy crest, black forehead and chin outlined white; upperparts green-grey, wing-coverts more rufous and strongly iridescent; breast black; remainder of underparts white; short reddish, black-tipped bill, and long, dark-red legs. Eye and orbital ring deep red. *In flight* greyish above, with large white area at bend of wing formed by secondary coverts; rump and uppertail-coverts white; feet extend beyond black tail, more so in race *cayennensis*. Sexes similar, with no seasonal variation; juvenile separable. Red carpal spur at bend of wing, visible in flight. *Feeds* in typical lapwing manner on insects, worms and other invertebrates, often in flocks when non-breeding.

Juvenile A dull, brownish version of the adult, with brown-black forehead and chin, buff-fringed crown and upperpart feathers, brown-black breast and short crest. Legs duller than adult.

First non-breeding/adult non-breeding Once body moult is complete ageing becomes difficult, though worn, brownish primaries may perhaps be seen in flight. Red of bill may be tinged greyish.

Adult breeding Age at first breeding not known. Plumage described under Identification.

Call Very noisy; a loud *kew, kew, kew,*...., becoming louder and more staccato when excited.

Status, habitat and distribution Common and widespread in lowland South America, south from Panama, northern South America, Trinidad and Barbados. In grasslands, particularly seasonally flooded ones. Has wandered north as far as Mexico.

◄ **61a. Juvenile**. Brazil, early March. Race *lampronotus* from location, but not identifiable as such at this age. Carlos Grande.

Racial variation Race *cayennensis* occurs in northern South America and is the race that has wandered to the northern hemisphere. Other races, which differ only in facial pattern, have occurred, but seem unlikely to have done so naturally and are assumed to have been escapes.

Similar species Northern Lapwing (62) shares the crest of Southern Lapwing but is easily separated by having a black-and-white (not largely grey) head pattern.

Reference Hilty & Brown (1986).

▲ **61b. Adult *cayennensis***. Venezuela, mid November. This is the race that has wandered to the northern hemisphere (to Mexico). RJC.

▼ **61c. Adult *cayennensis* in flight**. Venezuela, mid November. RJC.

62. NORTHERN LAPWING
Vanellus vanellus

The common lapwing of Europe and much of Asia, occurring in large flocks in the non-breeding season.

Identification L 29cm (11.5"); **WS** 66cm (26"). Medium-sized, with distinctive wispy crest, short black bill and medium-length pinkish or red legs. Overall appearance black-and-white, but upperparts have a dark green metallic sheen; crown and crest black, sides of head and nape generally white; variable broad black neck/breast-band extends to bill when breeding. Undertail-coverts rusty, rest of underparts white. *In flight* 'flappy', with broad rounded wings alternately flashing black and white; above black with pale tips to primaries, white rump and narrow white tail-tip (moulting adults often show white patches on wings); below, black flight feathers contrast with white underwing-coverts and body; feet do not extend beyond tail. Width of the outer wing depends on age and sex, with the adult male having particularly noticeable 'bulging' primaries and white at tips of four outermost primaries (three outermost in females and immatures). Plumage varies seasonally, and with age and sex; juvenile separable. Some webbing between the outer and middle toes. *Feeds* in typical lapwing manner.

Juvenile Short crest; obscure black markings on face. Dull metallic-green upperparts with narrow pale-buff fringes composed of closely spaced small spots. Throat white; feathers of breast-band tipped white. Legs dull pink.

▼ **62a. Juvenile**. Eastern England, early August. Aged by largely pale face, short crest, and upperparts, scapulars and tertials with narrow buff tips and spotting on feather sides. RJC.

First non-breeding/adult non-breeding As the juvenile, but upperparts have brighter metallic-green sheen with broad rich-buff tips (not fringes), which gradually wear away. Fairly short crest. First non-breeding birds may retain a few juvenile pale-buff fringes to upperpart feathers and wing-coverts, but the fringes wear quickly.

Adult breeding Some breed in second calendar year, remainder in third. Buff fringes on upperparts wear off by about early June. Black of breast-band extends upwards. In male forehead, lores, chin and throat all become black, and crest is longer; female generally retains some white feathers on foreneck and has slightly shorter crest. First-breeding probably not separable.

Call A plaintive *pee-wit*, with rising inflection.

Status, habitat and distribution A common, generally inland species, breeding in grassland and grassy wetlands throughout Europe south to France and, locally, Spain. A rare breeder in Faeroes and Iceland. Extensive westerly movements of adults occur from east and central Europe to east North Sea area, Britain and Ireland, typically from June; westerly and southerly movements also occur during hard weather. Non-breeding from Britain and Ireland south to Mediterranean and north-west Africa, east to north India, south China and southern Japan. Vagrant to Spitsbergen, eastern North America and several Caribbean islands.

Racial variation No races recognised.

Similar species The only other lapwings with crests are Southern Lapwing (61) of South America and Black-headed Lapwing (55) of Africa, neither of which are likely to occur with Northern Lapwing.

References Barter (2002), Ebels (2002), Shrubb (2007), Thorup (2006).

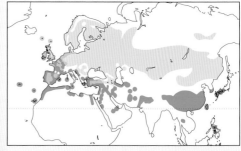

▲ **62b. Adult non-breeding**. Eastern England, early September. Moulting to non-breeding; has very short crest, more dark on face than juvenile, and new upperparts with broad buff-fringed tips, different from the narrower juvenile fringes. RJC.

▲ **62c. Adult non-breeding**. Eastern England, mid December. Much longer crest than 62b at this later date. RJC.

▲ **62d. Adult breeding male**. Eastern England, mid April. Long crest, forehead to throat entirely black, and has lost buff tips to many upperparts. RJC.

◀ **62e. Adult breeding male.**
Eastern England, early July.
Just commencing moult to
non-breeding, starting to acquire
white on forehead. Still has quite
long crest, but buff fringes on
upperparts have completely worn
away. RJC.

◀ **62f. Adult breeding female.**
Eastern England, mid April. Shorter
crest than male (compare with
62d), and has white on forehead
and throat. RJC.

▼ **62g. Non-breeding in flight.**
English midlands, mid November.
Adult males (front two) show white
on the outer four primaries (three
in females and immatures, lowest
bird), and the outer wing is rather
broader than in the females, with the
primaries 'bulging' slightly more at
the rear of the wing. RJC.

CALIDRIS AND RELATED SANDPIPERS

Twenty-four species in six genera worldwide, all of which breed in the northern hemisphere: Surfbird *Aphriza virgata*, Great Knot *Calidris tenuirostris*, Red Knot *C. canutus*, Sanderling *C. alba*, Semipalmated Sandpiper *C. pusilla*, Western Sandpiper *C. mauri*, Red-necked Stint *C. ruficollis*, Little Stint *C. minuta*, Temminck's Stint *C. temminckii*, Long-toed Stint *C. subminuta*, Least Sandpiper *C. minutilla*, White-rumped Sandpiper *C. fuscicollis*, Baird's Sandpiper *C. bairdii*, Pectoral Sandpiper *C. melanotos*, Sharp-tailed Sandpiper *C. acuminata*, Curlew Sandpiper *C. ferruginea*, Stilt Sandpiper *C. (Micropalama) himantopus*, Purple Sandpiper *C. maritima*, Rock Sandpiper *C. ptilocnemis*, Dunlin *C. alpina*, Spoon-billed Sandpiper *Eurynorhynchus pygmaeus*, Broad-billed Sandpiper *Limicola falcinellus*, Buff-breasted Sandpiper *Tryngites subruficollis* and Ruff *Philomachus pugnax*. These birds are part of the large family Scolopacidae, which also includes snipes, woodcocks, dowitchers, godwits, curlews, the *Tringa* sandpipers, turnstones and phalaropes.

Calidris sandpipers and their close relatives are a group of small to medium-sized shorebirds, including the smallest shorebird, Least Sandpiper, and the medium-sized Great Knot and Ruff. For the most part they are fairly compact, with medium-length bills and legs. All have distinctive juvenile, non-breeding and breeding plumages, while Great Knot and Ruff additionally have a supplemental breeding plumage. All *Calidris* sandpipers except Sanderling have a hind toe, and two (Semipalmated and Western Sandpiper) have webbing between the front toes. All these shorebirds are migratory, some very strongly so, with species such as Great Knot breeding in Siberia and migrating to Australia, Sanderling breeding in the high Arctic and migrating to the far south of South America, South Africa and Australasia, and Curlew Sandpiper breeding in northern Siberia and migrating to southern Africa and Australasia.

The seven smallest *Calidris* sandpipers, which are collectively known as 'stints' or 'peeps', can be difficult to identify, particularly in non-breeding plumage. Table 8 (p. 216) highlights some of the key structural and plumage features that may help with this problem.

Many *Calidris* species, particularly those which spend the non-breeding season in the southern hemisphere and which breed in their second calendar year, moult some or all of their primaries, starting at the end of their first calendar year. Details are given in the species accounts. Different races and populations of the same species may differ in their general moult strategy. One case is that both adults and juveniles of the European races of Dunlin moult either at a staging area during migration or in their non-breeding area, whereas adults of all the other Dunlin races commence primary moult when breeding, and the juveniles undergo body moult to first non-breeding before migration. A similar situation apparently applies to juvenile European Purple Sandpipers *vis-a-vis* other populations, the latter undergoing body moult before migration.

Male Ruff breeding plumages The most complex range of variation and sequence of plumages of any shorebird is shown by the Ruff. The nature and variety of supplemental breeding plumage of male Ruffs, particularly their extraordinary ruffs of extended neck feathers, has doubtless evolved as a consequence of their lekking system of mating. It is probable that no two male Ruffs are identical (see p. 287); the variations are innumerable. Some consistent plumage patterns can be identified, however, which are related to the role of individual males at the lek. Within the lek, some individuals hold and aggressively defend small territories; these are 'resident' males. Other males visit the lek periodically but are subservient to the resident males; known as 'marginal' males, they often have only partially developed plumage and seem rarely to mate successfully. Both resident and marginal males are together referred to as 'independent' males, and these are the largest birds, with darker ruffs, typically chestnut or black. 'Satellite' males, which always have a generally pale coloration and a conspicuous white ruff and/or white ear-tufts, seem to visit several different leks but wait at the margins. Generally smaller than the resident males, they are largely tolerated by them, and they seize the chance of mating while the independent males are involved elsewhere. It is thought that the very conspicuous satellites may help to attract females to the lek.

It has recently been discovered that a third mating strategy exists for males – the 'faeder' or 'sneaker', which number only about 1% of the Ruff population. The faeder has a female-like breeding plumage and is intermediate in size between typical males and females, but with body proportions closer to males than females. Their behaviour at the lek is a combination of that of male and female and, rather like the satellite males, they attempt to 'sneak' matings when the opportunity arises.

Table 7. Stints and peeps; comparative structural and plumage characteristics can be useful in separating the various species.

Species (leg colour)	Main structural features	Juvenile	Non-breeding	Breeding
67. Semipalmated Sandpiper (black legs)	Shortish blunt-tipped bill; short primary projection; feet partly webbed.	Relatively uniform dull brownish upperparts, fringed paler.	Coarsely streaked crown; breast streaking most prominent on sides.	Dullish upperparts; strongly streaked breast and flanks.
78. Western Sandpiper (black legs)	Longish slightly de-curved bill; very short primary projection; longish legs; feet partly webbed.	Rufous fringes on mantle and scapulars; coverts grey.	Finely streaked crown; fine streaking across breast.	Rufous on crown and ear-coverts; small dark chevrons on breast and flanks.
69. Red-necked Stint (black legs)	Shortish deep straight bill; long primary projection; elongate rear-end; legs shortish; feet unwebbed.	Rufous fringes on mantle and scapulars; indistinct pale mantle-V; coverts grey.	Uniformly grey upperparts with dark shaft-streaks.	Entire face and throat uniform brick-red bordered below with streaked 'necklace'; indistinct mantle-V.
70. Little Stint (black legs)	Shortish slightly decurved bill; long primary projection; feet unwebbed.	Split supercilium; rufous fringes on upperparts and coverts; clear pale mantle- and scapular-Vs.	Upperparts with darker centres; grey-washed breast-band.	Face and breast dark-streaked over orange-rufous wash; throat white; clear mantle-V.
71. Temminck's Stint (yellowish legs)	Slightly decurved bill; tail projects beyond primaries; white outer tail feathers.	Upperparts dull with dark submarginal lines and pale fringes; breast brown.	Dull uniform brownish upperparts; breast brown.	Random dark feathers (sometimes very few) in mantle and scapulars; breast brown.
72. Long-toed Stint (yellowish legs)	Slightly decurved bill; pale base to lower mandible; very short primary projection; upright stance; toes beyond tail in flight.	Split supercilium; mantle and scapulars fringed bright rufous; white covert fringes; breast-sides streaked.	Upperparts with dark centres and broad greyish fringes.	Dark of forehead reaches bill; broad rufous tertial fringes; pale mantle-V.
73. Least Sandpiper (yellowish legs)	Slightly decurved bill; very short primary projection; bent-legged crouching stance.	Diffuse supercilium; rufous-fringed upper-parts; indistinct pale mantle-V; breast uniformly streaked.	Upperparts with dark centres and broad greyish fringes.	Supercilia usually meet above bill; dull rufous sometimes scalloped fringes to tertials; breast coarsely streaked.

63. SURFBIRD
Aphriza virgata

A large, plump, short-billed calidrid, almost exclusively on rocky coasts when non-breeding.

Identification L 24cm (9.5"); **WS** 55cm (21.5"). Medium-sized, uniformly grey above when non-breeding, with plain head; short, stubby, dark-grey bill with extensive orange base to lower mandible; narrow white eye-ring and dark eye. Breast strongly spotted or streaked when breeding, uniform grey in non-breeding; remainder of underparts white, but flanks and belly have bold dark chevrons (adults) or are extensively spotted (juvenile); short to medium-length yellow legs. *In flight* dark grey above, with narrow white wing-bar, white uppertail-coverts and dark tail; toes just reach tail-tip. Plumages vary with age and season; females are typically very slightly larger but the difference is insufficient to allow sexes to be identified. No webbing, but toes thickened to improve grip on wet rocks. *Feeds* by picking on intertidal rocks, eating small shellfish (which may be pulled off quite vigorously); largely consumes insects on its inland breeding grounds.

Juvenile Upperparts, coverts and tertials pale brown-grey, with subterminal narrow black bar and narrow white fringe. Breast mottled brown-grey with much white flecking; lower breast and belly with small grey-brown spots.

First non-breeding/adult non-breeding Mantle,

scapulars and tertials plain grey (sometimes with broad sooty-grey fringes), breast uniform grey. First non-breeding retains some slightly browner juvenile wing-coverts (though by November these can be difficult to see), and averages less spotted beneath than adult non-breeding.

First-breeding Most probably breed in their second calendar year and are as adult breeding, though may have very worn primaries. A few remain in the non-breeding range, mostly acquiring adult-type non-breeding plumage, though apparently some develop partial breeding plumage.

Adult breeding Mantle, scapulars and some wing-coverts have dark tips and large, paired, orange spots (similar to Great Knot), which fade with age; breast with heavy dark streaks; flanks and belly with bold dark chevrons.

Calls Generally silent outside the breeding season but when disturbed may give a high-pitched, dry, irregular trill.

Status, habitat and distribution Breeds above tree-line in alpine tundra of inland Alaska and adjacent Yukon Territory; non-breeding exclusively on rocky coasts and jetties from south-central Alaska and Kodiak Island south down Pacific coast, including Galápagos, to Tierra del Fuego, but local in Central and South America. There are records of non-breeders during the breeding season from at least as far south as Panama. Scarce inland in North America, vagrant to Alberta, Pennsylvania, Gulf coast (Texas east to south-west Florida) in spring; not recorded outside Americas.

Racial variation None.

Similar species Great Knot (64) and Red Knot (65); both unlikely to be seen on rocky coasts, have

▼ **63a. First non-breeding**. California, early November. Note slightly greyer non-breeding upperparts compared to the browner juvenile coverts, which have a dark shaft and subterminal line. Juvenile upperparts have whitish fringes (see the wing-coverts on this bird), and the breast is paler and slightly mottled with pale spotting on the flanks. RJC

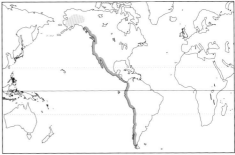

proportionally longer bills and in flight lack white uppertail-coverts. Rock Sandpiper (81) occurs in similar rocky habitat but is smaller and darker, has a proportionally longer, slightly decurved bill, and also lacks Surfbird's white uppertail-coverts. Black Turnstone (131) shares the rocky shore habitat and has a somewhat similar flight pattern, but is slightly smaller and very dark above, with a dark breast.

References Paulson (1993), Senner & McCaffery (1997), Tomkovich *et al.* (1998).

◀ **63b. Adult non-breeding**. California, early November. Adult wing-coverts, and marginally more spotting on flanks and belly than juvenile/first non-breeding. RJC.

◀ **63c. Adult breeding**. California, mid April. Not only acquiring breeding upperparts, but also a few wing-coverts of breeding-type; extensive spotting below. Note the similarity to the upperparts of breeding Great Knot (64e and 64f), to which the Surfbird is closely related. RJC.

◀ **63d. Adult non-breeding in flight**. California, early November. RJC.

64. GREAT KNOT
Calidris tenuirostris

Similar to Red Knot in proportions (but not in breeding plumage), this species is a long-distance migrant between Siberia and northern Australia.

Identification L 27cm (10.5"); **WS** 56cm (22"). Medium-sized with an indistinct broad supercilium and medium-length, slightly decurved, tapered, black bill; short to medium-length, usually grey, legs. *In flight* grey-brown above, with narrow white wing-bar formed by tips of greater coverts; darker primary coverts, white uppertail-coverts and dark tail; feet do not extend beyond tail. Male marginally smaller on all measurements; plumages vary with age and season. *Feeds* on mudflats by picking and probing, on invertebrates, and on berries on the breeding grounds.

Juvenile Head and neck streaked brown, crown darker; mantle and scapulars dark brown, wing-coverts and tertials pale brown with dark subterminal marks, all with off-white fringes. Foreneck heavily streaked brown, merging into spotted upper breast; flanks with large dark spots, rest of the underparts are white. Bill black with brownish or greyish base to lower mandible; legs yellowish or greenish grey.

First non-breeding/adult non-breeding Upperparts uniformly brown-grey with dark shaft-streaks, paler towards edges; upper breast heavily streaked grey; flanks with grey and dark grey spots and bars. Bill black, legs grey. First non-breeding retains juvenile wing-coverts and some tertials.

First/second breeding Probably do not breed until fourth calendar year. First-breeding acquire adult-type non-breeding plumage, though perhaps with some heavier black breast spots; may renew outer primaries. Second-breeding gains variable (none to 50%) extent of adult-type breeding mantle and scapular plumage, and has significant breast-spotting.

Adult breeding Head and neck heavily streaked black, crown darker; mantle largely black, scapulars with large paired bright chestnut spots and dark tips. Upper breast is heavily spotted, almost completely black; flanks have extensive black spots and bars. Bill black, legs dark greyish black. Amount of chestnut on the scapulars is variable but male averages brighter. Most fourth calendar year birds probably attain full breeding plumage. On departure from the non-breeding areas in March or April the proportion of red upperpart feathers is small. Once at the Yellow Sea (where it is thought that most Great Knots stage) the number of red feathers increases considerably as they acquire their supplemental breeding plumage.

Call Often silent, but sometimes calls *nyut-nyut*, recalling Red Knot.

Status, habitat and distribution Breeds in alpine tundra, above the tree-line, in eastern Siberia. Non-breeding on coastal mudflats and estuaries, in eastern Arabian Gulf, Pakistan and northern India to Bangladesh (scarce), south-east Asia, and Australia (most commonly); rarely New Zealand. Scarce migrant Middle East; vagrant to western

▼ **64a. Juvenile**. Japan, early September. Larger than Red Knot, with a longer, slightly downturned bill. Heavy underpart spotting makes juvenile remarkably similar to adult breeding, but upperpart and wing-covert patterns differ. Nobuhiro Hashimoto.

Europe north to Norway, Morocco, Djibouti, South Africa; rare vagrant North America.

Racial variation None.

Similar species Surfbird (63), which see; Red Knot (65) is slightly smaller, with proportionally shorter straight bill; Red Knot lacks the bold underpart spotting of Great Knot in most plumages, and in flight shows barred grey, not white, uppertail-coverts.

References Barter (2002), Battley *et al.* (2006), Lethaby & Gilligan (1992), Rogers (2006), Tomkovich (1996a).

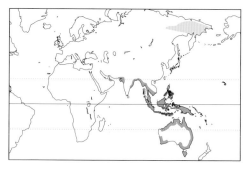

◄ **64b. First non-breeding.** Thailand, late November. In moult, showing fresh non-breeding upperparts and retained juvenile coverts. RJC.

◄ **64c. Immature, possibly first breeding.** Western Australia, late March. Second and third calendar-year birds do not breed, and remain in the non-breeding area. Lack of any breeding plumage above or below suggests this is in its second calendar year. RJC.

◄ **64d. Immature, possibly second-breeding.** Western Australia, late March. As with 64c, will not breed, but has attained some breeding-type plumage both above and below. RJC.

▲ **64e. Adult breeding**. Western Australia, late March. Has the restricted amount of red on the upperparts of the 'initial' breeding plumage acquired prior to migration. The black-centred scapulars are 'alternate' (not non-breeding) plumage; many will be replaced by chestnut-marked feathers while staging on migration. RJC.

◄ **64f. Adult breeding**. South Korea, early May. Though it may not leave the Yellow Sea for another two weeks, this bird has already acquired much of the red upperparts of its supplemental breeding plumage. RJC.

▼ **64g. Adult breeding Great Knots in flight**. South Korea, late April. RJC.

65. RED KNOT
Calidris canutus

A chunky, compact calidrid which is a circumpolar breeder and a long-distance migrant.

Identification L 24cm (9.5"); **WS** 51cm (20"). Medium-sized, shortish-necked, with medium-length dark grey or black bill, and medium-length, brownish-yellow to dark grey legs. *In flight* from above uniformly greyish, with narrow white wing-bar and pale grey rump and tail; feet do not project beyond tail. Sexes of similar size; plumages vary with age and season. *Feeds* by methodical picking, stitching or probing to shallow depth.

Juvenile Prominent whitish supercilium. Upperparts brownish grey; mantle, scapulars, wing-coverts and, to a lesser extent, tertials, have thin dark submarginal lines and narrow whitish fringes, giving neatly scalloped appearance. Underparts, especially breast, lightly spotted and barred over pale salmon-pink wash. Bill grey or dark grey with black tip; legs dull yellow.

First non-breeding/adult non-breeding Uniform grey upperparts, neatly fringed white when fresh. Bill usually black, legs dark grey. First non-breeding retains many juvenile wing-coverts and, depending on latitude of non-breeding area, may moult inner, outer or all primaries (South Africa), while those non-breeding in northern hemisphere do not commence primary moult until well into second calendar year; legs often with yellow tinge.

First-breeding Much as adult non-breeding.

Adult breeding May not breed until third or even fourth calendar year. Mantle and scapulars dark brownish-black with large rusty spots and whitish fringes. Supercilium, face, foreneck, breast and most of belly brick red. Females generally greyer above, with scattered whitish feathers (often with a dark terminal bar) on rufous underparts. Bill black, legs dark grey.

Call A subdued *knut*.

Status, habitat and distribution Locally common, breeding in drier upland tundra: Race *islandica* in northern Greenland and the Canadian high Arctic islands; nominate *canutus* on Taimyr Peninsula; *piersmai* on New Siberian islands; *rogersi* in Chukotka and eastern Siberia; *roselaari* on Wrangel Island and in north-west Alaska; and *rufa* in Canadian Arctic from Victoria Island to northern Hudson Bay. When non-breeding occurs almost exclusively on extensive coastal mudflats, often in very large flocks: *islandica* in western Europe, *canutus* in west and southern Africa, *piersmai* from India to Australasia (particularly north-west), *rogersi* in Australasia (particularly south-east Australia and New Zealand) and possibly the Pacific coast of the Americas, *roselaari* in south California to central America (apparently not Gulf of Mexico or Florida as sometimes suggested), and *rufa* from Massachusetts south to Florida and South America, especially Argentina. Some races share flyways: *islandica* occurs with

Table 8. Breeding male Red Knots: comparison of races, based on moderately worn skins taken in June. Females tend to be more similar and thus more difficult to assign to race. There can be considerable individual variability (particularly resulting from wear and fading), and caution is required before assigning any bird to a particular race, particularly in the field. Based on Tomkovich (1992, 2001).

Race (breeding area)	Breeding plumage (male)	
	underparts	upperparts
C. c. islandica (N Canada, Greenland)	all three races similar, darkish rufous below, *islandica* generally averaging paler than *canutus*; all with some rufous on lower belly and under tail.	all three races similar with relatively deep red markings, *islandica* generally averaging paler above (narrower black markings, paler and larger paired cinnamon spots) than *canutus*.
C. c. canutus (Taimyr)		
C. c. piersmai (New Siberian Islands)		
C. c. rogersi (Chukotka)	extensive white and restricted rufous on lower belly and under tail.	both similar, paler and more variegated than above three races, with slightly wider grey-white fringes, wider black markings, and larger buff spots.
C. c. roselaari (Wrangel Island)	paler rufous extending to lower belly.	
C. c. rufa (low Canadian arctic)	extensive white and restricted rufous on lower belly and under tail, as *rogersi*.	has palest upperparts; 'silvery' (grey/white) with restricted black and large buff spots.

canutus in the Wadden Sea (*canutus* is rare in the British Isles), *piersmai* and *rogersi* share the west Pacific flyway between Australasia and Japan, and *rogersi* and *roselaari* may occur together on Pacific coast of America, particularly when staging in west Alaska in May.

Racial variation The various races have been described by a combination of morphology and breeding plumage; some are separable but only in breeding plumage (notably *piersmai* and *rogersi* on the west Pacific flyway). Race *piersmai* typically has rufous-and-black upperparts and terracotta-red underparts; *rogersi* has paler upperparts owing to silvery fringes to scapulars and mantle feathers, lighter red underparts, and more white on the lower belly and undertail. Most other races are too similar to be safely identifiable in the field, and can only be separated on measurements taken in the hand; see Table 8 (p. 216). Juvenile *rogersi* and *piersmai* tend to have broader dark submarginal lines on the upperparts than the other races.

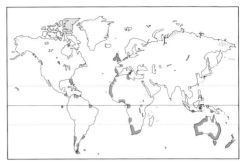

Similar species Great Knot (64), which see. Curlew Sandpiper (78) at all ages has similar plumages to Red Knot, but Red Knot is obviously larger and more bulky, lacking Curlew Sandpiper's longer decurved bill.

References Balachandran (1998), Barter (2002), Boyd & Piersma (2001), Engelmoer & Roselaar (1998), Harrington (2001), Harrington & Flowers (1996), Thorup (2006), Tomkovich (1992a, 2001).

▲ **65a. Juvenile, presumably *islandica* from location.** Eastern England, late August. Aged by neat upperparts and wing-coverts with pale fringes and dark submarginal lines; on younger individuals the breast and belly are pale peach. RJC.

◀ **65b. Juvenile, either *piersmai* or *rogersi* from location (more likely the latter).** Japan, late September. These two races have slightly wider dark submarginal fringes on the upperparts than other races, though as seen here the difference may be slight. Nobuhiro Hashimoto.

◀ **65c. First non-breeding, presumably** *icelandica*. Eastern England, late February. Non-breeding upperparts, but still has juvenile coverts and tertials. RJC.

◀ **65d. Adult non-breeding,** *rufa* **(from location)**. Florida, late October. The rings show that this bird was at least three years old when photographed. RJC.

▼ **65e. Adult breeding** *rufa* **(from location)**. Florida, mid May. This and the following photographs show something of the range of plumages exhibited by breeding Red Knots. The variability of individual birds, quite apart from differences resulting from sex and date, makes it extremely difficult to identify the different races in the field, and usually all that can be done is to infer the race from location. See also the discussion of the similar problem of racial identification of Dunlins (p. 271). RJC.

◀ **65f. Adult breeding *rufa* (from location)**. Florida, late August. Rings show that this bird was at least two years old when photographed. Commencing moult to non-breeding, with both upper- and underparts worn and faded; compare with the fresh-plumaged individual in 65e. RJC.

◀ **65g. Adult breeding, possibly *roselaari* from location**. California, mid April. RJC.

▼ **65h. Adult breeding *piersmai* (from location)**. Western Australia, early April. With Great Knots and Bar-tailed Godwits. RJC.

▲ **65i. Adult breeding Red Knots (three in flight), with Great Knots (in the water).** South Korea, early May. Races *piersmai* and *rogersi* occur together on the west Pacific flyway, with *rogersi* being much less strongly coloured below than *piersmai*; it is possible that the right-hand two are *rogersi* and the brighter left-hand bird is *piersmai*. RJC.

▲ **65j. Non-breeding Red Knots in flight.** Florida, late October. Two (centre left) still have outer primaries to replace. RJC.

66. SANDERLING
Calidris alba

A calidrid of sandy beaches, pearly grey above and white below for much of the year, and larger than the stints.

Identification L 20 (8"); **WS** 39 (15.5"). Small, with short, straight, heavyish black bill and short black legs. *In flight* from above has a prominent white wing-bar (emphasised by darker outer wing), rump and uppertail-coverts grey-centred with white sides; feet do not reach tail. Sexes differ marginally when breeding; plumages vary seasonally and with age. The only *Calidris* that lacks a hind toe. *Feeds* generally on beaches, running vigorously at the water's edge, pausing briefly to pick or probe, sometimes stitching.

Juvenile Darkish crown, mantle and scapulars have white marginal spots on black feathers, giving chequered appearance at close range; wing-coverts are variable, with dark subterminal marks, or plainer with pale fringes. White beneath.

First non-breeding/adult non-breeding Uniform pale grey above. Dark forewing may show at bend of wing, but adults have larger scapulars which often obscure both dark forewing and relatively unworn coverts. First non-breeding usually retains worn juvenile wing-coverts and tertials, often until February or March.

First-breeding Many probably do not breed until third calendar year. First-breeding acquires a variable amount of breeding plumage, often still retaining very worn juvenile coverts and/or tertials; those non-breeding in southern hemisphere have variable primary moult: some complete it, others suspend, leaving the outer primaries unmoulted; yet others replace outers only.

Adult breeding Whole of head, neck and upper breast is rufous, spotted black. Mantle and scapulars are black with extensive whitish fringes when fresh, becoming more rufous with wear; coverts are grey. Remainder of the underparts are white. Male more rufous on upperparts and upper breast than female, but sexes only safely separable in breeding pairs.

Calls A staccato *kip* or *quick*.

Status, habitat and distribution A common species, breeding in the high Arctic on stony, well-drained tundra. Nominate race *alba* from Greenland east to north-central Siberia; *rubidus* in northern Canada, Alaska and perhaps eastern Siberia. When non-breeding widespread coastally, typically on extensive sandy beaches, south from mid-latitude northern hemisphere around Africa, Asia, southern South America and Australasia.

Racial variation and hybridisation The two races differ slightly in measurements and breeding plumage, when *rubidus* averages greyer above owing to broader mantle fringes. Hybridisation reported between Sanderling and Dunlin; a bird in breeding plumage had bill shape and hind toe of Dunlin but lacked black belly and had Sanderling-type scapulars.

▼ **66a. Juvenile. Florida, mid September**. The neatly chequered upperparts are characteristic of juveniles. RJC.

Similar species Single Sanderlings can suggest a stint, the comparison in breeding plumage with Red-necked Stint (69) being particularly striking. The greater size of Sanderling – slightly larger than Ringed Plover (31), Semipalmated Plover (32) or Dunlin (82) – rules out this confusion, as does the Sanderling's lack of hind toe.

References Clark (1987), Engelmoer & Roselaar (1998), Gosbell & Minton (2001), MacWhirter *et al.* (2002), Thorup (2006).

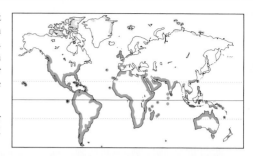

▲ **66b. First non-breeding**. California, early November. Aged by retained juvenile wing-coverts and tertials. RJC.

▼ **66c. Adult non-breeding**. Eastern England, early April. Plain wing-coverts and tertials; the dark area at the bend of the wing, a characteristic feature of Sanderlings when non-breeding, is often obscured by the breast feathers (as it also is in 66b). RJC.

◄ **66d. First breeding**. Florida, early May. Many remain in their non-breeding areas, gaining only partial breeding plumage; this individual retains very worn juvenile tertials. RJC.

▲ **66e. Adult breeding *alba***. Eastern England, early May. Confusion with breeding Red-necked Stint (69f), which also has a red neck and upper breast, is eliminated by Sanderling's larger size, different upper breast pattern and lack of a hind toe. RJC.

◄ **66f. Adult breeding *rubidus***. Florida, mid May. Separable in this plumage from the Palearctic race *alba* (66e) by wider grey mantle fringes. The differences from *alba* are not always as obvious as this! RJC.

▲ **66g. Adult breeding *alba*.** Wales, late July. Commencing moult to adult non-breeding. Darker above than in fresh breeding, the grey fringes having worn off; the paler red on face and breast results from a combination of fading and the growth of white non-breeding feathers. RJC.

▼ **66h. Non-breeding Sanderlings in flight**. Florida, late December. Pale grey, with a prominent white wing-bar and dark forewing. The three darker birds at the front of the flock are Dunlins. RJC.

67. SEMIPALMATED SANDPIPER
Calidris pusilla

A very small short-billed peep with partially webbed feet.

Identification L 14cm (5.5"); **WS** 30cm (12"). Forms a species pair with Western Sandpiper (68). Very small, with a medium-length, straight, blunt-tipped, black bill and medium-length black (occasionally dark olive) legs. Short primary projection beyond tertials. *In flight* from above shows narrow white wing-bar and white sides to dark-centred rump; feet do not reach tail tip. Sexes of similar size, females having marginally longer bills. Plumages vary with age and season. Partial webbing between all three toes. *Feeds* by picking from surface, sometimes wading.

Juvenile Whitish supercilium, accentuated by dark crown, loral line and ear-coverts. Crown uniformly dark and rather coarsely streaked; upperparts have dark grey-brown feather centres, with buffish fringes (July/August) which fade to off-white (September), largely lacking a mantle or scapular-V. Rear lower scapulars have dark anchor-shaped subterminal mark. Breast sides streaked dark, with buffish wash across upper breast.

First non-breeding/adult non-breeding Upperparts and wing-coverts uniformly brownish-grey, underparts white, with streaking on sides of breast sometimes continued across centre. Moult to first non-breeding occurs in the non-breeding area. First non-breeding retains juvenile wing-coverts, which are more worn, browner, and more broadly fringed than adult non-breeding; a few moult outer primaries.

First/second breeding Age of first breeding varies from second to fourth calendar year. Many do not breed and remain in non-breeding area, or move north to southern United States; many probably acquire non-breeding type upperparts, or just a scattering of darker breeding-type feathers.

Adult breeding Crown coarsely streaked. Rather dull upperparts have dark centres and grey or yellowish-buff fringes. Breast and upper flanks strongly streaked, the coarse breast-streaking usually forming a continuous band.

Calls Short, coarse *chrup* or *turp*.

Status, habitat and distribution Common, breeding in subarctic on northern edge of boreal forest, from far-eastern Siberia, Alaska to Baffin Island and Labrador. When non-breeding generally uses coastal estuarine habitats, the entire population leaving North America. Non-breeding season spent in Pacific central America, West Indies and northern and central South America. Vagrant to Galápagos, the Azores, and Europe east to Hungary.

Racial variation No races recognised but eastern Canadian birds on average have slightly longer bills.

Similar species In North America long-billed individuals are most likely to be confused with Western Sandpiper (68), the only other *Calidris* with partially webbed feet. Typically, Western Sandpiper has a larger, squarer, head, and a longer, more decurved, finely-tipped bill; juvenile Western has

◀ **67a. Juvenile**. Florida, mid September. In all plumages rather brownish and dull, usually with a rather short blunt-tipped bill; note olive tinge to the legs of this individual, which occurs on some birds. RJC.

bright chestnut-fringed upper scapulars, and moults earlier to first non-breeding than Semipalmated. Adult breeding Western has much bright chestnut on upperparts and diagnostic small chevrons on breast and upper flanks. Short-billed, non-breeding Western is particularly confusing, but note subtly decurved and finely-pointed bill, finer dark streaking on crown and breast. See also Table 7 (p. 210).

References Chandler (1990a), Donahue, P.K. (1996), Grant (1981), Gratto-Trevor (1992), Jonsson & Grant (1984), Kaufman (1990), Olah & Tar (2003), Phillips (1975), Veit & Jonsson (1984).

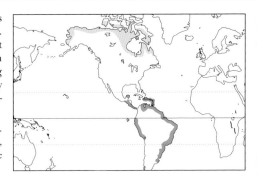

▲ **67b. First non-breeding**. Venezuela, mid November. Retains juvenile wing-coverts; virtually all move to Central and South America when non-breeding, and so are rarely seen in North America in this or adult non-breeding plumage. RJC.

▼ **67c. First breeding**. Florida, mid May. Semipalmated Sandpipers often do not breed in their second or third calendar year, and either remain in non-breeding areas or move north, but only to southern United States. This individual has worn coverts and some dark-centred grey-fringed first-breeding upperparts; compare with first-breeding Red-necked Stint (69e), which has acquired similar feathers. The slightly longer bill suggests this may be a female. RJC.

▲ **67d. Adult breeding**. Florida, mid May. Very dull in comparison to breeding Western Sandpiper, with dark-centred upperparts, though shows a rufous tinge on the ear-coverts. RJC.

▼ **67e. First or second breeding**. Florida mid-May. Both Semipalmated and Western Sandpiper show webbing between all three toes. This individual has a particularly long bill, and is presumably from the long-billed eastern Canadian population. The bill length invites comparison with Western Sandpiper, but note similarity with the plumage of both 67c and 67d, including a coarsely streaked crown, and the absence of Western Sandpiper's bright rufous tones on upperparts and chevrons on underparts. RJC.

▼ **67f. Juvenile in flight**. New Jersey, early October. A similar flight pattern is shown by many of the small *Calidris* sandpipers. Matthew Studebaker.

68. WESTERN SANDPIPER
Calidris mauri

Confusingly similar to Semipalmated Sandpiper in non-breeding plumages but with a longer, slightly decurved bill.

Identification L 15cm (6"); **WS** 30cm (12"). Forms a species pair with Semipalmated Sandpiper (67) Very small, with variable, but typically medium-length, slightly decurved, black bill (which gives the impression of a miniature Dunlin) and medium-length black legs. Primary projection beyond tertials is very short. *In flight* from above shows narrow white wing-bar and white sides to dark-centred rump; toes extend to tail-tip or marginally beyond. Sexes of similar size but males are often noticeably shorter-billed; plumage varies with season and age. Partial webbing between toes. *Feeds* by picking, probing and stitching, often wading quite deeply.

Juvenile Crown greyish, darker in centre; ear-coverts rather pale; whitish supercilium, contrasting with dark eye-stripe. Upperparts greyish but with strong chestnut fringes on mantle and, especially, on upper scapulars (often the last of the upperparts to be moulted); mantle and scapular-Vs faint; dark anchor-shaped marks on rear lower scapulars.

First non-breeding/adult non-breeding Uniform grey above, crown finely streaked; below, white with narrow band of fine streaking across upper breast. First non-breeding retains brownish, dark-centred, pale-fringed, juvenile wing-coverts; adult has uniform grey wing-coverts.

Adult breeding At least some breed in second calendar year. Bright rufous on crown, ear-coverts and at base of lower scapulars; extensive small chevron markings across breast and on flanks are diagnostic of this species among the stints, and may be retained into September after much of the breeding plumage has been lost.

Call Thin, high-pitched *jeet*.

Status, habitat and distribution Common, breeding on subarctic tundra in a relatively restricted area in western Alaska and adjacent north-east Siberia. Non-breeding from Washington (scarce), California, south to central America and Peru, and New Jersey south to South America. Rare in eastern Canada. Vagrant to the Azores, Madeira and western Europe.

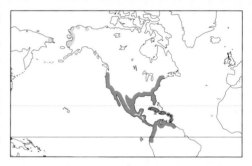

▼ **68a. Juvenile. Florida, early September**. Fairly heavy based, tapered and decurved bill separates from Semipalmated Sandpiper; longish bill suggests a female. Aged by neat upperparts with rufous fringes, and small, pale-fringed coverts. RJC.

Racial variation No races recognised.

Similar species Semipalmated Sandpiper (67), which see. Dunlin (82) is larger, has a similar shaped bill (but proportionally longer in American races) and appears smaller-headed and longer-bodied. See also Table 7 (p. 210).

References Jonsson & Grant (1984), Kaufman (1990), Veit & Jonsson (1984), Wilson (1994).

◄ **68b. First non-breeding.** Florida, late September. Juvenile Western Sandpipers moult earlier to first non-breeding than Semipalmated Sandpipers, but often retain a few of their rufous-fringed upper scapulars, as here. The wing-coverts and tertials are juvenile feathers. Though the bill is fairly short (suggesting a male) it is still tapered and decurved, unlike the usually shorter, straighter, blunter-tipped bill of Semipalmated Sandpiper. RJC.

◄ **68c. Adult male non-breeding.** California, mid November. A relatively high proportion of Westerns non-breeding in the southern United States are short-billed males; the females, as with many other shorebirds when non-breeding, tend to migrate further south. This predominance of males (see 68f) can lead to confusion with Semipalmated Sandpipers which rarely, if ever, occur in the United States when non-breeding. RJC.

◄ **68d. Adult female breeding.** Florida, mid May. Much bright rufous on head and upperparts, and many chevrons on the flanks are diagnostic of Western Sandpiper in this plumage. Male and female plumages are similar. Note webbing between toes, as with Semipalmated Sandpiper. RJC.

▲ **68e. Adult female non-breeding**. Florida, late August. Still moulting, retaining a few breeding-plumage upperparts and diagnostic chevrons on breast and flanks. The tapered, pointed appearance of the bill is lost when it is held very slightly open, as here. RJC.

▲ **68f. Non-breeding Western Sandpipers in flight**. California, early November. All are short-billed males, but have the typical slightly decurved, tapered bill of Western Sandpiper. RJC.

69. RED-NECKED STINT
Calidris ruficollis

A particularly long-winged stint with a distinctive breeding plumage.

Identification L 14.5cm (5.75"); **WS** 32cm (12.5"). Forms a species pair with Little Stint (70). Very small, with short to medium-length, usually straight and blunt-tipped black bill, and medium-length black legs. Primary projection beyond tertials long. Sexes are of similar size; plumage varies seasonally and with age. *In flight* from above shows narrow white wing-bar and white sides to dark-centred rump. *Feeds* actively, picking from surface, occasionally wading.

Juvenile Head pattern rather plain; supercilium diffuse, non-existent in front of eye and lacking obvious 'split' effect shown by Little Stint. Mantle and upper scapulars dark-centred with rufous fringes, indistinct (or absent) off-white mantle-V; lower scapulars have white fringes and lozenge-shaped dark centres (but less dark than Little Stint). Wing-coverts and tertials grey, with narrow, darker centres and whitish fringes, never as dark as Little Stint.

First non-breeding/adult non-breeding Generally similar to Little Stint but rather more uniform and paler grey above. Has dark area from lores through eye to ear-coverts. Lacks breast band. First non-breeding retains juvenile wing-coverts; those of adult are greyer and less worn.

First-breeding Many remain in non-breeding area; most individuals probably do not breed until their third calendar year. Generally acquire only partial breeding plumage; some start to replace outer primaries in April.

Adult breeding Very distinctive, with whitish or dull red supercilium behind eye (depending on progress of moult or wear); face, throat and upper breast brick-red, bordered with a necklace of small black spots across upper breast. Broad grey or rusty fringes to black-centred upper scapulars and some tertials; indistinct mantle-V.

Calls Differ from Little Stint: *kreet, kreep* or *chreek.*

Status, habitat and distribution Breeds in low mountain tundra in Siberia, from Taimyr to far north-east, and in north-west Alaska. When non-breeding found on coastal mudflats, but also on inland wetlands, in eastern India, south-east Asia, Indonesia and Philippines to Australia and New Zealand. Vagrant to Aleutians, both Pacific and Atlantic North American coasts (rare inland), Peru and Bermuda. Rare vagrant to western Europe, South Africa and Iran.

Racial variation No races recognised.

Similar species Combination of dark legs and

▼ **69a. Juvenile.** Japan, late August. In fresh plumage; some individuals at this age have pale buff on throat and upper breast. Separated from Little Stint by shorter, blunter bill, grey (not dark-centred) and buff-fringed coverts, narrow or absent rufous fringes to tertials and lack of 'braces' on mantle. Nobuhiro Hashimoto.

lack of webbing between toes eliminates all other stints, except Little Stint (70). In juvenile and adult breeding Little Stint can be separated by brownish coverts with rufous fringes, and when non-breeding may be doubtfully distinguished by paler centres to upperparts and, when present, grey-washed breast-band. See also Table 7 (p. 210).

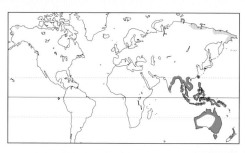

References Alström & Olsson (1989), Bamford *et al.* (2005), Barter (2002), Jonsson & Grant (1984), Kirwan (2007), Urban *et al.* (1986), Veit & Jonsson (1984).

▲ **69b. Juvenile**. Northern Territories, Australia, mid September. Has reached its non-breeding grounds; although in complete juvenile plumage it is worn and faded, but note the distinctive juvenile scapular pattern and pale rufous fringes to the tertials. RJC.

▲ **69c. Probable adult non-breeding**. Thailand, early December. A difficult bird to age; in spite of buffy fringes to tertials has adult-type non-breeding wing-coverts. RJC.

▲ **69d. First non-breeding/first breeding**. Western Australia, early April. Red-necked Stints do not breed until their third calendar year. Second calendar year birds remain in non-breeding areas and acquire limited breeding plumage, in this case one or two scapulars, though it may get more such feathers; otherwise as first non-breeding. RJC.

◄ **69e. First breeding/second non-breeding**. Victoria, Australia, late August. Many first-breeding birds have prominent dark centres to crown and upperpart feathers. RJC.

◄ **69f. Adult breeding**. South Korea, late May. Fresh breeding plumage. Compare with breeding Sanderlings (66e and 66f). RJC.

◄ **69g. Adult breeding wing-stretching**. Japan, late May. Note grey wing-coverts; as with juvenile, helpful for separation from Little Stint. Nobuhiro Hashimoto.

70. LITTLE STINT
Calidris minuta

A common Eurasian stint wintering in Africa and South Asia, readily identified in most of its range by black legs.

Identification L 13cm (5"); **WS** 30cm (12"). Forms a species pair with Red-necked Stint (69). Very small, with medium-length, finely-tipped, slightly decurved black bill, and medium-length black legs (but see juvenile below). Primary projection beyond tertials long. *In flight* from above shows narrow white wing-bar and white sides to dark-centred rump. Sexes are of similar size; plumage varies seasonally, and with age. *Feeds* actively, picking from surface, occasionally wading.

Juvenile Crown has darker central area and narrow, white, lateral stripe behind eye, usually giving a prominent split-supercilium effect. Dark-centred mantle and scapular feathers fringed rufous, except for prominent white mantle- and scapular-Vs; lower rear scapulars have solid dark centres but occasionally they resemble those of Red-necked Stint. Sides of upper breast neatly streaked over orange-buff wash, forming a distinctive patch. Some individuals have brownish-black legs.

First non-breeding/adult non-breeding Grey above but generally with feather centres darker, giving scalloped appearance at distance. Usually has uniform greyish wash, sometimes streaked, across upper breast. First non-breeding retains some worn juvenile wing-coverts, which appear browner than the greyer, more uniform, newly-grown upperparts; some moult all primaries from December, others outer ones only.

▼ **70a. Juvenile**. Eastern England, late August. In fresh plumage, showing double supercilium and the 'braces' (formed by alignment of the pale feather edges) that are a prominent feature of juvenile plumage. RJC.

Adult breeding Age at first breeding not known but probably second calendar year. Head and breast are variably orange-rufous with dark streaks and mottling; throat always white. Mantle, scapulars and wing-coverts dark brown, with broad orange to orange-rufous fringes; prominent mantle-V.

Call High pitched staccato *stit*.

Status, habitat and distribution A breeder in high Arctic coastal tundra in northern Scandinavia in small numbers, and more commonly eastward in central Siberia, irregular to far north-eastern Siberia, where it overlaps with Red-necked Stint. On migration and when non-breeding, uses muddy coastal and estuarine localities, also inland wetlands. Non-breeding mainly in Africa south of Sahara, but small numbers found around Mediterranean and Arabia to India. Vagrant to Aleutians, and both Pacific (south to Mexico) and Atlantic coasts of North America, Barbados, the Maldives and Australia.

Racial variation and hybridisation No races recognised but occasional 'grey-morph' juveniles occur, with less rufous, more uniformly coloured, upperpart fringes, and mantle- and scapular-Vs indistinct or absent. A presumed juvenile hybrid Little x Temminck's Stint had upperparts, particularly scapulars, with a pattern intermediate between the two species, an intermediate length primary-projection, a pale Little Stint-like mantle-V, and yellow-brown legs.

Similar species Red-necked Stint (69), which see. See also Table 7, p. 210.

References Cox (1988b), Ebels (2002), Iliff *et al.* (2004), Jonsson (1996), Jonsson & Grant (1984), Prince & Croxall (1996), Thorup (2006), Tomkovich (1996b), Veit & Jonsson (1984).

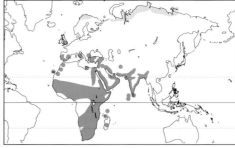

▲ **70b. Juvenile**. Eastern England, mid October. An older individual; note that if the upper scapulars were raised the white edges would line up as a scapular-V. Some have brownish, not black legs, as this bird, but even so the leg colour is still much darker than any of the pale-legged stints. RJC.

◀ **70c. Juvenile wing-stretching**. Eastern England, mid October. Showing dark-centred, rufous fringed coverts; the same individual as 70b. RJC.

◀ **70d. Juvenile**. Eastern England, mid September. A 'grey morph' individual; an unusual colour phase where the rufous upperpart fringes are replaced by grey. Birds like this clearly differ in colour from the norm, but are otherwise similar in size, structure and behaviour. Also shows the double supercilium. This bird is in the same plumage state as 70b and 70c. RJC.

▲ **70e. Adult breeding**. Cyprus, late April. The rich, rufous-fringed feathers are newly acquired breeding plumage. Some birds have a little more red on the face, which this one may still acquire. RJC.

▼ **70f. Adult breeding/non-breeding**. Eastern England, mid August. Moulting to non-breeding; adults at this plumage stage can be very confusing. RJC.

▼ **70g. Juvenile/first non-breeding in flight**. Oman, early November. See also 78g for Little Stints in flight. Markus Varesvuo.

71. TEMMINCK'S STINT
Calidris temmincki

A rather plain stint with yellowish or greenish legs and white outer tail feathers.

Identification L 14cm (5.5"); **WS** 31cm (12"). Very small, with short, finely-tipped slightly decurved black bill, and medium-length yellowish or greenish legs. Tail often extends beyond folded wing. *In flight* from above shows narrow white wing-bar, white sides to dark-centred rump and, diagnostically among stints, white outer tail feathers (best seen on take-off and landing); feet do not extend beyond tail. Sexes of similar size; plumage varies seasonally and with age. *Feeds* in freshwater wetlands by picking, often in crouched bent-legged stance, usually amongst vegetation at edge of open mud.

Juvenile Uniformly plain buffish-grey upperparts, with narrow, dark subterminal lines and pale buff fringes, a pattern unique amongst the stints.

First non-breeding/adult non-breeding Uniformly greyish-brown upperparts; breast (sometimes sides only) light brown. First non-breeding retains some worn juvenile wing-coverts; outer primaries replaced from December.

Adult breeding First breeds in second calendar year. Mantle and scapulars irregularly blotched darker owing to random mixture of uniform greyish- and blackish-centred pale-fringed feathers. Prominence of supercilium varies with individual, those with an indistinct supercilium appearing plain-faced.

Call Distinctive thin, high-pitched trilling, often continuously repeated.

Status, habitat and distribution Breeds in Arctic and subarctic, often near or amongst low vegetation, from Scotland (very rare) and Norway continuously to far-eastern Siberia. An uncommon migrant in western Europe, usually occurring singly or in small numbers. Non-breeding in muddy freshwater environments, both near the coast and inland, locally around the Mediterranean, Africa from Sahara to the equator, Arabia and the Indian subcontinent to south-east Asia and Borneo. Vagrant to Ireland, Aleutians and British Columbia.

Racial variation and hybridisation No races recognised. For hybrid Little x Temminck's Stint, see Little Stint.

Similar species Among stints, yellowish legs are shared only with Long-toed Stint (72) and Least Sandpiper (73); but white outer tail feathers, juvenile upperpart feathers and call are all diagnostic. In all plumages, Temminck's has a rather uniform brownish breast, which is white or sharply streaked in the other two yellow-legged species. See also Table 7 (p. 210).

References Hendenström (2004), Jonsson (1996), Jonsson & Grant (1984), Mudge & Dennis (1995), Thorup (2006), Veit & Jonsson (1984).

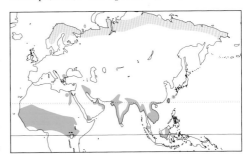

◄ **71a. Juvenile**. Central England, late September. Among the pale-legged stints the dark submarginal lines on the upperparts are diagnostic of both age and species. The dark breast, which is common to all plumages, gives the appearance of a miniature Common Sandpiper. RJC.

▲ **71b. Adult non-breeding**. India, late December. Uniform grey-brown upperparts, but still with dark breast. RJC.

▲ **71c. Adult breeding**. Cyprus, mid April. Breeding adults gain a variable proportion of dark-centred breeding upperparts; this one has very few. RJC.

▲ **71d. Adult breeding**. Finland, June. Has a large proportion of breeding upperparts. Gordon Langsbury.

▲ **71e. Adult breeding in flight**. Finland, early July. The only stint with white outer tail feathers. Markus Varesvuo

72. LONG-TOED STINT
Calidris subminuta

An Asian species with long yellowish legs, and longer neck and toes than other stints.

Identification L 13cm (5.5"); **WS** 29cm (11.5"). Forms a species pair with Least Sandpiper (73). Very small, with short to medium-length, very slightly decurved black bill, with pale base to lower mandible (but see below; bill entirely black in Least Sandpiper), and medium-length yellowish legs. Bill proportionally shorter than Least Sandpiper. Has a distinctive facial pattern (best shown on juvenile) with split supercilium, pale oval on lores connected to supercilium framed by broken (or waisted) loral line and darker forehead; supercilium diffuse behind eye. Longer neck, longer legs and longer toes than other stints; toes (including claws) slightly longer than tarsus, bill noticeably shorter than tarsus. Primaries project marginally beyond tertials. *In flight* from above has narrow white wing-bar and white sides to dark-centred rump, as does Least Sandpiper. Toes extend beyond tail, unlike Least Sandpiper. Sexes of similar size; plumages vary with age and season. *Feeds* by picking, generally at freshwater margins; usually singly or in small loose groups. Has rather upright stance, not crouching as Least Sandpiper; regularly extends neck vertically when disturbed ('giraffe' pose).

Juvenile Crown rusty, streaked; contrasting, off-white supercilia do not meet on forehead (usually do so in Least Sandpiper); may have lateral crown stripes giving a split-supercilium effect. Upperparts have dark feather centres with bright rufous fringes and obvious creamy mantle-V; innerwing-coverts have white tips and fringes. Breast and flanks finely streaked over brownish wash, often paler at centre (compare Least Sandpiper); remainder of underparts white. Pale bill base usually fairly obvious.

First non-breeding/adult non-breeding Head pattern as juvenile but less contrasting. Upperparts generally brownish-grey, with distinctive blackish feather centres and broad grey-brown fringes to scapulars and wing-coverts. Pale bill-base may be restricted or absent. First non-breeding retains worn juvenile wing-coverts; some (Australia) moult outer primaries from March.

Adult breeding The majority probably breed in their second calendar year. Head pattern much as juvenile. Mantle and scapulars, innerwing-coverts and tertials blackish with broad rufous fringes, the latter never scalloped, unlike Least Sandpiper; creamy mantle-V. Breast finely streaked dark brown, with abrupt lower margin (recalling Pectoral Sandpiper). Pale bill-base may be restricted or absent.

Calls Soft, liquid *chree* or disyllabic *chuilp*.

Status, habitat and distribution Breeds near tree-line in marshy, sparsely vegetated subarctic, discontinuously from south-west to far-east Siberia and Kurile Islands. Non-breeding from eastern

◄ **72a. Juvenile.** Japan, late August. Note oval pale area on lores connecting with supercilium, and the near-broken dark loral-line, which, together with the split supercilium, gives Long-toed Stint a characteristic facial pattern. The mantle 'braces' are shown by both juvenile and adult breeding, but the pale base to the lower mandible is usually obvious only on juveniles. Nobuhiro Hashimoto.

India to south-east Asia, Taiwan and Australia. Vagrant Aleutians, Alaska. Rare vagrant to Pacific North America and western Europe.

Racial variation No races recognised.

Similar species Temminck's Stint (71), which see. For separation from Least Sandpiper (73) see Identification above, and Table 7 (p. 210).

References Alström & Olsson (1989), Higgins & Davies (1996), Jonsson & Grant (1984), Tomkovich (1996b), Veit & Jonsson (1984).

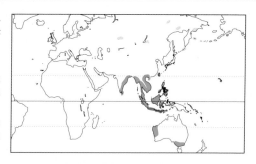

▲ **72b. Adult non-breeding**. Thailand, early December. RJC.

▼ **72c. Adult non-breeding**. Western Australia, mid March. Toe length is important for separation from Least Sandpiper; the longest toe is slightly longer than tarsus (well shown here) and the bill is shorter than tarsus, which is often easier to judge than the length of the toe. RJC.

◄ **72d. Adult non-breeding**. Western Australia, mid March. Just commencing moult to adult breeding, with rufous-fringed breeding-type tertials. This bird is in the 'giraffe' pose, frequently shown by Long-toed Stint. RJC.

▼ **72e. Adult breeding**. Western Australia, mid March. Bright rufous fringes and pale mantle lines. RJC.

◄ **72f. Non-breeding wing-stretching**. Japan, late November. Nobuhiro Hashimoto.

73. LEAST SANDPIPER
Calidris minutilla

The smallest of the peeps, this species can usually be distinguished by its yellowish legs.

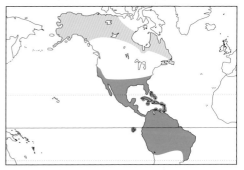

Identification L 13cm (5.5"); **WS** 28cm (11"). Forms a species pair with Long-toed Stint (72). Very small, with medium-length, tapered, slightly decurved, all-black bill and medium-length yellowish legs. Toes (including claws), bill and tarsus all very similar in length. Primary tips barely extend beyond tertials. *In flight* from above has narrow white wing-bar and white sides to the dark-centred rump; toes just reach end of tail. Sexes similar; plumages vary seasonally and with age. *Feeds* typically by picking on mud or at water's edge, with crouched bent-legged stance, which can make it appear smaller than other stints, even though its body size is much the same; more tolerant of brackish or saline conditions than Long-toed Stint. Only rarely shows the 'giraffe' pose characteristic of Long-toed Stint.

Juvenile Crown rusty, streaked; dull supercilia are wider in front of eye and often, but not always, meet above bill. Upperparts have dark feather centres and bright rusty-brown fringes, and an inconspicuous creamy mantle-V. Breast uniformly streaked and washed buff; remainder of underparts white. Some individuals can be very dull, particularly when worn.

First non-breeding/adult non-breeding Head pattern as juvenile but less contrasting. Upperparts brownish-grey with darker feather-centres; breast streaked or washed brownish-grey, remainder of underparts white. First non-breeding loses some or all juvenile wing-coverts, so only those retaining juvenile buff (fading to off-white) covert fringes can be aged with confidence. Some first non-breeding replace outer primaries from January.

Adult breeding Many breed in their second calendar year. Heavily streaked crown, rusty ear-coverts; mantle and scapulars have blackish-brown centres with broad, pale grey-brown fringes when fresh; the fringes quickly wear, revealing rusty feather bases. Fringes of inner coverts and tertials often scalloped. Creamy mantle-V. Breast heavily streaked dark brown.

Call Very high-pitched *kreet*, with rising inflection.

Status, habitat and distribution Common, breeding in wettish tundra or grassland near the tree-line, continuously from western Alaska to Newfoundland. Non-breeding birds occur

◄ **73a. Juvenile**. Florida, late August. A bright individual; on many (as here) the supercilia meet above the bill. Bill slightly longer than tarsus, which separates from Long-toed Stint. RJC.

both coastally and on inland wetlands in southern United States, the Caribbean and Central and South America south to Chile and Brazil. Vagrant to Iceland, the Azores and western Europe.

Racial variation No races recognised.

Similar species Temminck's Stint (71) and Long-toed Stint (72), which see. See also Table 7 (p. 210).

References Cooper (1994), Jonsson & Grant (1984), Veit & Jonsson (1984).

◄ **73b. Juvenile.** Southwest England, mid September. On this individual the supercilia do not meet above the bill, a feature more typical of Long-toed Stint. RJC.

◄ **73c. Juvenile.** Florida, late August. A duller, more worn individual than 73a. RJC.

◄ **73d. First non-breeding.** Florida, early November. Grey, non-breeding upperparts but rufous-fringed juvenile coverts and tertials. RJC.

▲ **73e. First non-breeding**. Venezuela, mid November. Similar plumage to 73d, but more worn. The 'giraffe' pose shown here is typical of Long-toed Stint, but is only occasionally shown by Least Sandpiper. RJC.

▼ **73f. Adult non-breeding**. Florida, early January. RJC.

▲ **73g. Adult breeding**. Florida, early May. In fresh plumage, with broad rufous upperpart fringes; note the slightly scalloped upper tertial, a feature not shown by adult breeding Long-toed Stint. RJC.

▼ **73h. Non-breeding Least Sandpiper in flight**. California, early November. Has grey outer tail feathers, like all stints except Temminck's. RJC.

74. WHITE-RUMPED SANDPIPER
Calidris fuscicollis

Larger than the stints but smaller than a Dunlin, this is a common breeding species of the North American tundra.

Identification L 16.5cm (6.5"); **WS** 38cm (15"). Forms a species pair with Baird's Sandpiper (75). Small with medium-length, slightly decurved, black bill (often pale at base of lower mandible), and medium-length blackish legs (Baird's Sandpiper's bill is typically all black, proportionally shorter, straighter and less tapered). Breast and flanks streaked over white ground in all plumages (Baird's Sandpiper flanks are virtually unstreaked, but a streaked breast-band overlies brownish wash). Long-winged, with wing-tip extending beyond tail; white of uppertail-coverts can sometimes be seen between incompletely folded wings. *In flight* from above shows narrow white wing-bar, uppertail-coverts form a narrow white patch (which eliminates Baird's Sandpiper), and darker tail; feet do not extend beyond tail. Sexes of similar size; plumages vary with season and age. *Feeds* with body held horizontally, picking from surface.

Juvenile Rusty crown, broad white supercilium; mantle and upper scapulars fringed rusty but with whitish mantle and scapular-Vs; wing-coverts tipped whitish. Breast and flanks finely streaked grey, remainder of underparts white.

First non-breeding/adult non-breeding Adults have brownish-grey upperparts, with darker centres to mantle and upper scapulars. First non-breeding retains some juvenile scapulars and wing-coverts until at least January. Has variable pre-breeding primary moult, some moulting all primaries, some outer ones only, others none.

Adult breeding Probably breeds in second calendar year. White supercilium, streaked crown, ear-coverts and mantle, all with rusty tinge; dark-centred scapulars have grey, rusty and whitish fringes. Breast and flanks with bold dark streaks and chevrons. When breeding, the male can be distinguished by the extended throat, which is used in display.

Call A distinctive, thin, high-pitched *tzeet*.

Status, habitat and distribution Common, breeding in wet, near-coastal tundra from northern Alaska east through northern Canada to Southampton Island; possible recent breeding on Bear Island, Barents Sea, so might possibly be breeding elsewhere in the Eurasian high Arctic. Non-breeding birds occur on near-coastal and inland muddy freshwater margins, in south-eastern South America and Falkland Islands. Vagrant to western Europe, the Azores, West Africa, Australasia and Antarctica.

Racial variation and hybridisation No races recognised. Has apparently hybridised with Pectoral

▼ **74a. Juvenile/first non-breeding**. Eastern England, mid October. Long wings give an attenuated rear profile shared by Baird's Sandpiper, but has streaked flanks in all plumages and a pale-base to the lower mandible, though this is not very evident in this individual. Has replaced a few juvenile scapulars. RJC.

Sandpiper (76) on several occasions, and also with Dunlin (82) and Buff-breasted Sandpiper (85).

Similar species Slightly larger than all the stints. Similar to Baird's Sandpiper (75; see Identification above). The only other small white-rumped sandpipers are Curlew Sandpiper (78) and Stilt Sandpiper (79), both of which are slightly larger, longer-necked and longer-legged.

References Cox (1987), Fishpool & Demey (1991), Golley (1990), Goodwin (1981), Grønningsaeter (2005), McLaughlin & Wormington (2000), Parmelee (1992), Prince & Croxall (1996).

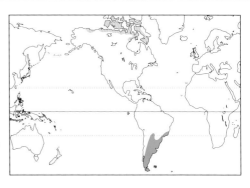

◀ **74b. Non-breeding**. Falkland Islands, January. The grey wash on the flanks is a useful separation from stints and Curlew Sandpiper in this plumage. Gordon Langsbury.

◀ **74c. Adult breeding**. Indiana, late May. Dark upperparts, heavy breast-streaking and chevrons on flanks are apparent in this plumage. J. D. Phillips.

▲ **74d. Adult breeding/non-breeding**. Florida, late September. Still retains worn wing-coverts, and has a few dark upperparts and flank chevrons from breeding plumage, but has many new plain-grey non-breeding upperparts. RJC.

▼ **74e. Adult breeding/non-breeding**. Florida, late September. The same bird as 74d, preening, showing the white uppertail-coverts, which are the best separation from Baird's Sandpiper. RJC.

▼ **74f. Adult breeding wing-lifting**. Rhode Island, mid June. The flight pattern, with a narrow white wing-bar and white uppertail-coverts is shown well. This photo was taken well south of the breeding grounds, so this individual is unlikely to breed! Michelle L. St.Sauveur.

75. BAIRD'S SANDPIPER
Calidris bairdii

An elegant, long-winged calidrid that migrates between the high Arctic and southern South America.

Identification L 15cm (6"); **WS** 39cm (15.5"). Forms a species pair with White-rumped Sandpiper (74). Very small, with straight, rather slender, medium-length blackish bill, and medium-length blackish legs. Long-winged, with wing-tip extending beyond tail. *In flight* from above shows a very narrow white wing-bar, dark-centred grey rump with some white at sides, and dark tail; feet do not extend beyond tail. Sexes similar; plumages vary seasonally and with age. *Feeds* with horizontal carriage, often on dry sparsely vegetated areas, less often at water's edge, picking from surface.

Juvenile General appearance brown, with rusty tinge. Streaked brown crown, indistinct supercilium, upperparts and wing-coverts brown, strongly scalloped with buffish or whitish fringes; breast streaked brown over buff wash, usually forming clear-cut breast-band. Remainder of underparts white.

First non-breeding/adult non-breeding Uniform mid-brown above, with indistinct supercilium, brown across breast. First non-breeding generally retains juvenile coverts and has variable pre-breeding primary moult; some moult all primaries, some outer ones only, others none.

Adult breeding Breeds in second calendar year. Crown strongly streaked dark brown; mantle and scapulars dark brown or blackish with broad buff fringes. Breast heavily streaked dark brown.

Call A low *treep*.

Status, habitat and distribution Breeds in high Arctic dry tundra, both coastally and inland, in north-eastern Siberia, and from northern Alaska east to Baffin Island, Ellesmere Island and north-west Greenland, and at higher altitudes in southern Alaska. Non-breeding period spent in relatively dry, sparsely vegetated areas in central and southern South America. Vagrant to Hawaii, Japan, Galápagos, Falkland Islands, Iceland, the Azores, western Europe, western and southern Africa, Indonesia, Australasia.

Racial variation and hybridisation No races recognised, but some rare individuals of all ages may be noticeably greyer than usual. Possible hybridisations include Baird's Sandpiper x Buff-breasted Sandpiper, Baird's Sandpiper x Pectoral Sandpiper and Baird's Sandpiper x Dunlin.

Similar species White-rumped Sandpiper (74), which see. From Dunlin (82) by smaller size, finer bill, primary extension beyond tail, and less conspicuous wing-bar; browner and more scaly juvenile; lacks adult breeding Dunlin's black belly.

References Gill & Tomkovich (2004), Laux (1994), Moskoff & Montgomerie (2002), O'Brien *et al.* (2006), Prince & Croxall (1996), Tomkovich (1996b).

▼ **75a. Juvenile**. Southwest England, mid September. Neat, pale-fringed juvenile plumage; all-black bill and lack of flank streaking help separate Baird's from White-rumped Sandpiper. RJC.

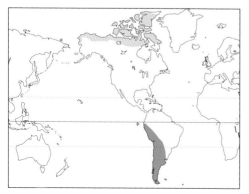

▲ **75b. Adult non-breeding**. Chile, late December. Upperparts uniformly brown-grey when non-breeding. Note lack of flank streaking and brown wash under breast-streaking, which helps separate from White-rumped Sandpiper. Hanne and Jens Eriksen.

◄ **75c. Adult breeding**. Kansas, late April. Dark upperparts with broad buff fringes and breast heavily streaked dark brown. RJC.

◄ **75d. Baird's Sandpiper in flight**. Colorado, August. Flight pattern, with narrow white wing-bar and dark-centred rump, is similar to most of the other small *Calidris* shorebirds. Bill Schmoker.

76. PECTORAL SANDPIPER
Calidris melanotos

With its sharply demarcated breast, this is one of the easiest calidrids to identify, in all plumages.

Identification Male **L** 23cm (9") **WS** 44cm (17.5"); female **L** 21cm (8") **WS** 40cm (16"). Forms a species pair with Sharp-tailed Sandpiper (77). Sexually dimorphic; medium-sized (male) or small (female), with long neck, medium-length slightly decurved brownish bill, darker at tip; medium-length yellowish legs. Streaked breast has abrupt junction with white belly, unlike Sharp-tailed Sandpiper. Male 5–10% larger than female. *In flight* from above uniform except for a narrow, indistinct, white wing-bar, and white sides to dark-centred rump and uppertail-coverts; toes just reach tail-tip. *Feeds* among vegetation and at muddy freshwater margins, picking from surface and occasionally probing.

Juvenile Crown streaked; whitish supercilium; dark-brown upperparts with rusty fringes, except for whitish mantle and scapular-Vs, and off-white wing-covert fringes. Rusty fringes give a bright appearance, but becomes greyer as fringes fade and wear. Brown-streaked foreneck and breast; remainder of underparts white.

First non-breeding/adult non-breeding Head pattern much as juvenile, though less rusty; upperparts greyish, with darker feather-centres and broad, rather diffuse, paler fringes. Once juvenile mantle and scapulars are lost, first non-breeding and adult non-breeding are doubtfully separable in the field. Juveniles moult only in non-breeding area, adults mostly so. Some first non-breeding may moult primaries (Australia).

First-breeding Most apparently return to breeding grounds and attempt to breed. At least some show a mixture of adult-type non-breeding and breeding upperparts.

Adult breeding Like juvenile, but duller, with inconspicuous supercilium; mantle and scapulars dark, with broader buff or off-white fringes. Breast of male is dark, mottled or spotted; that of the female streaked brown over brownish wash. Male inflates throat sac in breeding display, revealing extensive black at bases of pale-fringed throat and breast feathers.

Call A low, slightly rasping *krrick*.

Status, habitat and distribution Breeds in low-lying, poorly drained Arctic coastal tundra, from Yamal Peninsula in northern Siberia eastwards to northern Alaska and across northern Canada east to Hudson Bay; perhaps bred in Scotland in 2004 and in Spitsbergen in 2005. Non-breeding in grasslands and marshy wetlands, mainly in south-ern South America; small numbers in south-east Australia and New Zealand. Since breeding grounds extend to western Siberia, it is possible that a few may migrate via Europe, and perhaps spend the non-breeding season in Africa. Commonest North American vagrant (or possibly scarce migrant) to Europe, found less frequently throughout Africa, east to western India.

Racial variation and hybridisation No racial variation. Has apparently hybridised with White-rumped Sandpiper (74, which see), Baird's Sandpiper (75), Curlew Sandpiper (78; hybrid known as 'Cox's Sandpiper') and Dunlin (82).

▼ **76a. Juvenile**. California, early November. Still in near-complete juvenile plumage despite the late date, having lost only a few scapulars. Compare with Figure 7 (p. 12), which is in much fresher plumage. RJC.

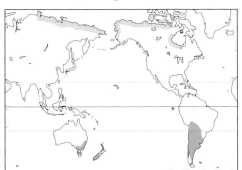

Similar species Sharp-tailed Sandpiper (77) has slightly shorter, straighter bill, slightly shorter legs and is 'pot-bellied'; in all plumages has diffuse (not abrupt as in Pectoral) lower margin to streaked breast, and has whiter supercilium. Juvenile Sharp-tailed is generally more rufous, with rusty cap, rufous wash on breast and white eye-ring; breeding adults have bold chevrons on breast and flanks.

References Alexander-Marrack (1992), Britton (1980), Cox (1990), Holling *et al.* (2007), Fishpool & Demey (1991), Gantlett & Grant (1989), Grønningsaeter (2007), Higgins & Davies (1996), Hjort (2005), Holmes & Pitelka (1998), Lees & Gilroy (2004), Prince & Croxall (1996), Sangster (1996), Schepers *et al.* (1991), Undeland & Sangha (2002).

◄ **76b. Adult non-breeding.** Texas, mid April. Well on its way back north to its breeding grounds, but still in complete non-breeding plumage. Difficult to assign to sex in this plumage unless size comparison is possible. RJC.

◄ **76c. Adult male non-breeding.** Florida, mid March. Many of the upperpart feathers are non-breeding with broad, greyish, diffuse fringes. The expanded breast feathers show the black bases of breeding male. Many individuals seem to be quite late to moult to breeding plumage. RJC.

◄ **76d. Adult male breeding.** Alaska, mid June. Lower margin of breast is swollen and heavily streaked in breeding male; upperparts have dark centres and narrow, even rufous fringes. René Pop.

◀ **76e. Adult female first breeding/breeding**. Alaska, early June. The breast-streaking is much less extensive than in male; upperparts, particularly scapulars, are non-breeding type with diffuse fringes, suggesting this may be a second calendar year bird. René Pop

▲ **76f. Adult female breeding**. Texas, mid August. The relatively restricted breast-streaking indicates a female. In very worn but near complete breeding plumage. RJC

◀ **76g. Juvenile in flight**. Massachusetts, early September. Note the thin central wing-bar. Glen Tepke.

77. SHARP-TAILED SANDPIPER
Calidris acuminata

A long-distance migrant that breeds in northern Siberia and spends the non-breeding season in Australia.

Identification Male **L** 20cm (8"), **WS** 43cm (17"); female **L** 18cm (7"), **WS** 41cm (16"). Slightly smaller than Pectoral Sandpiper (76), with which it forms a species pair. Compared to Pectoral Sandpiper has slightly shorter legs and a 'pot-belly', a diffuse (not abrupt) lower margin to streaked breast, whiter supercilium and eye-ring. Small, with medium-length, slightly decurved, blackish bill (though often with pale base, and pale cutting edge to lower mandible), prominent white eye-ring, and medium-length brown-yellow legs; breeding adults have bold chevrons on breast and flanks. *In flight* all primaries have white shafts; narrow white central wing-bar, and white sides to dark-centred rump and uppertail-coverts; toes just extend beyond tail-tip (similar to Pectoral Sandpiper, but sides of rump more streaked). Sexes similar, but male 5–10% larger; plumages vary seasonally and with age. Frequently raises wings on landing, a habit that aids identification. *Feeds* among vegetation and at muddy freshwater margins; picks from surface and occasionally probes.

Juvenile Crown strongly rusty, prominent pale supercilium, particularly behind eye. Upperparts, tertials, and wing-coverts dark with bright rusty fringes; throat white, necklace of fine streaking on buff ground across upper breast, remainder of breast washed buff-brown. Very bright when fresh, gradually fading and becoming duller with age. Legs often brighter yellow than adult.

First non-breeding/adult non-breeding Rather dull, with greyish-brown crown, white supercilium,

▼ **77a. Juvenile.** Japan, late October. Very bright rufous in this plumage; breast washed bright rufous, with far less streaking than Pectoral Sandpiper. Nobuhiro Hashimoto.

upperparts dark brown with broad diffuse paler fringes. Underparts largely white, with some fine dark streaking on upper breast and flanks. With adults, heavy streaking on breast and chevrons on breast sides, often retained into September. First non-breeding acquired by October–January, but retains at least some bright-buff fringed juvenile wing-coverts; the majority moult outer primaries during December–March.

First-breeding/adult breeding Many birds probably acquire breeding plumage and breed in their second calendar year. Dark overall, with supercilium less prominent owing to fine streaking. Upperparts dark brown with bright rusty fringes, pale mantle-V; throat and breast with much fine dark spotting; numerous dark chevrons on remainder of underparts, often from February.

Call Repeated short soft *pleep* or similar, when flushed.

Status, habitat and distribution Fairly common, breeding in hummocky, damp Arctic tundra in northern and north-eastern Siberia. When non-breeding, widely distributed in freshwater wetlands, intertidal marshes and mudflats from southern New Guinea and islands of the south-west Pacific to Australia (particularly in south-east), small numbers in New Zealand. Regular, mainly juveniles, in small numbers in the Aleutians and western Alaska; vagrant throughout North America, particularly Pacific coast; rare vagrant western Europe (west to Ireland), and north to northern Scandinavia; also India, Sri Lanka and south-east Asia.

Racial variation and hybridisation No racial variation. Apparent hybrids ('Cooper's Sandpiper') between Sharp-tailed Sandpiper (77) and Curlew Sandpiper (78) have occurred in south-east Australia.

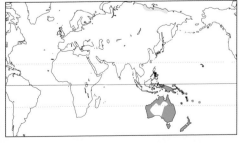

Similar species Pectoral Sandpiper (76), which see. Purple Sandpiper (80) is of similar size, has yellowish legs, and could be confused away from its usual rocky coastal habitat, but has proportionally longer bill and shorter legs, more chunky shape, is darker grey and lacks prominent pale supercilium.

References Barter (2002), Britton (1980), Kaufman (1990), Mlodinow (2001), Rosenband (1982), Smiddy & O'Sullivan (1995).

▲ **77b. Adults moulting to non-breeding**. Queensland, Australia, early September. These have newly acquired non-breeding dark-centred, grey-fringed upperparts, but both birds still have some breeding-plumage chevrons on flanks. RJC.

◄ **77c. Adult breeding**. South Korea, late April. Upperparts have broad rufous fringes, underparts heavily streaked with extensive chevrons on flanks. RJC.

◄ **77d. Adult moulting to breeding, wing-stretching**. Western Australia, late March. Showing wing-pattern. Rufous fringes on some scapulars; outermost primary is still growing. RJC.

78. CURLEW SANDPIPER
Calidris ferruginea

A small, elegant calidrid, larger and longer-billed than Dunlin, with a white rump in flight.

Identification L 20cm (8"); **WS** 41cm (16"). Small, with longish neck, long decurved black bill and medium-length black legs. *In flight* from above has narrow white wing-bar, and white lower rump and uppertail-coverts (with some barring when breeding); feet extend just beyond tail. Female is marginally larger, but this is not usually evident in the field; plumage varies seasonally, and with age. *Feeds* by picking from mud, but also often wades quite deeply; very occasionally feeds while swimming.

Juvenile Streaked darkish crown, prominent whitish supercilium; greyish-brown upperparts (mantle and upper scapulars dark brown) and wing-coverts, with dark subterminal lines and narrow whitish fringes. Breast lightly streaked and washed pale salmon-buff; remainder of underparts white.

First non-breeding/adult non-breeding Grey crown, prominent white supercilium. Grey upperparts and wing-coverts, with dark shaft-streaks and narrow white fringes; grey flecks across upper breast, otherwise underparts white. First non-breeding retains worn juvenile wing-coverts; most replace outer primaries, particularly those non-breeding in the more southern areas.

First-breeding Remains in non-breeding area, and attains adult non-breeding-type plumage or has a scattering of adult breeding mantle, scapular or underparts feathers.

Adult breeding Many probably breed from their third calendar year. Crown streaked; supercilium faint; mantle and scapulars blackish with red lateral spots and fairly broad greyish tips when fresh. Face (with small but conspicuous white area at bill base when fresh), nape, neck and remainder of underparts brick-red, though vent and undertail-coverts can be brick-red or white. Female has variable dark barring on underparts, and in breeding pairs may perhaps be distinguished by slightly longer bill and generally slightly duller, less deep-red coloration. Active body moult to or from breeding plumage makes the latter feature valueless for sexing migrants.

Call A quiet *chirrup*.

Status, habitat and distribution Breeds in tundra with scattered ponds in northern and north-eastern Siberia between the Taimyr Peninsula and Chukotka. Non-breeding season is spent in Africa, south and south-east Asia, and Australasia. A widespread vagrant.

Racial variation No races recognised but eastern breeding females show less dark barring below than do western females.

Similar species Dunlin (82) is more compact, shorter-legged, has distinctive juvenile and adult breeding plumage (the latter with a black belly), and lacks Curlew Sandpiper's white rump in flight; main potential confusion is with non-breeding North American and east Asian Dunlins, which

◀ **78a. Juvenile**. Wales, early September. In juvenile plumage; pale fringes with dark submarginal lines are a feature shared only with Red Knot and Temminck's Stint, neither of which have Curlew Sandpiper's decurved bill. RJC.

have longer, more decurved bills than other races; *hudsonia* (but not *pacifica*) Dunlin has streaked flanks (not shown by Curlew Sandpiper); but check for white rump. Stilt Sandpiper (79) is similarly proportioned and also has a white rump, but has blunter-tipped, straighter bill and longer yellow or greenish legs, and in flight shows no wing-bar.

References Barter (2002), Carmona *et al.* (2003), Ebels (2002), Engelmoer & Roselaar (1998), Minton (2000), Tomkovich (1996b).

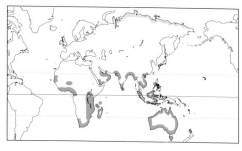

▲ **78b. First non-breeding.** Namibia, mid December. Uniformly grey adult-type upperparts, but retains worn juvenile wing-coverts. RJC.

▲ **78c. Non-breeding.** Western Australia, early April. Curlew Sandpipers in their second calendar year remain in non-breeding areas and have no or limited breeding-type plumage, so this may be a first-breeding individual. RJC.

▼ **78d. Adult non-breeding/breeding.** Cyprus, late April. Some breeding females are quite grey; alternatively, this individual may simply be delayed in moulting. RJC.

▲ **78e. Adult breeding**. Cyprus, late April. Difficult to sex except on the breeding grounds, but depth of colour suggests a male. Curlew Sandpipers frequently wade belly-deep, as here. RJC.

◄ **78f. Adult breeding wing-stretching**. Western Australia, late March. RJC.

◄ **78g. Adult breeding in flight**. Puglia, Italy, early May. The dark barring on the uppertail-coverts is a feature of breeding plumage and is not shown by juveniles or non-breeding birds, when both have completely white uppertail-coverts. The two lowest birds are Little Stints. Mathias Schäf.

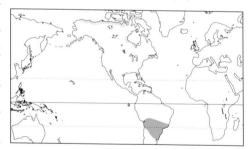

79. STILT SANDPIPER
Calidris himantopus

Elegant long-legged and long-necked calidrid with distinctive rufous ear-coverts when breeding.

Identification L 20cm (8"); **WS** 41cm (16"). Similar to Curlew Sandpiper but with longer, yellowish, legs. Small, with long neck and long, slightly decurved black bill and long yellowish or greenish legs. Brownish or greyish above; underparts pale, except adult breeding which is heavily barred beneath. *In flight* from above uniform apart from white rump and tail-coverts (with some barring when breeding); feet extend beyond tail. Sexes are of similar size, plumage varies seasonally and with age. Small webs between all three front toes. *Feeds* by vertical probing in both mud and water, often wading to belly.

Juvenile Crown streaked, tinged rusty, whitish supercilium; mantle and scapulars dark with rufous and whitish fringes. Wing-coverts and tertials darkish, with darker tips and whitish fringes. Neck and breast lightly streaked brown; remainder of underparts white. Legs often yellow.

First non-breeding/adult non-breeding Generally greyish, including crown; white supercilium. Upperparts narrowly fringed whitish. First non-breeding retains some worn juvenile dark-tipped tertials and wing-coverts, which are replaced only from November; renews some outer primaries.

First-breeding Some remain in non-breeding area, and are much as adult non-breeding; many, however, apparently acquire near-complete breeding plumage. Curiously, few apparently return to breed until their third calendar year, which is inconsistent with the acquisition of breeding-type plumage in the second calendar year.

Adult breeding Crown rufous with dark streaks, whitish supercilium; lores and ear-coverts bright rufous. Mantle and scapulars blackish, with broad buffish and whitish fringes; neck and upper breast heavily streaked blackish, underparts heavily barred blackish from lower breast to undertail-coverts. Legs usually olive green.

Call A monosyllabic *querp.*

Status, habitat and distribution Breeds in sub-arctic damp tundra just north of tree-line in a variety of habitats, from northern Alaska east to Hudson Bay. A few non-breeding birds occur in freshwater wetlands, particularly ponds, in Florida, on Gulf coast and southern California, but most winter in central South America. Vagrant to Iceland, Europe, Japan, Taiwan and Australia, mainly July–October.

◄ **79a. Juvenile**. Ontario, mid August. Superficially similar to Curlew Sandpiper, but has longer, yellow legs, straighter bill, and juvenile has darker-centred upperparts. Steve Pike.

Racial variation No races recognised.

Similar species Curlew Sandpiper (78), which see. Grey-tailed Tattler (124) and Wandering Tattler (125) of Asia and the Pacific respectively are the only comparable species with barred underparts when breeding, but both are larger (medium-sized), grey above, have medium-length straight bills, much shorter yellow legs, lack the white rump and occur in different habitats.

Reference Klima & Jehl (1998).

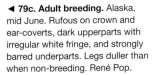

▲ **79b. Non-breeding**. Florida, late March. Uniformly grey upperparts, white underparts. Like Curlew Sandpiper, regularly wades belly-deep. RJC

◀ **79c. Adult breeding**. Alaska, mid June. Rufous on crown and ear-coverts, dark upperparts with irregular white fringe, and strongly barred underparts. Legs duller than when non-breeding. René Pop.

◀ **79d. Adult breeding Stilt Sandpipers in flight**. New York, early August. Unlike Curlew Sandpiper has no wing-bar, but shares the dark barring on the uppertail-coverts when in breeding plumage. Feet extend much further beyond the tail than in Curlew Sandpiper. Jim Gilbert.

80. PURPLE SANDPIPER
Calidris maritima

A dark, dumpy shorebird that breeds on the Arctic tundra and winters on rocky coasts of northern Europe and North America.

Identification L 21cm (8"); WS 41cm (16"). Forms a species pair with Rock Sandpiper (81). Small, plump and short-necked, with uniformly grey head and neck, lacks supercilium; narrow white eye-ring, medium-length, slightly decurved, blackish bill with dull orange base, and short yellow-orange legs. *In flight* from above is rather dark, with white wing-bar and narrow white sides to dark-centred rump; feet do not project beyond tail. Sexes are of similar size; plumages vary seasonally and with age. *Feeds* in rather leisurely manner on intertidal rocks; uncommon (usually immatures on southward migration) in other habitats.

Juvenile Crown has rusty streaks, narrow white eye-ring; faint supercilium quickly lost. Mantle and scapulars dark grey, fringed rufous and whitish; wing-coverts and tertials dark brown-grey with neat off-white fringes. Upper breast and flanks heavily streaked dark grey, which is quickly replaced by paler spotting with commencement of body moult. In Europe may migrate before replacing juvenile body feathers but the reverse is often true in North America.

First non-breeding/adult non-breeding Head, mantle and scapulars uniformly dark grey, the latter having a purplish sheen; throat and upper breast with fine dark streaks; lower breast, and particularly flanks, extensively spotted light brown-grey. Adult wing-coverts have diffuse paler fringes; first non-breeding retains juvenile contrasting covert fringes.

First-breeding At least some breed in second calendar year. As adult breeding, but may replace fewer upperparts; retains very worn juvenile wing-coverts and tertials.

Adult breeding Crown dark brown; sometimes with indistinct buff supercilium. Mantle and scapulars blackish, with broad buff fringes; breast, upper belly and flanks heavily marked greyish-brown. Bill largely black, with restricted yellow-orange at base; legs dull orange to brown, darker than in other plumages.

Call *Weet, wit*, given in flight.

Status, habitat and distribution Fairly common, breeding in lichen-rich tundra, on Canadian arctic islands, Greenland, Iceland, Scotland (very rare), Spitsbergen, Norway and northern Russia to the Taimyr Peninsula. Non-breeding season spent on rocky coasts of eastern North America from Newfoundland to Carolinas (rarely Florida), Great Lakes and, in Europe, Iceland and northern Finland south to Portugal. Probably non-breeding further north than any other shorebird except the closely related Rock Sandpiper. Some non-breeding birds in Britain are believed to come from Canadian populations. Vagrant to Portuguese and Spanish Atlantic islands, Mediterranean, eastern Europe; also Alaska.

Racial variation and hybridisation Various races proposed on morphological grounds (not separable in the field), though Scandinavian, Spitsbergen and northern Russian breeders average paler than the brighter Greenland and Iceland populations. Presumed hybrid Purple Sandpiper x Dunlin has been recorded in England.

Similar species Dumpy shape and short, yellowish legs are distinctive, even away from usual rocky habitat. Rock Sandpiper (81) of the northern Pacific (not recorded in the range of Purple Sandpiper) has black area on lower breast when breeding, but is very similar to Purple Sandpiper in juvenile and non-breeding plumages when, except perhaps for nominate Rock race *ptilocnemis*, the two species are perhaps not safely separable. Juvenile Rock Sandpipers, however, have a buff wash on the breast (probably lost on the breeding grounds), which Purple Sandpipers lack. Except when breeding, Purple Sandpipers usually have more orange legs.

References Dennis (1983), Engelmoer & Roselaar (1998), Gibson & Kessel (1992), Holling *et al.* (2007), Millington (1994), Payne & Pierce (2002), Smith & Summers (2005), Thorup (2006).

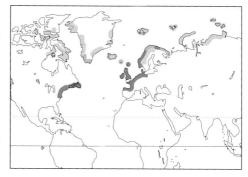

▼ **80a. Juvenile**. Eastern England, early September. Has migrated south without losing much of its juvenile plumage (except perhaps its faint supercilium), unlike North American Purple Sandpipers, which moult to first non-breeding then migrate. RJC.

▼ **80b. First non-breeding**. Eastern England, late September. Has acquired adult-type upperparts and breast/flanks, but retains juvenile wing-coverts and tertials. RJC.

▼ **80c. First non-breeding/first breeding**. Eastern England, early April. The juvenile wing-coverts are retained until the moult to second non-breeding. RJC.

▼ **80d. First non-breeding/first breeding, wing-stretching**. Eastern England, early April. The same individual as 80c, showing wing pattern. RJC.

◄ **80e. Adult non-breeding.** Eastern England, early April. Has diffuse pale grey fringes to the wing-coverts, rather than the sharper narrow white fringes of the juvenile coverts. RJC.

◄ **80f. Adult breeding.** Iceland, late June. RJC.

▼ **80g. Adult breeding.** Wales, early August. In complete but very worn plumage. RJC.

81. ROCK SANDPIPER
Calidris ptilocnemis

A true Arctic specialist, breeding on upland tundra in summer; the northernmost shorebird in winter, occurring on north Pacific coasts.

Identification L 22cm (8.5"); **WS** 40cm (15.5"). Forms a species pair with Purple Sandpiper (80), to which it is very similar, differing most in nominate race *ptilocnemis* (see below) and when breeding. Small, plump and short-necked, with uniformly grey head and neck, medium-length, slightly decurved, black bill with orange base, and short, yellow or (when breeding) dark-olive legs. *In flight* from above is rather dark, with white wing-bar (extent depends on race, see below) and narrow white sides to dark-centred rump; feet do not project beyond tail. Females have slightly longer bills; plumages vary seasonally and with age. *Feeds* in breeding season on insects; when non-breeding on molluscs and other marine invertebrates, on intertidal rocks, and by picking on intertidal mudflats.

Juvenile All races except nominate *ptilocnemis* have uniformly grey head and neck, and inconspicuous paler supercilium. Upperparts, wing-coverts and tertials dark grey with prominent off-white fringes (rusty fringes to some upper scapulars); breast and flanks with fine dark streaks. Race *ptilocnemis* is paler than other races, with head and neck streaked white, not solidly grey, and has a white area below and behind eye. Many probably commence moult of upperparts before migration.

First non-breeding/adult non-breeding All races except *ptilocnemis* have entire head, neck and upperparts uniformly darkish grey, with extensive large grey spots on breast and flanks. Race *ptilocnemis* is paler, with a light grey head, neck and mantle contrasting with medium grey scapulars, wing-coverts and tertials; smaller, less extensive, breast and flank spots and streaks make underparts appear paler. First non-breeding retains pale fringed juvenile wing-coverts.

First-breeding Most breed in their second calendar year, so as adult breeding but retains worn juvenile wing-coverts.

Adult breeding Varies with race; see Table 9 (p. 265). Upperparts dark with broad, buff, rusty or chestnut fringes, some races with white terminal fringes. Crown dark with fine white streaks, white or obscure supercilium, white throat; upper breast streaked dark on buff ground; lower breast black, variably mottled white (sometimes just heavily spotted black); remainder of underparts white, streaked and spotted grey or dark grey on flanks. Legs dark olive. Males of all races probably separable, with paler nape, brighter upperparts, and more extensive black on lower breast.

Calls Abrupt *chreet, cheet* or *cheerrt*, similar to Purple Sandpiper.

Status, habitat and distribution Fairly common, breeding in upland tundra, non-breeding on rocky coasts and intertidal mudflats. Nominate race *ptilocnemis* on Bering Sea islands (Pribilof, Hall and St Matthew Islands, non-breeding coastal Alaska south of Bering Sea); *tschuktschorum* coastal west Alaska, St Lawrence Island and Chukotski Peninsula, non-breeding central and southern Alaska south to northern California (less frequently south of Alaska in recent years); *couesi* largely sedentary, from the Aleutian Islands to Alaska Peninsula and Kodiak Island; *quarta* is also largely sedentary, on Commander and Kurile Islands. A further putative race, *kurilensis*, quite similar to *quarta*, breeds in southern Kamchatka, non-breeding in Hokkaido and Honshu, Japan.

Racial variation Four subspecies usually recognised (but see Status above). The races differ both in size (*ptilocnemis* largest, *tschuktschorum* and *couesi* intermediate, *quarta* smallest), and in breeding plumage (see Table 9, p. 265, but beware individual variability); *ptilocnemis* is paler than other races in all plumages (including non-breeding, see above), and in flight has a broader and more extensive wing-bar.

Similar species Purple Sandpiper (80) (which see) is extremely rare in the Pacific. Dunlin (82; races *pacifica, arcticola* and *sakhalina* in range of Rock Sandpiper) is very similar in breeding plumage but has whiter head, lacking the dark ear-covert patch. Dunlin has a longer, more decurved and entirely black bill, and more extensive black belly-patch that reaches the all-black legs.

References Buchanan (1999), Gill *et al.* (2002), Lehman (2006), Tomkovich (1996b).

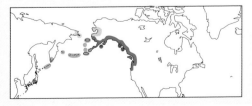

Table 9. Typical breeding plumages of the races of Rock Sandpiper. Individual birds may show some variability, in part owing to sexual differences.

Race (breeding area)	*ptilocnemis* (Pribilof Islands)	*tschuktschorum* (northern Bering Sea)	*couesi* (Aleutian chain)	*quarta* (Commander Islands)
Upperparts	Fringes rusty, some off-white terminally	Fringes rich chestnut, white terminally	Broad cinnamon-rufous fringes, white terminally	Broad tawny fringes, many lack white terminally
Ear-coverts	Well-defined dark spot	Well-defined dark spot	Uniformly dusky, sometimes darker spot	Darker spot
Breast patch	Variable in extent	Extensive, mottled white	Distinct, but mottled white	Poorly developed, often just heavily spotted
Flanks	Fine streaks	Streaks and/or spots	Streaks and spots	Heavily spotted

◀ **81a. First non-breeding.** Grays Harbor County, Washington, early January. Michael Woodruff.

▼ **81b. Adult non-breeding, probably *tschuktschorum*.** Alaska, early May. Rock Sandpipers, as with Purple Sandpipers, can sometimes be quite late acquiring breeding plumage. This individual has one or two breeding scapulars, darker legs, but is otherwise in non-breeding plumage. Race *ptilocnemis* is paler than the other races in all plumages, and when non-breeding the head and mantle are obviously paler than the scapulars and folded wing, not similarly toned as here. Richard Crossley.

▲ **81c. Adult breeding *ptilocnemis*.** St George, Pribilof Islands, early June. This race has rusty upperpart fringes and a variable breast-patch, some having much more black than this individual. Nikolay Konyukhov.

◀ **81d. Adult breeding *tschuktschorum*.** St Lawrence Island, Alaska, June. Upperparts have chestnut fringes and white tips, and the large belly patch has white mottling. Martin Hale.

◀ **81e. Adult breeding *couesi*.** Alaska, June. Wide cinnamon-rufous white-tipped upperpart fringes and distinct, but mottled breast-patch. Richard Sale.

▲ **81f. Adult breeding *quarta*.** Commander Islands. Tawny upperpart fringes, breast-patch often just heavily spotted.

▲ **81g. Adult breeding *ptilocnemis* in flight.** Pribilof Islands, late June. This race is paler in all plumages than the others, and has a more prominent wing-bar. Glen Tepke.

▼ **81h. Non-breeding Rock Sandpipers in flight**. Alaska, late February. South of Bering Sea; probably a mix of races, but the paler individuals are likely to be *ptilocnemis*. Jim Zipp.

82. DUNLIN
Calidris alpina

The most widespread small shorebird, common throughout the northern hemisphere, but one that can present stern identification challenges.

Identification L 19cm (7.5"); **WS** 34cm (13.5"). Small (though can be variable in size), with medium (European races) or long (east Siberian and North American races) slightly decurved, black bill, and medium-length black legs; folded primaries typically just reach tail-tip and extend a short distance beyond tertials. **In flight** from above brown-grey with narrow white wing-bar and dark-centred white rump; feet just reach tail-tip. Sexes similar in size, but males have shorter bills than females on average (by 3–5mm; all races); plumage varies seasonally, and with age. **Feeds** with hunched, short-necked stance, by picking, probing to shallow depth and stitching; commonly wades.

Juvenile Rather rufous-toned head and breast with brownish crown. Upperparts darkish brown with rusty-fringes, often with pale mantle-V; breast and belly strongly streaked or spotted, remainder of underparts white. Only European races migrate south in this plumage; other races have body moult on or near breeding grounds.

First non-breeding/adult non-breeding Uniform brownish-grey upperparts; crown darker. Foreneck and breast diffusely streaked grey, remainder of underparts white. First non-breeding retains darker, pale-fringed and dark-tipped juvenile wing-coverts and tertials; those of adults are paler and more uniform.

First-breeding/adult breeding Most birds probably breed in second calendar year. Bright, rusty or chestnut fringes to black-centred feathers of crown and upperparts; brownish streaked neck and breast, black belly, and white vent and undertail-coverts. Belly patch, particularly in females (both sexes when worn), can show some white centrally. Males of most races have whiter hindneck than females but this can only be used to sex birds not in breeding pairs in European races. First-breeding can be distinguished by well-worn wing-coverts; sometimes acquires only partial breeding plumage.

Call *Treep*, often given at take-off, and in flight.

Status, habitat and distribution Common, breeding in tussocky, hummocky tundra. Non-breeding occurs coastally and on extensive inland wetlands. Ten races recognised here. In Europe, nominate *arctica* breeds in north-east Greenland, migrating through Iceland, Britain, Ireland and western France, probably non-breeding in north-west Africa; *schinzii* breeds south Greenland, Iceland, Britain and Ireland and the Baltic, non-breeding in north-west Africa; *alpina* breeds Norway and northern Russia, migrating via Norway and the Baltic, non-breeding in the British Isles and France to western Mediterranean. In Siberia *centralis* breeds in central Siberia (perhaps intergrading with *alpina* in the west), non-breeding is probably from the Mediterranean through the Middle East to India; *actites* breeds in northern Sakhalin; *kistchinski* breeds in Kamchatka; *sakhalina* breeds in Chukotka. These latter three Siberian races occur in non-breeding plumage from Japan and Korea to south-east China. In North America, *hudsonia* breeds in Arctic Canada west of Hudson Bay, non-breeding along the east coast of the United States from Massachusetts to Florida and the Gulf Coast; *arcticola* breeds in north Alaska, non-breeding in Japan, Korea to south-east China; *pacifica* breeds in western Alaska, non-breeding along the Pacific American coast from south Alaska to Mexico. Vagrants resembling *hudsonia* have been reported (mainly September–October) from Britain and Ireland, and small individuals, perhaps *arctica*, from eastern North America.

Racial variation and hybridisation Identification to race may be possible in favourable circumstances, most readily in breeding plumage. At the end of the breeding season separation will often be difficult, as feather wear renders the distinctions less obvious and upperparts become dull and blackish. The three European races occur together on migration; *alpina* attains breeding plumage four to six weeks later than *arctica* and *schinzii*, so the latter two races are duller and more worn when *alpina* is in fresh plumage; Table 10 (p. 269) shows distinctions between these races. The seven Siberian and North American races are all very similar, though with small differences in size and plumage, and timing of moult. Siberian races *actites, kistchinski, sakhalina* (the latter moulting both in and out of breeding plumage relatively early), together with the Alaskan race *arcticola*, may all occur together on migration, and also when non-breeding; see Table 11 (p. 269). The three North American races each have different migration routes and non-breeding areas, though it is possible that *pacifica* and *arcticola* might occur together in west Alaska. When breeding, *pacifica* is much less heavily streaked than *hudsonia*, with streaking virtually absent immediately above the black belly. Non-breeding *hudsonia* generally has neat lines of streaks down flanks, though occasional individuals of all races may show some flank-streaking. Non-breeding *pacifica* often appear rather browner, less grey above, than other races. There are reports of possible hybrids with Sanderling (see 66), White-rumped Sandpiper, Baird's Sandpiper, Pectoral Sandpiper and Purple Sandpiper.

Similar species Rock Sandpiper (81), Western Sandpiper (68), Baird's Sandpiper (75), Curlew Sandpiper (78), all of which see. Broad-billed Sandpiper (84) differs from Dunlin in all plumages by double supercilium, bill-shape, and short, grey legs; in juvenile and adult breeding lacks Dunlin's breast spots or belly patch, respectively.

References Barter (2002), Browning (1991), Curry *et al.* (2003), Engelmoer & Roselaar (1998), Ferns (1981), Gantlett & Grant (1989), Greenwood (1984, 1986), Harrison & Harrison (1971, 1972), McLaughlin & Wormington (2000), Millington (1994), Stoddart (2007), Thorup (2006), Tomkovich (1998), Warnock & Gill (1996), Wallace *et al.* (2001), Wenink & Baker (1996), Wenink *et al.* (1996).

Table 10. Distinctions between European (*arctica*, *schinzii* and *alpina*) and North American Dunlin races (*pacifica* and *hudsonia*) in breeding plumage; first-breeding may be attained later than adult breeding. In this plumage the European races may generally be sexed by the paler hindneck of the male. Bill lengths are the range for both sexes combined (data largely from Ferns 1981 and Prater *et al.* 1977).

Race	Bill length (mm)	Mantle/scapular fringes	Breast streaking	Belly patch
arctica (smallest)	24.0–31.2	Narrow, cinnamon	Fine	Small
schinzii	24.0–36.3	Broad, yellowish-red; narrow grey terminal fringes	Moderate, spotted near belly patch	Small
alpina	27.7–36.3	Rich chestnut; wide grey terminal fringes when fresh	Heavy; separated from belly patch by white band	Large
pacifica	31.7–43.7	Very bright chestnut	Very faint	Large
hudsonia	33.0–42.4	Very bright chestnut	Faint	Large

Table 11. Distinctions between Siberian and Alaskan Dunlin races in breeding plumage. Data largely from Browning (1991); bill lengths are the range for both sexes combined.

Race	Bill length (mm)	Back colour	Extent of black on crown and mantle	Extent of breast streaking
centralis	30.1–37.9	Race *centralis* is perhaps best regarded as intermediate between *alpina* and *sakhalina*		
actitis (smaller)	28.4–33.0	Pale rusty	Greatest, especially crown	Fairly heavy
kistchinski	30.4–40.6	Rusty, yellowish fringes	Intermediate	Intermediate
sakhalina	29.7–43.0	Rusty	Least	Fine
arcticola	29.8–39.8	Bright rusty	Least	Fine

▶ **82a. Juvenile.** Eastern England, early September. Neatly fringed upperparts and strongly spotted underparts, recalling black belly of breeding adult. This individual already has one or two adult-type non-breeding scapulars; see Figure 2 (p. 8) for another juvenile Dunlin in similar plumage. The underpart spotting is quickly lost, as seen in 82b and 82c. The long bill of this bird might suggest this individual is of a North American or far-eastern race, but this is unlikely given the almost complete juvenile plumage. North American Dunlins largely moult to first non-breeding before southward migration, unlike European populations. This is probably a female of the race *alpina*, which can have bill lengths in the range of the longer-billed races. RJC.

◀ **82b. Juvenile/first non-breeding**. Southeast England, mid September. Has lost some juvenile upperparts and much of its underpart spotting. RJC.

◀ **82c. First non-breeding**. Eastern England, mid October. Has adult-type plain brown-grey upperparts and has lost all juvenile underpart spotting, but still retains dark-tipped juvenile wing-coverts and tertials. RJC.

◀ **82d. First non-breeding**. Eastern England, mid December. Plumage fluffed-up owing to cold weather, obscuring all but the rearmost juvenile wing-coverts; the tertials are also juvenile. RJC.

▲ **82e. First non-breeding** *hudsonia*. Florida, early January. Ageing becomes more difficult as the dark tips of the coverts and tertials wear off. Bill averages longer and slightly more decurved in the North American races, and can cause confusion with Curlew Sandpiper. All non-breeding Dunlins can have some breast/flank streaking, but *hudsonia* typically has neat lines of streaks on breast-sides and flanks, as here. RJC.

▲ **82f. Adult non-breeding** *pacifica*. California, early November. Plain adult-type feathers on upperparts, coverts and tertials. In this plumage *pacifica* has far less flank streaking than *hudsonia* (82e). RJC.

It is not easy to identify any of the races of Dunlin in breeding plumage away from the breeding grounds, and the following images provide examples of both migrants and birds on the breeding grounds in a range of plumage states. In April and May, birds have bright, fresh plumage (as 82g), but on the breeding grounds in June they are already showing fading and wear, and their appearance has changed. Since races are established largely on the basis of the average characters of specimens taken on the breeding grounds, variation between individuals and differences between sexes, as well as variation with time owing to fading and wear as the season progresses, result in caution being needed when attempting to assign race to any particular individual. Consequently, there will always be more than an element of speculation with the racial identification of migrant Dunlins in breeding plumage. A similar caveat applies, for similar reasons, to the racial identification of many shorebirds, in particular to migrant Red Knots in breeding plumage.

▲ **82g. Adult breeding** *arctica*. Wales, mid April. The smallest race. Narrow grey and cinnamon upperpart fringes, fine breast-streaking, and small breast-patch. RJC.

▲ **82h. Adult breeding** *schinzii*. Eastern England, early May. In fresh plumage, upperparts with broad yellow-red fringes and grey tips, breast streaked, with spotting near rather small belly patch. RJC.

▲ **82i. Adult breeding *schinzii*.** Iceland, late June. Already rather worn, and much duller than 82h. Probably a female from bill length and from streaked, rather brownish nape. RJC.

◄ **82j. Adult breeding, probably race *schinzii*.** Wales, late July. Race *schinzii* has smallish belly patch, but upperparts now very dark as fringes are largely worn away, leaving only the black feather centres, so difficult to assign racial identity. Longish bill and brown nape suggest a female. RJC.

◄ **82k. Adult breeding, possibly race *alpina*.** Southeast England, mid September. At this date most adult Dunlins are well into their moult to non-breeding, and show little or no trace of any breeding plumage. This exceptionally late-moulting individual shows extreme wear, and is so black both above and below that it gives the impression of being oiled, though it is not. The large belly patch with largely white upper border suggests this may be race *alpina*. RJC.

▲ **82l. Adult breeding, probably race *sakhalina*.** South Korea, early May. Rusty upperparts, whitish nape, and fine breast-streaking are all features of *sakhalina*. The combination of leg flags and (part of) the ring number showed that, in spite of the very worn wing-coverts, the bird is an adult and not first-breeding. RJC.

▲ **82m. Adult breeding *sakhalina*.** Chukotka, Siberia, early July. All non-European races commence primary moult on the breeding grounds, an interesting parallel with juveniles of the same races, which similarly commence moult to first non-breeding prior to migration. This individual has started wing-moult, and has shed several of its inner primaries. Augustino Faustino.

▲ **82n. Adult breeding *actites*.** Sakhalin, mid June. The smallest Siberian race. Pale rusty upperpart fringes, extensive black on crown and upperparts; most show heavier breast-streaking than this individual. Andrey Blokhin.

◄ **82o. Adult breeding, probably race *kistchinski*.** Japan, mid May. This race is a rare migrant in Japan. Yellowish-rusty upperpart fringes and darkish nape; long bill, so probably female. Nobuhiro Hashimoto.

▼ **82p. Adult breeding *kistchinski*.** Kamchatka, late July. Worn, so plumage rather dark. Short bill suggests a male. Alexandre Andreeva.

◄ **82q. Adult breeding, possibly race *arcticola*.** South Korea, early May. Very similar to *sakhalina*, but averages brighter and more red above. RJC.

▲ **82r. Adult breeding *pacifica*.** California, late April. Peter LaTourrette.

◄ **82s. Adult breeding *hudsonia*.** Florida, mid May. Relatively slight breast-streaking; *hudsonia* is the only Dunlin race whose distribution does not overlap with other races, either when breeding, on migration or when non-breeding. RJC.

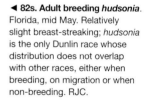

◄ **82t. Non-breeding *pacifica* in flight.** California, early November. Flight pattern, with central white wing-bar, and rump with dark centre and white sides, is similar to many of the other small *Calidris* sandpipers. RJC.

83. SPOON-BILLED SANDPIPER
Eurynorhynchus pygmaeus

A small east Asian sandpiper with a unique spatulate-tipped bill. Endangered.

Identification L 15cm (6"); **WS** 32cm (12.5"). Very small, with rather large head, short neck, broad supercilium particularly in front of eye; spatulate-tipped medium-length black bill, and black legs of medium length. The bill is best seen when bird is head-on; from the side it can appear slightly uptilted. *In flight* from above, narrow white wing-bar, white primary shafts, white sides to dark-centred rump and tail-coverts, tail grey; toes just extend beyond tail. Males average smaller than females, but there is much overlap; plumage varies seasonally and with age. *Feeds* on biofilm, small invertebrates and insects, by rapid stitching on wet mud adjacent to surface water with bill held at 60° to horizontal, by picking from surface and, in shallow water, by sweeping bill sideways.

Juvenile Forehead and lores white with some dark streaking; crown and patch from eye to ear coverts streaked dark brown and rusty. Upperparts with black centres, fringed rusty and white, white mantle-V. Underparts completely white but upper breast washed buff when fresh; brown-streaked upper breast-side patches. Rusty buff tones either fade or are off-white in some individuals. Migrates south in this plumage.

▼ **83a. Juvenile/first non-breeding**. South Korea, late September. One of the smallest shorebirds. Has acquired some grey first non-breeding upperparts, but the brown-streaked crown and upper breast-side patches are juvenile, as are the retained dark-centred upperparts and wing-coverts. Kim Shin-Hwan.

First non-breeding/adult non-breeding Forehead and lores white, crown grey with darker streaks, patch from eye to ear coverts grey; uniformly grey-brown above, initially with pale fringes. Grey-streaked upper breast-side patches; remainder of underparts white. First non-breeding retains juvenile wing-coverts, and possibly some tertials.

Adult breeding Age when first breeds uncertain. Head and neck entirely orange-red to deep brick red, largely obscuring supercilium; crown has heavy dark streaks; upperparts have dark centres and reddish fringes. Breast and remainder of underparts white, with brown streaking or spotting in centre of upper breast, spotting at sides. Male is generally brighter than female.

Calls A quiet *preep*, or sharp *wheet*.

Status, habitat and distribution Rare (numbers declining), breeding in coastal arctic tundra in north-east Siberia. Non-breeding on coastal and estuarine mudflats and wetlands, in southern and south-eastern Asia, from Vietnam, Thailand and Burma west to Bangladesh and India. On migration Japan, Korea, south-east China; vagrant west to Sri Lanka, east to Aleutian Islands and coastal north-west Pacific south to British Columbia and inland Alberta.

Racial variation None.

Similar species Small size and similar plumages at all ages invite comparison with, particularly, Red-necked Stint (69), and also Little Stint (70), but spatulate bill-tip seen in a good (head-on) view is immediately diagnostic.

References Alström & Olsson (1991), Baicich (2000), Barter (2002), McWhirter (1987), Shama (2003), Tomkovich (1992b).

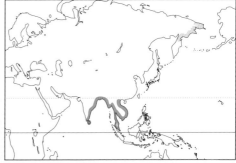

▲ **83b. Non-breeding**. Thailand, late January. Both non-breeding and breeding plumage recall a miniature Sanderling at the equivalent plumage stage, as do the short runs these sandpipers often make between feeding bouts. Hanne and Jens Eriksen.

▼ **83d. Adult breeding**. South Korea, late May. Same individual as 83c. RJC.

▲ **83c. Adult breeding**. South Korea, late May. From the side the bill appears slightly upturned; the head is a deeper red than in Red-necked Stint, with which it often loosely associates. RJC.

▼ **83e. Juvenile in flight**. Japan, early September. Has a narrow white wing bar, broadening on the primaries (not visible here), and a dark centre to a largely white tail. Akito Teramae.

84. BROAD-BILLED SANDPIPER
Limicola falcinellus

A small, rather dark Eurasian shorebird
with a prominent double supercilium
that joins at the forehead.

Identification L 17cm (7"); **WS** 33cm (13"). Small,
short-necked, with long, slightly decurved blackish bill
that is flattened horizontally and kinked downwards
at the tip; short dark-grey legs. In all plumages has
a double supercilium which joins at forehead. *In
flight* from above leading edge of wing blackish,
narrow white wing-bar, white at sides of blackish-
centred rump; feet do not extend beyond tail. Sexes
of similar size; plumages vary seasonally and with
age. *Feeds* by gentle, vertical probing in mud or
shallow water.

Juvenile Crown brown with a creamy double
supercilium. Mantle and scapulars are dark brown
with rufous-buff fringes and prominent cream
mantle and scapular-Vs; wing-coverts dark brown,
fringed buff. Upper breast and flanks streaked
brown, remainder of underparts white.

First non-breeding/adult non-breeding Crown
greyish, whitish double supercilium. The upperparts are dull grey with diffuse pale fringes (paler
at the tips), and darker shaft-streaks; wing-coverts
and tertials of adult are as upperparts. First non-
breeding retains the dark juvenile wing-coverts
but moults the outer primaries. Dark forewing
may show at the bend of the folded wing, as in
non-breeding Sanderling.

First-breeding Much as adult breeding but many
probably remain in non-breeding area and do not
breed until their third calendar year.

Adult breeding Rather similar to juvenile, initially
with whitish fringes to upperparts giving 'frosty'
appearance. Wear then reveals bright rufous-buff
fringes to otherwise blackish mantle and scapulars,
but further wear renders upperparts very dark;
mantle and scapular-Vs generally obvious. Upper
breast and flanks heavily spotted; remainder of
underparts white.

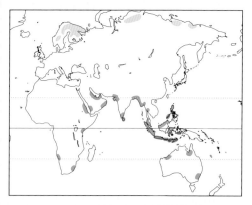

▼ **84a. Juvenile *falcinellus*.** Southern European Russia, September. The straight bill of this species has a downturned 'kink' at the tip. There is a double supercilium in all plumages. Evgeny Kotelevsky

Call A trilled, almost buzzing *chrreet*.

Status, habitat and distribution Uncommon, breeding in extensive marshland in boreal forest. Non-breeding in muddy wetlands and intertidal mudflats. Nominate race *falcinellus* breeds from Norway to north-west Russia, and migrates south-east, non-breeding from East Africa to India, and so is rare in western Europe; race *sibirica* breeds discontinuously through northern Siberia, non-breeding in south and south-east Asia, Australia and (small numbers) New Zealand. Vagrant western Europe west to Ireland, also Morocco and western Africa.

Racial variation Breeding *falcinellus* has near-black upperparts; *sibirica* is brighter, more rufous, with rufous fringes and pale tips. Upperparts of fresh juvenile *sibirica* have broader buff fringes; races indistinguishable when non-breeding.

Similar species Dunlin (82), which see. From Jack Snipe (87) by latter's straight bill, more rufous coloration with strong, creamy lines on head and body, and pale legs.

References Buckley (1980), Higgins & Davies (1996).

▲ **84b. Juvenile *sibirica***. Japan, late August. Upperpart fringes in this race are chestnut, not whitish. Nobuhiro Hashimoto.

▼ **84c. Adult non-breeding**. Thailand, mid November. RJC.

▲ **84d. Adult breeding *falcinellus*.**
Finland, early June. Markus
Varesvuo.

◄ **84e. Adult breeding *falcinellus*
wing-stretching.** Finland, early
June. The same individual as 84d.
Markus Varesvuo.

◄ **84f. Adult breeding *sibirica*.**
Western Australia, early April.
As with juvenile has chestnut
upperpart fringes in this race. RJC.

85. BUFF-BREASTED SANDPIPER
Tryngites subruficollis

A small-headed buffish shorebird with an exceptionally long migration. Near Threatened.

Identification L 19cm (7.5"); **WS** 41cm (16"). Small, long-necked, with a small rounded head; short, fine, straight black bill and medium-length yellow or yellow-ochre legs. Fine dark streaking on crown, plain buff face with prominent dark eye; upperparts scaly, underparts uniform buff. *In flight* from above uniform, with primaries and secondaries slightly darker; below, striking white underwing contrasts with buff body, toes just extend beyond tail. Males are 5% or so larger than females and can, with care, be separated in the field. Adults are the same year-round, juvenile is separable. *Feeds* by picking, usually in short grassland, with rather hunched stance.

Juvenile Mantle and scapulars dark brown or blackish, with contrasting narrow whitish fringes; wing-coverts have brown centres and dark subterminal marks, and broader, less contrasting buff fringes. Apparently a few may have white lower breast and belly.

First non-breeding Probably has a complete post-juvenile moult, including primaries, but since this is not completed until spring, can be aged while still retains juvenile upperparts and wing-coverts.

First-breeding/adult breeding/adult non-breeding Probably breeds in second calendar year. Once juvenile plumage is lost there is apparently no significant seasonal variation. Head and underparts much as juvenile; mantle and scapulars are dark-centred with fairly broad, buff fringes; wing-coverts similar, but with less-dark centres.

Call A quiet, trilled *pr-r-r-reet* but generally rather silent.

Status, habitat and distribution Breeds on well-drained grassy tundra, locally in north-east Siberia, northern Alaska and northern Canada east to King William Island. Non-breeding on sparsely vegetated wet grasslands, primarily in Argentina. Regular vagrant to Europe, scarcer Africa, Sri Lanka, Japan, Pacific islands and Australia; most mid August to mid October.

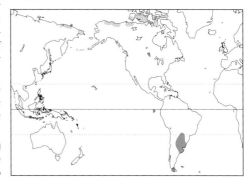

◄ **85a. Juvenile. South-west England, late September**. Pale fringes to the dark-centred upperparts give a strongly scalloped appearance. RJC.

Racial variation and hybridisation No races recognised. There is a report of a possible hybrid Buff-breasted Sandpiper x Baird's Sandpiper.

Similar species Most likely confusion is with juvenile (female) Ruff (86) but Buff-breasted Sandpiper is always smaller than the smallest female Ruff has shorter, straighter and finer bill, yellow legs, and is buff beneath as far as the undertail-coverts.

References Alexander-Marrack (1992), Ali & Ripley (1983), Buden (2006), del Nevo (1984), Lanctot & Laredo (1994), Laux (1994), Robbins & Alderfer (2000).

▲ **85b. Adult**. Nebraska, mid May. Much as juvenile, but upperparts have brown-buff fringes; once adult plumage is acquired there is little change in appearance year-round. Carlos Grande.

◄ **85c. Adult**. Nebraska, mid May. Showing upperwing in display; flight pattern is uniform above. Carlos Grande.

86. RUFF
Philomachus pugnax

Males in breeding plumage are the most variable of all shorebirds; females are considerably smaller and more uniform.

Identification Male **L** 29cm (11.5") **WS** 61cm (24"); female **L** 22cm (9") **WS** 47cm (18.5"). Medium-sized with smallish head; medium-length and slightly decurved bill, long neck and medium-length legs. Colours of plumage and bare-parts vary (but generally systematically) with age and sex, adult breeding male having ear-tufts and uniquely long neck feathers ('ruff'); both bill and legs are variably dark and/or yellow or orange. *In flight* looks 'hump-backed', has white wing-bar of variable extent and white oval areas either side of dark-centred rump; feet project beyond tail. Males are usually about 20% larger than females (but see 'faeder' below), with proportionally slightly shorter bills and longer legs. For comparison, Common Redshank is slightly smaller and shorter legged than most males, but larger than females, and Lesser Yellowlegs is very slightly larger than female Ruff. *Feeds* by picking in muddy areas, often amongst vegetation; sometimes wades.

Juvenile Plain head with streaked crown and indistinct buff supercilium, upperparts very scaly, with blackish-brown feather centres and narrow pale-buff fringes; often shows mantle-V. Wing-coverts and tertials have variable feather centres, solidly dark-brown or dark-brown patterned buff, but always have buff fringes. Breast dull buff, remainder of underparts white. Bill blackish, with brownish base; legs dull yellowish.

First non-breeding Much as juvenile but crown and hindneck greyish-brown; upperparts as adult non-breeding, though a variable number of worn juvenile wing-coverts and tertials are retained. Legs as juvenile, or greenish, or mottled yellow/orange. Some southerly non-breeders, usually females, may moult a few outer primaries.

First-breeding May have similar plumage to adult breeding and breed in second calendar year or, particularly those remaining in non-breeding area, may have only scattered breeding feathers on upperparts; a few show white at base of bill. Many have a mixture of retained juvenile and fresh non-breeding type wing-coverts. Leg colour probably as first non-breeding.

Adult non-breeding Crown and hindneck dull brown; upperparts greyish-brown, with darker centres and diffuse pale fringes. Incomplete, narrow white eye-ring. Often have white at base of bill (which can also be seen on breeders of both sexes). Throat and neck white, greyish wash across breast, remainder of underparts white. Bill dark, base may be orange; legs often yellowish or dull orange. A few (more frequently males) have head and neck white.

Adult breeding Both sexes replace a large proportion of body feathers, scapulars, wing-coverts and tertials from January: the 'true' breeding plumage. *Male*: commences moult to breeding January or February; by late April or May acquires ruff and ear-tufts ('supplemental' plumage), and a week or so later feathering of forehead and around eye is lost to reveal orange area of warty skin. Supplemental plumage is lost quite quickly, often by the end of June, though remainder of breeding plumage is retained longer. Individuals are very variable, with any two rarely the same. Ruff, ear-tufts, mantle, scapulars and tertials may be black, white, rufous or chestnut, uniformly coloured, spotted or barred white, buff or rusty; belly is often black but this is variable. Birds that are largely white, including the ruff ('satellites'), are often obviously smaller than darker ('independent') males, though still larger than the females. Bill is usually dark-tipped and orange- or pink-based, but may be wholly dark, or wholly orange or pink; legs orange. About 50% of independent males can be identified as such by black ear-tufts, and 50% of satellites by combination of white or pale-patterned ruffs and/or ear-tufts; other ruff/tuft combinations may be either independents or satellites. Individual males apparently have similar plumage in successive years. The ruff and ear-tufts are less well-developed in younger males. A very small proportion of males ('faeders') have plumage and bare-part colours that mimic breeding females (thus lacking ruff and ear-tufts); on all measurements they are intermediate, only a little larger than females, but have the structural proportions of males. *Female*: typically head and hindneck streaked dark brown; mantle, scapulars and tertials brown, plain or patterned (or barred) rusty, fringed buff. Foreneck and breast generally brown, but may be patterned as upperparts. However, individuals are variable; some have extensive pale areas on head and neck. Bill usually dark, but may have orange at base; legs orange.

Call Usually silent.

Status, habitat and distribution Common, breeding in damp areas of low vegetation discontinuously from England, The Netherlands, north to northern Scandinavia, northern Russia to eastern Siberia.

When non-breeding generally frequents muddy freshwater areas, locally in Europe but the majority in Africa south of the Sahara. Females move further south than males when non-breeding. Vagrant worldwide beyond its normal range, perhaps rare migrant through much of North America.

Racial variation No races recognised.

Similar species Buff-breasted Sandpiper (85), which see.

References Chandler (1990b), Gill *et al.* (1995), Jukema & Piersma (2000, 2006), Karlionova *et al.* (2007), Meissner & Scebba (2005), Rohde (1998), Thorup (2006), van Rhijn (1991).

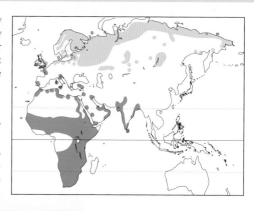

◄ **86a. Juvenile**. Eastern England, early August. Aged by neat, pale-fringed upperparts and wing-coverts, peach-buff head, neck and breast. Proportionally smaller body and more rounded head suggest a female. RJC.

◄ **86b. Juvenile**. Spain, early September. Aged as 86a, but an individual with markings on the feather centres. The feather patterns of juvenile Ruffs show much variability; rather bulky proportions suggest a male. RJC.

◄ **86c. First-breeding female**. Cyprus, late April. Proportions are subtly different from the male, more elegant, with a less bulky body. Aged by its dull greenish legs and worn plain brown-grey wing-coverts retained from juvenile plumage; full adults replace most coverts before breeding. RJC.

▲ **86d. Adult male**. East England, early April. In breeding plumage (note barred and dark-centred upperparts, wing-coverts and tertials), but still has to acquire its ruff, which is its 'supplemental' breeding plumage. RJC.

▲ **86e. Adult male**. Poland, late April. Has acquired its ruff and 'ear-tufts', though both these are surprisingly inconspicuous except when the bird is actually displaying. In a few days it will get an area of bare, orange warty skin at the base of the bill. Note the white band of feathers at the bill base, which is shown by many Ruffs, and, at least in females, is a feature more often seen in older birds. The dark plumage suggests that at the lek this is an 'independent' male. Dark-headed birds often seem to have all-dark bills. RJC.

▲ **86f. Adult male**. Poland, late April. The white head and ruff shows this to be a 'satellite' male (slightly smaller than an 'independent' male). Note the brighter orange legs compared to non-breeding birds. RJC.

▲ **86g. Adult male 'faeder'**. Cyprus, late April. Bearing the plumage of a breeding female, this bird is slightly larger than a true female Ruff and has the structure of a male (smaller head in proportion to its rather bulky body), though it is smaller than even the smallest satellite male. RJC.

◀ **86h. Adult breeding female**. Cyprus, late April. Very similar to 86c, but note freshly acquired wing-coverts, and orange legs; the white feathering at the bill base is also an indication of maturity. RJC.

◀ **86i. Adult breeding female in flight**. Cyprus, late April. Identified in flight by 'hump-backed' shape, narrow white wing-bar, white ovals at sides of rump, and feet extending well beyond tail. RJC.

◀ **86j. Adult breeding females in flight.** Cyprus, late April. As with 86i, note feet beyond tail; also variable leg colours, and the white head of the leading bird, a feature occasionally shown by females and also by non-breeding males. RJC.

▲ **86k. Breeding male Ruffs**. Poland, late April. This series shows something of the extraordinary range of plumages seen in these birds. RJC.

SNIPES

There are nine species of snipe (in two genera) in the northern hemisphere: Jack Snipe *Lymnocryptes minimus*, Solitary Snipe *G. solitaria*, Latham's Snipe *G. hardwickii*, Wood Snipe *G. nemoricola*, Pintail Snipe *G. stenura*, Swinhoe's Snipe *G. megala*, Common Snipe *Gallinago gallinago*, Wilson's Snipe *G. delicata* and Great Snipe *G. media*. There are 18 species (in two genera) worldwide. Wilson's Snipe is treated here as a separate species from Common Snipe, on account of its larger number of tail feathers and different display flight, in which the two outer tail feathers are used when drumming, rather than the single outermost feather used by Common Snipe. The majority of northern hemisphere snipe species are migratory, though the Solitary and Wood Snipes of Central Asia are only either altitudinal or short-distance migrants.

The snipes are chunky, small or medium-sized shorebirds, with short or very short legs. They generally appear short-necked but, apart from Jack Snipe, are in fact fairly long-necked, and they may show this when alarmed. All have cryptic brown plumage – in the case of Jack Snipe, this is practically identical irrespective of age or season. In the *Gallinago* snipes, there is a distinctive juvenile plumage which is quite quickly lost, after which the plumage remains the same throughout the year. Snipe sexes are indistinguishable.

Snipe identification The very similar colours and plumages of snipes make specific identification difficult. Apart from Jack Snipe, the head pattern is basically the same in all northern hemisphere species: a thin pale central crown-stripe, dark lateral crown-stripes, a usually prominent pale supercilium, a dark line from bill through the eye to the ear-coverts, and a dark line (the width of which varies with species) on the cheek below the eye ('cheek-line'). Jack Snipe lacks a crown-stripe but has a double supercilium (divided by a dark line) above the eye. See also the section on plumage terminology (p. 8).

Three of the larger species, the Solitary, Wood and Great Snipes, and the smallest, the Jack Snipe, are the more straightforward to identify, though none is easy. The most difficult are the snipes of intermediate size – Latham's, and the closely related species pairs Swinhoe's and Pintail Snipe, and Common and Wilson's Snipe. In some circumstances, examination of the bird in the hand (or a lucky, close-range photograph of the spread tail) may be needed to be sure of the number and shape of the tail feathers, which are diagnostic of species. The display flights (which use these feathers) are also diagnostic of species, but these are only seen on the breeding grounds.

Range will often eliminate at least some of the possible confusion species, and so can habitat. There are a number of other pointers that can be useful. Comparative bill length can be helpful. There are subtle differences in the head patterns, particularly the widening of the supercilium in front of the eye in Swinhoe's and Pintail Snipes, which is the consequence of the narrowness of the line from bill to eye in these species. Swinhoe's tends to have a darker lateral crown-stripe with less pale streaking than Pintail. Unfortunately, none of these features is entirely constant. On the upperparts, the extent and detail of the pale buff or creamy edges to mantle and scapulars, and the pattern of the wing-coverts should be noted. Structure, too, can be helpful. On a standing bird the relatively shorter tail and wings of Pintail Snipe are such that the wing-tips are cloaked by, or barely extend beyond the tertials, and both tertials and wing-tips almost reach the tail tip. In Swinhoe's, Common and Wilson's Snipe the wing-tips usually extend a short distance beyond the tertials and fall well short of the tail-tip, but again these differences are not always constant.

Snipes are often more easily identified in flight; many are first seen when flushed, so it is fortunate that both structure and jizz in flight can be helpful. Features to note are the distance at which the bird flushed, the call (if any), height and distance of flight, the angle at which the bill is held (near horizontal in Swinhoe's and Pintail, downward in Common and Wilson's), and the presence (Common Snipe) or absence (Swinhoe's, Pintail, and Wilson's Snipes) of a clear, though narrow, white trailing edge to the secondaries. Note also if the feet extend beyond the tail (Pintail Snipe), or if they just reach the tail-tip (the other three species). A rather diffuse, pale, centre-wing panel on the upperwing contrasts with the darker flight feathers on both Latham's and Common Snipes.

All *Gallinago* snipe except Great Snipe have aerial displays in which the males spread their outer tail feathers to make a 'winnowing' or 'drumming' sound that is remarkably far-carrying. They use a switch-back flight, and drum when in a steep dive. The pitch and nature of the drumming depends on the shape, stiffness and number of tail feathers used, which differ with species, so that the sound produced is species-specific. Two species (Solitary and Swinhoe's Snipes) have a communal aerial display, which is sometimes referred to as a 'tok'. Great Snipe is unique in having a lekking breeding system.

▼ Common Snipe in 'parachute' descent from display flight. Common Snipe has a roller-coaster display flight which takes place at a height of 80 metres or so. It features alternating ascents and descents of about 20 metres, with drumming on the descent. At the end of this display the bird will sometimes 'parachute' to the top of an isolated tree, or to the ground, with wings raised. Poland, late April. RJC.

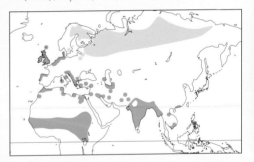

87. JACK SNIPE
Lymnocryptes minimus

The smallest of the snipe, with a relatively short bill and bold, striped upperparts.

Identification L 18cm (7"); **WS** 34cm (13.5"). Small, with long, pale-based, dark-tipped bill and short yellowish legs. All-dark crown, lacking central crown-stripe of other snipes, creamy double supercilium, dark eye-stripe; upperparts brownish, with greenish purple gloss (rarely seen in the field), and prominent creamy scapular 'lines'. Breast and flanks streaked, remainder of underparts white. *In flight* from above dark, with prominent scapular lines, white trailing-edge to secondaries and dark pointed tail; darkish beneath, with paler underwing-coverts, legs folded under body. Often flushes at very close range, without calling, dropping down quite quickly and typically aligning itself with the immediate vegetation. Sexes are of similar size, with similar plumages at all ages. The wedge-shaped tail has 12 feathers. *Feeds* in muddy areas with a horizontal stance and bobbing action, flexing legs repeatedly.

Juvenile Separable only in the hand by finer and paler streaks on white undertail-coverts.

Adult Age at first breeding not known but probably second calendar year. Described under Identification; there appear to be no characters by which this species may be aged in the field.

Call Usually silent when flushed but may, very rarely, give a single call similar to Common Snipe, but quieter and less rasping.

Status, habitat and distribution Uncommon, but probably often overlooked owing to secretive behaviour, breeding in swamps in northern Europe to central Siberia. Non-breeding spent in freshwater wetlands in Britain and Ireland, Denmark, south to central Africa, west to southern Asia. Rare vagrant to Alaska, California, Labrador and the Caribbean.

Racial variation No races recognised.

Similar species Broad-billed Sandpiper (84), which see.

References Ebels (2002), Hollyer (1984), Svažas *et al.* (2002), Taylor (1977), Tuck (1972).

▼ **87a. Central England, late November**. These birds cannot be aged in the field. They often flush at very short range without calling, and settle again quite quickly, usually (but not here!) aligned with vegetation, using the camouflage provided by their upperpart stripes. RJC.

▼ **87b. Sweden, early April**. White trailing edge to wing as Common Snipe, but smaller, with a much shorter bill. Dan Mangsbo.

88. SOLITARY SNIPE
Gallinago solitaria

The largest northern hemisphere species, this is a bulky snipe with a long bill.

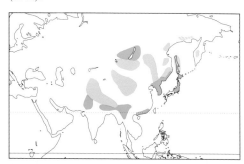

Identification **L** 30cm (12"); **WS** 52cm (20.5"). Medium-sized, has generally pale rusty plumage, with typical snipe markings that generally lack contrast, particularly on the head. Paler areas on head whitish, upperparts and wing-coverts with many small white spots; mantle, and particularly scapulars, have broad white outer edges. Very long, brown bill with blackish distal half, and short, sturdy, yellowish legs. Throat and upper breast closely mottled grey-brown, lower breast and flanks strongly barred, centre of belly white. On ground, tertials cover (or almost cover) tips of primaries; chestnut tail extends well beyond tertials. As name suggests, usually solitary. *In flight* bill held just below horizontal. Above has wide paler panel formed by secondary coverts but lacks white trailing edge to secondaries, or any other significant feature apart from chestnut tail; below much of wings and underparts dark owing to extensive barring; toes do not reach tail tip. When flushed, rises with zig-zag flight, slower and heavier than Common Snipe, usually dropping down after short distance. Female slightly larger, plumage similar at all seasons and ages. Tail usually has 20 feathers, but may have 22 or 24; outer five narrow, about 3mm wide. *Feeds* apparently in similar manner to other larger snipe, largely on invertebrates.

Juvenile No known differences from adult.

Adult Age at first breeding not known; adult is described under Identification.

Calls When flushed *scaap* or *pench*, similar to Common Snipe but coarser and deeper.

Status, habitat and distribution Scarce, breeding just below the tree-line in mountains at around 2,500–4,500m, favouring streams, especially those with associated marshland. Found discontinuously in south and eastern Siberia and north-east China to Kamchatka; non-breeding spent in similar habitat (often with Ibisbill) but at lower altitudes, in east Pakistan, northern India, Burma, Korea and Japan.

Racial variation Two races generally recognised but probably not separable in the field. Nominate *solitaria* breeds in Central Asia, non-breeding south to India; *japonica* occurs further east, non-breeding in south-east Asia, particularly Japan.

Similar species Wood Snipe (90) is almost as large but is a woodland species, with darker and heavily streaked breast (mottled in Solitary), and completely barred belly; in flight has broader wings, recalling Eurasian Woodcock (99).

References Ali & Ripley (1983), Ernst (2004), Tuck (1972).

◄ **88a. Adult**. Kazakhstan, early January. Pale markings in this species are near-white; the tertials cloak the primary tips, and the chestnut tail is prominent. Askar Isabekov.

▲ **89b. Adult**. South Korea, early January. Pale areas rather whiter than other snipe, with strongly barred underparts. Robert Newlin.

89. LATHAM'S SNIPE
Gallinago hardwickii

A fairly large species that breeds in The Russian Far East and Japan before migrating to Australasia. Alternative name: Japanese Snipe.

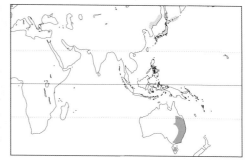

Identification L 29cm (11.5"); **WS** 51cm (20"). Medium-sized, with very long, brown, dark-tipped bill and medium-length dull yellowish legs. Typical snipe head pattern, with narrow, near-white, central crown-stripe, wide black side-crown stripes and prominent creamy supercilium, wider in front of eye. On ground, tertials usually cover tips of primaries but sometimes extend just beyond; long tail extension beyond the wing-tips. Combination of long wings and long tail give Latham's a more smoothly attenuated rear than both Swinhoe's and, particularly, Pintail Snipe. Females are larger and longer-billed, but have slightly shorter tails and fewer tail feathers than males. *In flight* from above outer primary has white shaft (brown in Pintail and Latham's Snipes), has pale panel formed by median coverts and lacks white trailing edge to secondaries; underwing dark owing to strong barring, toes do not reach tail tip. When flushed has heavier, more direct flight than Common Snipe, often dropping to ground quite quickly. Sexes are of similar size, adult plumage similar year-round, juveniles separable. Tail has 18 feathers, outer five spread when drumming, the outer-most -narrow (5mm). Feeds in typical snipe manner, by deep probing in soft ground for invertebrates.

Juvenile Mantle and scapular lines narrower and paler than adult, wing-coverts with pale buff fringes, giving overall neater appearance than adult.

First non-breeding/adult Age at first breeding not known. Adult described under Identification; aged by paired, darkish buff spots at tips of wing-coverts. Post-juvenile moult probably includes mantle, scapulars and wing-coverts; once these are replaced, first non-breeding and adult become indistinguishable.

Call Usually calls when flushed; a repeated harsh *krek* or *squaak*.

Status, habitat and distribution Fairly common, breeding in freshwater wet grassland in

▼ **89a. Juvenile**. Queensland, Australia, mid September. Aged by the scaly buff-fringed wing-coverts. Separable from both Swinhoe's and Pintail Snipes by the long tail, which gives an attenuated rear-end. RJC.

southern Kuril Islands, southern Sakhalin and Japan (Hokkaido and northern Honshu); non-breeding in east and south-east Australia, also in Tasmania, where it is the only snipe. On passage Korea, Taiwan, Philippines, New Guinea; vagrant New Zealand and Tasman Sea islands.

Racial variation None.

Similar species Pintail (91) and Swinhoe's Snipes (92) share Latham's head pattern but are smaller, and both have some foot projection in flight; Latham's has none. Pintail Snipe has proportionally shorter tail. See also Common Snipe (93).

References Takano *et al.* (1985), Tuck (1972), Ura *et al.* (2005).

◄ **89b. Adult**. Victoria, Australia, late October. Aged by large paired buff spots at tips to wing-coverts, and worn, faded primaries. Same bird as 89c. Rohan Clarke.

▼ **89c. Adult in flight**. Victoria, Australia, late October. An adult in wing moult, replacing the outer primaries. Identification features include the white shaft of the outer primary, the pale median-covert panel, lack of a pale trailing edge to the wing, and toes not projecting beyond the tail. Rohan Clarke.

90. WOOD SNIPE
Gallinago nemoricola

A large, dark snipe that breeds at up to 4,000m in and around the Himalayas. Vulnerable.

Identification L 30cm (12"); **WS** 46cm (18"). Medium-sized with rather deep, long, pinkish-brown, dark-tipped bill, and short greyish legs. Typical snipe head pattern; upperparts brown-black, with prominent pale buff fringes to mantle and scapulars forming stripes down back; wing-coverts have dark-brown and buff barring; below, there are heavy black streaks on brown breast, and strong dark barring on paler belly. *In flight* dark above and below, with broad, rounded wings (recalling Eurasian Woodcock) and short tail, lacking any white; toes just project beyond the tail. When flushed has slow, wavering flight with bill held pointing downwards, often dropping into cover after 50–100m. Sexes are of similar size and with similar plumage year-round; juvenile separable. Tail has 18 feathers, the outer four narrow, about 4mm wide. *Feeds* probably much as Eurasian Woodcock by probing and picking on earthworms and aquatic invertebrates.

Juvenile Narrow buff fringes to scapulars and wing-coverts give neater, more scaly appearance than adult.

Adult Age at first breeding not known. Adult described under Identification.

Call Often silent but occasionally gives a low, croaking *tok-tok*.

Status, habitat and distribution Uncommon, very scattered, breeding in high-altitude meadows and stony, shrubby alpine grassland in northern India, south Tibet, Nepal and Bhutan. Non-breeding at lower altitudes in hilly country with marshy areas and thick cover, south to southern India, east to Burma. Vagrant to Sri Lanka and northern Thailand.

Racial variation None.

Similar species Solitary Snipe (88), which see. Eurasian Woodcock (99) has transverse bars on crown and lacks Wood Snipe's prominent pale mantle and scapular lines, though both species use similar habitats. Great Snipe (95) has extensive white on the wing-coverts, and prominent white at the corners of the tail.

References Ali & Ripley (1983), Tuck (1972).

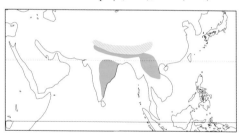

▼ **90a. Adult**. Nepal, mid May. Large and dark, with a relatively short bill for a snipe. Graham Deacon.

▼ **90b. Adult**. Nepal, mid May. A typical snipe head pattern. Pete Morris.

91. PINTAIL SNIPE
Gallinago stenura

A small, strongly migratory Asian snipe with stiff, very narrow outer tail feathers.

Identification L 25cm (10"); **WS** 42cm (16.5"). Forms a species pair with Swinhoe's Snipe (92), whose plumage is very similar but averages darker; separation from Swinhoe's is extremely difficult, particularly on the ground. Marginally the smallest of four very similar-sized snipes (other three are Swinhoe's, Common and Wilson's Snipe), Pintail is medium-sized with a very long bill (but relatively shorter and deeper than Common Snipe) and medium-length legs. Supercilium wider in front of eye, same colour as area above dark cheek-line, as in Swinhoe's. On ground, short tail gives blunt appearance to rear, and tips of primaries almost reach tail tip. No (or very small) primary projection beyond tertials. Generally more uniform brown, less contrasted than Common Snipe, with dull buff mantle and scapular stripes, and pale brown fringes to wing-coverts. *In flight*, compared to Common and Wilson's, bill is held near horizontal (as in Swinhoe's), has blunter, more rounded wing-tips, has paler, browner upperwing, and poorly-defined, paler-brown trailing-edge to secondaries (not white as Common Snipe); shows little white in tail, barred underwing (axillaries with black bars wider than white) gives overall dark appearance; toes extend well beyond short tail. Usually flies only a short distance when flushed, often rising at fairly close range, without zig-zags and often, but by no means always, calling. Sexes are of similar size; similar plumage year-round, juvenile separable. Tail usually has 26 feathers, but may have 24 or 28; the outer six to (usually) eight are stiff and very narrow ('pins'), with the outermost about 1mm wide. *Feeds* in mud and soft ground by repeated, vertical probing; less energetic than Common or Wilson's Snipe. See also the general discussion of snipe identification (p. 288.)

Juvenile Much as adult but fringes of upperparts, including mantle-lines and particularly the scapulars, are narrow and off-white. Wing-coverts very similar to adult, with wide, pale-buff fringes and dark subterminal lines, giving the coverts a scalloped appearance.

First non-breeding/adult Age at first breeding not known. Adult described under Identification; best separated from juvenile by broader, darker buff mantle and scapular fringes. Post-juvenile moult probably includes mantle, scapulars and wing-coverts; once these are replaced, first non-breeding and adult become indistinguishable.

Calls and song Often calls when flushed; a short, nasal *squik* or *ketch*, often compared to the quack of a duck; similar to Swinhoe's, less harsh than Common or Wilson's Snipe; also a short, high pitched call. Has unique mass aerial breeding display (a tok), with song composed of accelerating *chvin* calls; drumming (a higher pitched, more monotone humming than in Swinhoe's) is performed individually.

Status, habitat and distribution Fairly common, breeding in Arctic wetlands and scrubby tundra, in less wet areas than Common Snipe, in northern Russia east of Urals to eastern Siberia; non-breeding in eastern Arabia, East Africa, India, Sri Lanka, throughout south-east Asia to south-west China and Japan, south-east to Philippines and Borneo to East Timor, and (in small numbers), north-west Australia. Vagrant elsewhere in Australia, the Aleutian Islands, Hawaii and west to Israel and Sicily.

Racial variation None.

Similar species Within Pintail Snipe's range, particularly Latham's (89), Swinhoe's (92), and Common Snipe (93), which see.

References Bundy (1983), Carey & Olsson (1995), Corso (1998), Granit *et al.* (1999), Green & Overfield (1995), Leader & Carey (2003), Olsson (1987), Shirihai (1988), Taylor (1984), Tuck (1972), Wallace (2003).

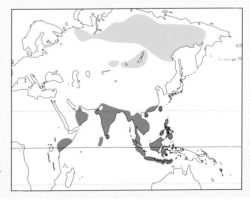

▲ **91a. Adult**. South Korea, late April. Very 'blunt' rear-end, with tertials and primary tips reaching close to tip of short tail. Difficult to separate from Swinhoe's Snipe when on the ground. For separation from Common Snipe note fringes of lower scapulars, which have buff on outer fringe running round to inner side; Common Snipe has a wider pale outer fringe, and narrower rusty inner fringe (compare with 93d). Aurélien Audevard.

▼ **91b. Adult**. Malaysia, late December. Compared to Common Snipe the bill averages shorter and the supercilium is particularly wide in front of the eye, though these are characters are shared with Swinhoe's Snipe. The primary tips only just project beyond tertials. Choo Tse Chien.

92. SWINHOE'S SNIPE
Gallinago megala

A migratory Siberian snipe with a very long bill, wintering south to Australia.

Identification L 28cm (11"); **WS** 45cm (17.5"). Forms a species pair with Pintail Snipe (91), from which can be separated only with difficulty. Largest of four very similar-sized snipes (the other three are Pintail, Common and Wilson's Snipe); Swinhoe's is medium-sized, pale and dull above, with very long bill (but relatively shorter and deeper than Common Snipe), and medium-length legs. Supercilium wider in front of eye, same colour as area above dark cheek-line, as Pintail Snipe. On ground, tail extends beyond primary tips; no (or short) primary projection beyond tertials. *In flight,* bill held near horizontal; from above as Pintail Snipe, lacking pale trailing edge to secondaries, underwing dark (coverts with black bars wider than white). Toes just reach beyond end of tail. When flushed, usually rises at close range, often without calling, and keeps rather low, flying only a short distance. Sexes similar in size and plumage with no seasonal variation; juvenile is separable. Tail usually has 20 feathers, but can have 18, 22 or even up to 26; the outer five or (usually) six on each side are

▲ **92a. Juvenile**. Japan, late August. When on ground very similar to Pintail Snipe. Aged by narrow white outer fringes to scapulars (narrower and whiter than juvenile Common Snipe), and wing-coverts with uniform pale fringes. Common Snipe of all ages shows much more white at tips of secondaries. Tadao Shimba.

stiff and are progressively narrower towards the outermost (which are 2mm wide). *Feeds* as other snipe, by deep, vertical probing. See also see general discussion of snipe identification (p. 288.)

Juvenile Much as adult, but fringes of upperparts, including mantle-lines and particularly scapulars, narrow, off-white. Wing-coverts very similar to adult, with wide pale-buff fringes and dark subterminal lines, giving coverts scalloped appearance.

First non-breeding/adult Age when first breeds not known. Adult plumage described under Identification; best separated from the juvenile by broader, more buff mantle and scapular fringes. Post-juvenile moult is not known in detail but probably much as in Pintail Snipe, so first non-breeding and adult similar.

Calls and song Calls only occasionally when flushed, usually a single, short nasal grunting *scaak*, similar to Pintail Snipe, less harsh than Common or Wilson's Snipe. In breeding display the male sings a repeated *grikka, grikka, grikka, ...* When drumming the spread outer tail feathers emit a low humming.

Status, habitat and distribution Breeds in grassy, lightly wooded wetland margins, in eastern Russia and south-central Siberia; non-breeding birds occur in south and east India, south-east China,

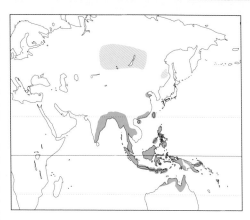

Taiwan, Philippines, Indonesia, New Guinea and, in small numbers, northern Australia. Vagrant to Sri Lanka.

Racial variation None.

Similar species Within Swinhoe's Snipe range particularly Latham's Snipe (89), Pintail Snipe (91) and Common Snipe (93), which see.

References Carey & Olsson (1995), Leader & Carey (2003), Madge (1989), Morozov (2004), Tuck (1972), Wallace (1989, 2003).

▼ **92b. Adult**. Taiwan, early September. Ming-Li Pan.

▼ **92c. Adult in display**. Finland, late June. A vagrant well west of its normal range. Toes extend beyond tail. Note use by this species of six outer tail feathers for drumming. Tom Lindroos.

93. COMMON SNIPE
Gallinago gallinago

The most widespread of the snipes, with a very long, straight bill.

Identification L 26cm (10"); **WS** 41cm (16.5"). One of four similarly sized snipes (the other three are Swinhoe's, Pintail and Wilson's Snipes), but particularly similar in plumage and behaviour to Wilson's Snipe (94), with which it forms a species pair. Medium-sized, bulky, with a very long, straight, black-tipped, brown bill and short, dull-yellow legs. Dark-brown crown, creamy crown-stripe and broad supercilium, latter only marginally wider in front of eye; whitish area above cheek-line, dark eye-stripe. Upperparts relatively bright and contrasting, dark brown, with longitudinal creamy stripes on mantle and scapulars, tertials barred dark brown and rusty. Wing-coverts duller and paler, tipped or fringed creamy-buff. Foreneck and breast dark brown with buff wash; flanks barred, remainder of underparts white. On ground, tail extends beyond primary tips; short (or no) primary projection beyond tertials. *In flight* bill is held tilted downwards; from above dark with narrow but obvious white trailing-edge to secondaries and indistinct mid-covert panel. Underwing relatively pale (barred axillaries have white bars wider than black), the palest of these four snipe; toes just project beyond tail. When flushed usually calls, rising with zig-zag flight at long range (typically up to 50m away), usually flying at some height away from immediate area. Sexes similar in size, plumage similar year-round, juvenile separable. Tail usually has 14 feathers, occasionally 12, 16 or, very rarely, 18; width of outermost feather 9.5–12.5mm at 20mm from tip; in display flight spreads only the outermost tail feather (outer two spread in Wilson's), and has lower pitched drumming than Wilson's Snipe. *Feeds* in mud by rapid, repeated, vertical probing; often wades up to belly. See also general discussion of snipe identification (p. 288.)

Juvenile Mantle- and scapular-lines narrower than adult; wing-coverts (particularly inner median coverts) uniformly fringed pale buff with black subterminal line.

First non-breeding/adult Breeds in second calendar year. Tips of wing-coverts have dark shaft-streak separating two oval cream spots, giving coverts overall spotted appearance when fresh. Post-juvenile moult includes mantle, most scapulars and wing-coverts; once these are replaced first non-breeding and adult become indistinguishable.

Call A rasping *scaap* when flushed.

Status, habitat and distribution Common, breeding in freshwater marshes; nominate *gallinago* in Europe from Portugal and Spain to northern Scandinavia, east through Russia and Siberia to western Aleutians (where scarce); race *faeroeensis* in Orkneys, Shetland, Faeroes and Iceland. Both races when non-breeding occur in freshwater wetlands from Denmark, Britain and Ireland (probably including most Icelandic *faeroeensis*), south to sub-Saharan Africa and east through the Mediterranean, Arabia, south and south-east Asia to southern China, tending to move further south in hard weather.

Racial variation Compared with *gallinago*, *faeroeensis* averages darker, with slightly more extensive rufous upperparts and narrower mantle and scapular stripes, but the two races are doubtfully separable in the field. A very rare black morph or melanistic form, 'Sabine's Snipe', has been recorded in both Common and Wilson's Snipe, apparently more frequently in the past.

Similar species Within Common Snipe's range, particularly Latham's (89), Pintail (91) and Swinhoe's (92), which see. Wilson's Snipe (94) is very similar in size, structure and plumage but is darker (even darker than race *faeroeensis*), less rufous, with breast and underwing more extensively barred; in flight has little or no white on trailing edge of secondaries and has a much darker underwing. Great Snipe (95) is bulkier, extensively barred beneath and has proportionally shorter bill; in flight has dark, white-bordered panel on secondaries and white outer tail. Eurasian Woodcock (99) also has greater bulk, transverse bars on crown and all-dark underparts; in flight is darker with more rounded wings.

Reference Tuck (1972).

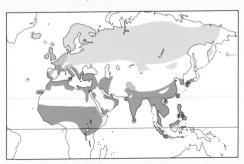

◄ **93a. Juvenile**. Eastern England, mid September. Aged by neat wing-coverts, with wide pale buff fringes and darker submarginal line. RJC.

◄ **93b. Juvenile**. Eastern England, mid September. Wing-lifting, showing narrow dark underwing barring characteristic of Common Snipe. RJC.

◄ **93c. Juvenile**. Eastern England, mid September. Wing-stretching, showing white trailing-edge to secondaries, which is much narrower or largely absent in Wilson's Snipe. RJC.

◀ **93d. Adult *gallinago*.** East England, late March. Adult wing-coverts have large paired creamy spots separated by a black line which reaches the feather tip. The scapulars, with pale, wider outer fringe and rusty, narrower inner fringe, help separate from Pintail Snipe (in which fringes are narrower, and the buff colour runs around a short distance to inner side). RJC.

◀ **93e. Adult *faeroeensis*.** Iceland, late June. The wing-coverts are more worn than 93d; the narrower upperpart stripes of this race characteristically fade to white, but feather centres are more extensively rufous. RJC.

▼ **93f. Adult *gallinago* in flight.** Poland, late April. As with 93c, note the white trailing edge to secondaries. RJC.

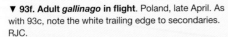

▼ **93g. Adult *gallinago* in flight.** Poland, late April. When displaying, only the outermost tail-feather is spread, unlike Wilson's Snipe, which extends the outer two. Underwing is much paler and the barred pattern more obvious than in other similarly sized snipes. RJC.

94. WILSON'S SNIPE
Gallinago delicata

The Nearctic equivalent of Common Snipe, with which it was formerly considered conspecific.

Identification L 26cm (10"); **WS** 40cm (15.5"). One of a group of four very similar sized snipes (the other three are Swinhoe's, Pintail and Common Snipe), particularly similar in plumage and behaviour to Common Snipe (93), with which it forms a species pair. Medium-sized, bulky, with very long, straight, black-tipped, brown bill and short, dull-yellow legs. Dark-brown crown, creamy crown-stripe and broad supercilium, latter only marginally wider in front of eye, whitish area above cheek-line; dark eye-stripe. Upperparts dark brownish black, with longitudinal creamy stripes on mantle and scapulars; tertials barred dark brown and rusty. Wing-coverts duller and paler, tipped or fringed buff. Foreneck and breast dark brown with buff wash, flanks barred; remainder of underparts white. On ground, tail extends beyond primary tips; short (or no) primary projection beyond tertials. *In flight* bill is held tilted downwards. From above dark, lacking Common Snipe's narrow white trailing-edge to secondaries, underwing heavily barred with black of axillary barring wider than white; toes just project beyond tail. When flushed usually calls, rising with zig-zag flight often at long range, usually flying at some height from immediate area. Tail has constant 16 feathers, width of outermost feather 4–9 mm at 20mm from tip; in display flight spreads outer two tail feathers (not just outermost feather as Common Snipe), and has higher pitched drumming than Common Snipe. *Feeds* in mud by rapid, repeated, vertical probing; often wades up to belly. See also general discussion of snipe identification (p. 288).

Juvenile Mantle- and scapular-lines narrower than adult; wing-coverts (particularly the inner median coverts) uniformly fringed pale buff with black subterminal line.

First non-breeding/adult First breeds in second calendar year. Tips of wing-coverts have dark line along shaft separating two oval whitish spots, giving coverts overall spotted appearance until worn. Post-juvenile moult includes wing-coverts; once these are replaced, first non-breeding and adult become indistinguishable.

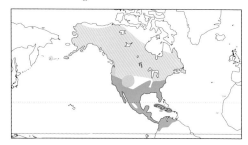

▼ **94a. Adult**. Colorado, late April. Averages blacker on upperparts than Common Snipe and has stronger, darker flank barring, but there is much overlap between the two species. Juvenile has wing-coverts with patterning similar to juvenile Common Snipe (93a). RJC.

Call A rasping *scaap* when flushed, very similar to Common Snipe; drumming see Identification.

Status, habitat and distribution Fairly common, breeding in freshwater marshes throughout North America at latitudes above about 43°N, non-breeding from 40°N south to northern South America. Vagrant to western Europe.

Racial variation None, but a very rare black morph,

'Sabine's Snipe' (also recorded for Common Snipe), apparently observed more frequently in the past.

Similar species Common Snipe (93), which see. American Woodcock (101) is of similar size to Wilson's Snipe but has transverse bars on crown and is rufous beneath.

References Bland (1998, 1999), Carey & Olsson (1995), Legrand (2005), Mueller (2005), Tuck (1972).

◄ **94b. Adult**. Colorado, late April. Wing-lifting, showing overall darker underwing, with wider dark barring than Common Snipe. RJC.

◄ **94c. Adult**. Colorado, late April. Wing-stretching, showing absence of white trailing-edge on secondaries, which aids separation from Common Snipe. RJC.

▼ **94d. Adult in flight**. New York, early May. Uses the outer two tail feathers in display; compare with Common Snipe (93g), which only uses the single outermost tail feather. Richard Crossley.

95. GREAT SNIPE
Gallinago media

A bulky Eurasian snipe with a distinctive wing-covert pattern and barred underparts. Near Threatened.

Identification L 28cm (11"); **WS** 45cm (18"). Medium-sized with a bulky body, very long dark-brown bill with pale yellow base, and short greenish or yellowish legs. Dark-brown crown, creamy crown-stripe and broad supercilium; dark eye-stripe obvious on lores but indistinct on ear-coverts. Upperparts dark brown, with longitudinal creamy stripes on mantle and scapulars; wing-coverts broadly tipped white, forming two prominent bars on folded wing. Throat and breast heavily spotted, most of belly, vent and undertail-coverts barred. *In flight*, bill is held near horizontal. From above dark, with dark white-bordered panel on greater coverts; tail has prominent white sides and tip (most obvious at take-off and landing); the feet do not extend beyond the tail. When flushed Great Snipe usually fly low and direct, dropping down after flying 100m or so. Sexes are of similar size; plumage is similar year-round; the juvenile is separable. Tail usually has 16 feathers, occasionally 14 or 18; unlike all other *Gallinago* snipe does not have an aerial display, but has a lekking breeding system. *Feeds* by picking as well as vertical probing for invertebrate prey, usually in damp areas but also on drier ground.

Juvenile More rufous above than the adult, but soon fades; wing-coverts are tipped off-white or buff, less broadly than in adult; brown markings on white outer tail.

First non-breeding/adult Age at first breeding not known. Adult described under Identification. Post-juvenile moult includes wing-coverts and tail; once these are replaced, first non-breeding and adult become indistinguishable.

Call Generally silent, but may give a low croak when flushed.

Status, habitat and distribution Breeds in damp boreal woodland in Scandinavia through western Siberia to about 90°E; probably moves south through central Europe, as is a scarce migrant in western Europe. Most non-breeding occur in Africa south of the Sahara.

Racial variation No races recognised.

Similar species Within range Common Snipe (93), which see.

References Svažas *et al.* (2002), Tuck (1972), Wallace (1976, 1977).

▼ **95a. Juvenile**. Finland, early September. Bill deeper (but shorter) than Common Snipe. Pale tips to wing-coverts form lines across folded wing. Antti Below.

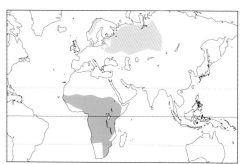

▲ **95b. Juvenile wing-stretching**. Finland, early September. The underpart chevrons are narrower and less dark than on adult. The same individual as 95a. Antti Below.

▼ **95c. Adult**. Kenya, mid September. Pale tips to wing-coverts form wider lines than on juvenile, and chevrons on underparts are also wider. RJC.

DOWITCHERS

Three species (in a single genus) worldwide, all breeding in the northern hemisphere: Short-billed Dowitcher *Limnodromus griseus*, Long-billed Dowitcher *L. scolopaceus* and Asian Dowitcher *L. semipalmatus*.

All three dowitchers are migratory. The three races of Short-billed Dowitcher breed in separate areas in northern Canada, while the monotypic Long-billed Dowitcher breeds in eastern Siberia, Alaska and north-west Canada. Some Long-billed Dowitchers migrate south-east across Canada to the North American Atlantic seaboard, and it is presumably some of these, that either continue eastward or are blown eastward across the Atlantic, that annually appear as vagrants in Europe. Short-billed Dowitchers, in contrast, migrate largely within North America, overland or coastally, which probably explains their much greater rarity as vagrants to Europe. Asian Dowitcher is larger than the two other species, and breeds in central Siberia, migrating south to its non-breeding areas, the most important of which are Java and Sumatra.

Both Short-billed and Long-billed Dowitchers, particularly the former, structurally resemble snipe with their long, straight, snipe-like bills, a resemblance that is enhanced by their 'sewing-machine' deep-probing feeding style. Unlike snipe, which appear similar year-round, dowitchers have distinctive juvenile, non-breeding and breeding plumages. Adult Short- and Long-billed Dowitchers are very similar, and care is required in making a specific identification, especially in non-breeding plumage. Asian Dowitchers are larger, with proportionally longer legs, and although they also exhibit the 'sewing-machine' style of feeding, they recall godwits rather than snipe. Indeed, when non-breeding they often associate with godwits, particularly Black-tailed Godwit.

Separation of Short-billed and Long-billed Dowitchers Detailed information on separating these two species is given in the individual species accounts, particularly under Short-billed Dowitcher (96), but there are other useful pointers. Female dowitchers have longer bills than the males. Since the bill lengths of these two species overlap considerably, only the shortest-billed male Short-billed and the longest-billed female Long-billed can be identified on this character alone.

The wing-moult strategy differs. Short-billed Dowitchers only moult their primaries once they reach their coastal non-breeding grounds, while Long-billed Dowitchers commence wing-moult while still on southward migration, often at favoured inland stop-over sites where quite large numbers gather. Thus any inland North American dowitcher seen in wing-moult during July–September is likely to be a Long-billed, since an inland Short-billed Dowitcher will still be on migration. Another useful, though by no means absolute, distinction between these two species is Short-billed's preference for brackish or saline coastal wetlands, and Long-billed's preference for freshwater habitats .

96. SHORT-BILLED DOWITCHER
Limnodromus griseus

A North American dowitcher with a preference for coastal or brackish wetlands.

Identification L 27cm (10.5"); **WS** 45cm (17.5"). Forms a species pair with Long-billed Dowitcher (97). Medium-sized, with long (male) or very long (female) straight, pale-based, blackish-tipped bill and medium-length, dull yellow or yellowish-green legs. Darkish crown, prominent supercilium and dark eye-stripe; both upperparts and underparts show considerable seasonal variation. Barred tail often has white bars as wide as, or wider than, dark bars (not the tail-coverts, which are more evenly barred). Wing-tips of adults (not juveniles which have shorter wings) usually extend at least to tail-tip, typically 2–3 mm beyond, with two primary tips exposed beyond tertials. *In flight* from above fairly uniformly darkish (including rump), white extending from above rump in V up back; toes extend beyond tail. Sexes of similar size, female with slightly longer bill; plumages vary seasonally and with age. *Feeds*, usually in water, and often in flocks, by rapidly probing bill vertically in mud. Larger web between outer toes, smaller between inner. Prefers brackish or saline water, where any dowitcher is much more likely to be a Short-billed than a Long-billed Dowitcher.

Juvenile Dark-brown crown, broad off-white supercilium; mantle, scapulars and tertials dark brown with bright buff fringes and internal markings ('tiger-stripes'), particularly on tertials. Breast buffish, spotted brown at sides; remainder of underparts largely white.

First non-breeding/adult non-breeding Grey crown, off-white supercilium; upperparts uniformly grey; grey upper breast, often streaked and spotted, and with irregular, spotted, lower margin; some darker grey spots and bars on breast sides and flanks. First non-breeding retains some distinctive juvenile scapulars and tertials, and many worn wing-coverts; can often be distinguished by these throughout the non-breeding season.

First-breeding Many do not breed until their third calendar year and remain well south of the breeding grounds. Such individuals show non-breeding or only partial breeding plumage.

Adult breeding Adult breeding varies with race (see Table 12, p. 309). Crown dark brown; reddish supercilium. Mantle feathers black-centred with rufous or pinkish-brown fringes; scapulars and tertials black, barred buff with paler, uniformly rusty fringes; underparts have orange-red or brownish-red base colour; breast spotted dark brown, flanks with brown spots and bars, belly variably orange-red to white.

Call A mellow, quick *tu-tu-tu*, recalling Ruddy Turnstone's 'rattle'.

Status, habitat and distribution Common, breeding in three distinct areas in subarctic coniferous forest and muskeg, in southern Alaska and adjacent Canada ('Western' Short-billed Dowitcher, race *caurinus*), between about 55°N and 62°N from British Columbia east to Hudson Bay ('Inland' Short-billed Dowitcher, race *hendersoni*), and in central Quebec and Labrador ('Eastern' Short-billed Dowitcher, nominate race *griseus*). Races *hendersoni* and *griseus* intergrade in northern Ontario, and occur together along Atlantic coast on migration (*griseus* being the more numerous, particularly north of New Jersey), non-breeding coastally from North Carolina south to Brazil (with *griseus* having a more easterly distribution than *hendersoni*); *caurinus* migrates south through the Pacific-coast states, non-breeding California south to Peru. Very rare vagrant to Europe, presumably *griseus* or *hendersoni*.

Racial variation Juvenile *caurinus* has narrower fringes to upperpart feathers, thus appearing darker than the other two races. Otherwise only separable in breeding plumage, when racial identification depends on width and colour of upperpart fringes, extent of internal light markings on scapulars and tertials, density of spotting and barring on breast and flanks, and extent of white on lower belly and vent. See Table 12 (p. 309). In general, *caurinus* is intermediate between the other two races.

Similar species Structurally, Long-billed Dowitcher (97) is very slightly larger, more bulky (humpbacked, with pot-belly), and has marginally longer legs, though these differences are small and not always well shown. Dowitchers with longest bills are female Long-billed. In all but juvenile plumage, folded wings of Long-billed Dowitcher typically fall marginally short of tail (usually with little or no extension beyond tertials); in Short-billed wing-tips just reach end of tail or extend marginally beyond (typically two primary tips showing). Juvenile Long-billed lacks the pale internal patterning of upperpart feathers, appearing plain and rather dark in comparison to Short-billed. Non-breeding Long-billed has more extensive grey on breast, less spotted at margin; breeding Long-billed most resembles Short-billed *hendersoni* (particularly when both are worn in late June or July), but has underparts entirely darkish red, a densely spotted foreneck and barred upper breast. Except when diagnostic juvenile feathers can be seen (some of which are usually retained at least to end of year in both species), identification is best confirmed by call.

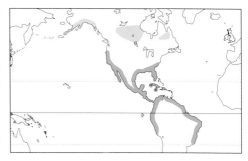

References Chandler (1998), Jaramillo & Henshaw (1995), Jaramillo *et al.* (1991), Jehl e*t al.* (2001), Kaufman (1990), Loftin (1962), Putnam (2005).

Table 12. Distinctions between Short-billed Dowitcher races when breeding; note that intergrades between nominate *griseus* and *hendersoni* occur.

Race	Mantle and scapulars	Breast ground-colour and markings	Belly
caurinus	Narrow fringes; scapulars may have whitish tips	Dull brown-orange; fairly heavily spotted, flanks sometimes barred	Small white area
griseus	Narrow fringes	Dull brown-orange; densely spotted and barred, flanks barred	Mainly white, usually with dark spots
hendersoni	Broad fringes, often with pale internal markings	Orange-peach; moderately to lightly spotted, flanks usually with limited barring	White area limited or absent

◄ **96a. Juvenile *griseus* or *hendersoni***. Florida, late August. These two races are inseperable in this plumage. The yellow internal markings ('tiger stripes') on the tertials and larger rear coverts are diagnostic of juvenile Short-billed Dowitcher. RJC.

◄ **96b. Juvenile *caurinus***. California, mid September. Moulting to first non-breeding. Has narrower upperpart fringes than juveniles of race *griseus* or *hendersoni*. RJC.

▲ **96c. First non-breeding**. Florida, late December. Many, as with this individual, retain at least some juvenile pale-marked tertials and wing-coverts until the end of the year. RJC.

▲ **96d. Adult non-breeding**. Florida, early January. Wing-tips of adults extend to or (more usually) 2–3mm beyond tail, exposing two primary tips beyond the tertials. RJC.

▲ **96e. First-breeding**. Florida, late June. Many second calendar year Short-billed Dowitchers do not return to the breeding grounds; some acquire near-complete breeding-type plumage, others (as the bird on the left) virtually none. The bill of Short-billed Dowitcher has a slight down-turn at the distal third, unlike the straight bill of Long-billed Dowitcher. RJC.

▲ **96f. Adult breeding *griseus***. Florida, late May. Narrow upperpart fringes, strongly spotted beneath with dense flank barring, and white on belly are characteristics of *griseus*. RJC.

▼ **96g. Adult breeding *hendersoni***. Texas, late April. Upperparts have broad fringes and pale internal markings, underparts orange-peach with fewer dark markings, mainly spots. RJC.

▲ **96h. Adult breeding intergrade**. Texas, late April. Intergrades between *griseus* and *hendersoni* occur, as this individual with fairly broad upperpart fringes and more spots and bars below than the typical sparsely marked *hendersoni*. RJC.

◄ **96i. Adult breeding *caurinus*.** California, mid April. Narrow upperpart fringes and rear scapulars with whitish tips. RJC.

▼ **96j. Non-breeding Short-billed Dowitchers in flight**. Florida, mid October. Very similar in flight to Long-billed Dowitcher (compare with 97g), but feet project slightly less far beyond tail. RJC.

97. LONG-BILLED DOWITCHER
Limnodromus scolopaceus

Marginally the larger of the two North American dowitchers, with a preference for freshwater habitats.

Identification L 29cm (11"); **WS** 46cm (18"). Very similar in all plumages to Short-billed Dowitcher (96), with which it forms a species pair, but slightly larger. Medium-sized, with long (male) or very long (female), straight, pale-based and dark-tipped bill, and medium-length, dull yellow or yellowish-green legs. Darkish crown, prominent supercilium and dark eye-stripe. Barred tail (not tail-coverts) has dark bars wider than white; at all ages folded wings fall just short of tail-tip and are cloaked by tertials. *In flight* as Short-billed Dowitcher, though feet project marginally further beyond tail. Sexes of similar size but female has longer bill; those dowitchers with longest bills are female Long-billed. Plumages vary seasonally and with age. Larger web between outer toes, smaller between inner. *Feeds* as Short-billed Dowitcher but usually found in freshwater habitats.

Juvenile Brown crown, off-white supercilium. Mantle, scapulars and tertials dull brown, usually lacking internal markings; mantle and scapulars have narrow, rufous, slightly scalloped fringes; greater coverts and tertials have dark tips and sometimes two paler subterminal patches, narrowly fringed off-white when fresh. Breast washed buff; remainder of underparts largely white. Overall, considerably darker above than Short-billed Dowitcher.

First non-breeding/adult non-breeding Very much as Short-billed Dowitcher, particularly adult non-breeding; grey on breast slightly darker and more extensive, with more abrupt lower boundary, with less spotting, but often with strong barring on lower flanks. Post-juvenile moult as Short-billed, so first non-breeding can be aged on worn wing-coverts, and specifically identified from retained plain, worn dark-tipped juvenile coverts or tertials, often to February or March.

First-breeding/adult breeding Unlike Short-billed Dowitcher, most attain breeding plumage in second calendar year and thus probably breed then. Black upperpart feathers have narrow rusty fringes, with white tips to scapulars; underparts entirely chestnut-red, throat and upper breast spotted, sides of breast and much of underparts strongly barred black, fringed white when fresh.

Call When disturbed typically gives a double or triple high, thin *keek*, then often repeated singly.

Status, habitat and distribution Fairly common, breeding in Arctic beyond tree limit, usually near fresh water; in eastern Siberia (where extending west) from the Yana east to western Alaska and north-west Canada. Both Siberian and North American birds move south down Pacific coast or south-east through Canada, non-breeding in freshwater wetlands south from California, Virginia, Gulf coast and Central America to Panama; rarely Hawaii. Regular vagrant to Europe, and much less

◄ **97a. Juvenile**. California, early November. Moulting to first non-breeding; has replaced the upper scapulars with adult-type non-breeding feathers, but the remaining upperpart feathers, typically with darkish tips and irregular narrow pale fringes, are all juvenile. Bill lacks slight down-turn at distal third of Short-billed Dowitcher. RJC.

commonly as far east as India, also Japan (where perhaps non-breeding), Korea, to south-east Asia; mainly during mid-September to early November.

Racial variation No races recognised.

Similar species Short-billed Dowitcher (96), which see.

References Chandler (1998), Holt (1999), Jaramillo & Henshaw (1995), Jaramillo *et al.* (1991), Kaufman (1990), Putnam (2005), Takekawa & Warnock (2000).

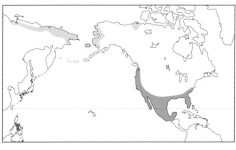

◀ **97b. First non-breeding.** Florida, late December. Still has many dark-tipped juvenile coverts, scapulars and tertials. Long bill suggests a female. RJC.

◀ **97c. Adult non-breeding.** California, mid November. Note that wing-tips do not reach tail tip and are completely covered by the tertials; strong flank barring, and black bars on tail that are much wider than white bars, help to separate from Short-billed Dowitcher. Shortish bill suggests a male. RJC.

◀ **97d. First non-breeding/first-breeding.** Florida, late March. Identified as Long-billed Dowitcher by broad black tail bars, and aged by single retained brownish dark-tipped juvenile tertial. RJC.

◀ **97e. Adult breeding**. California, late April. In fresh breeding plumage sides of breast have dark bars with broad white fringes. RJC.

◀ **97f. Adult breeding.** Alaska, mid June. White fringes of underparts have largely worn off, though dark bars on flanks remain. Very like *hendersoni* Short-billed Dowitcher at this stage, but underparts are deeper red and foreneck is spotted. René Pop.

◀ **97g. Non-breeding Long-billed Dowitchers in flight**. California, mid November. Feet extend marginally further beyond tail than in Short-billed Dowitcher; compare with 96j. RJC.

98. ASIAN DOWITCHER
Limnodromus semipalmatus

The largest of the dowitchers, with a heavy, very long, straight black bill. Near Threatened.

Identification L 35cm (14"); **WS** 57cm (22"). Large, with very long, blunt-tipped, straight, dark bill of constant width; dark crown, pale supercilium, dark loral line and long blackish legs. Godwit-like; when non-breeding often associates with both Black-tailed Godwits (race *melanuroides*) and Bar-tailed Godwits (race *baueri*). Similar to the latter in all plumages and easily overlooked when with godwits, but smaller than these godwit taxa. Best identified by its very long bill, often held at about 30° to horizontal. *In flight* from above, plain brown-grey, lacks wing-bar (though has paler area across wing-coverts and secondaries), tail, rump and V up back obscurely paler owing to extensive dark barring; feet extend beyond tail-tip. Sexes of similar size, female's bill averages marginally longer. Plumages vary seasonally and with sex; juvenile distinctive. Webbing between all three toes, more extensive than other dowitchers. *Feeds* like other dowitchers in soft mud, with sewing-machine action of rapid vertical probing, head often plunged below water surface.

Juvenile All upperparts and wing-coverts dark with broad pale fringes, recalling juvenile Bar-tailed Godwit, but lacking latter's pale-notched, lower-rear scapulars and tertials. May have pink area on bill base.

First non-breeding/adult non-breeding Upperparts and wing-coverts brown-grey with broad, diffuse pale fringes; neck and breast streaked and mottled brown, remainder of underparts white. Bill and legs entirely dark grey. First non-breeding acquires adult-type upperparts but retains many worn and bleached juvenile wing-coverts; bill base may retain some pink.

First-breeding Does not breed in second calendar year, and some perhaps not in third; first-breeding probably remains on non-breeding grounds, where attains a variable amount of breeding plumage, from none to about 25%. Bill all-dark.

Adult breeding Upperparts with black centres and broad rufous fringes; head, neck, breast and upper belly chestnut-red, with dark barring (particularly on lower breast and flanks) and white fringes when fresh. Bill and legs grey-black. Female probably less bright than male.

Call Mostly silent away from breeding grounds but gives a soft *eouw*, particularly in flight.

Status, habitat and distribution Uncommon, breeding discontinuously in freshwater wetlands through mid-latitude Siberia; on migration in coastal Japan and China; non-breeding on coastal

▼ **98a. Juvenile**. Hong Kong, mid August. Somewhat larger than the other two dowitcher species, with a sturdy straight bill and a different pattern on upperparts and coverts. Lacks pink base to bill shown by some juveniles. Ming-Li Pan.

and estuarine mudflats, from eastern India, Gulf of Thailand, Sumatra and Java south to north-western Australia. Vagrant in Korea, and east to Oman (where may be scarce migrant or non-breeder), Kenya and Australia other than the north-west.

Racial variation None.

Similar species From Black-tailed (102) and Bar-tailed Godwits (104) by Asian Dowitcher's smaller size, straight, all-dark blunt-tipped bill and, in flight, by barred rump and lack of wing-bar.

References Barter (2002), Chandler (1998), Higgins & Davies (1996), Mengel (1948), Paige (1965), Park & Kim (1994), Silvius (1988), Silvius & Erftemeijer (1989), Smart & Forbes-Watson (1971).

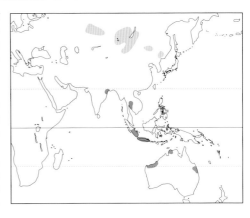

◄ **98b. Juvenile**. Queensland, Australia, late October. Much more worn and faded by this date. David Fisher.

▼ **98c. First non-breeding**. Western Australia, early April. Worn wing-coverts and total lack of any breeding plumage suggest first non-breeding/first breeding. The other shorebirds are Bar-tailed Godwits; Asian Dowitchers often associate with both Bar-tailed and Black-tailed Godwits. RJC.

◄ 98d. Adult breeding. Western Australia, early April. Bill length suggests a male. RJC.

◄ 98e. Adult breeding. Western Australia, early April. Bill length suggests a female; the shorebird immediately behind is a non-breeding Great Knot. RJC.

◄ 98f. Asian Dowitchers in flight. Hong Kong, early May. Note that the toes of dowitchers extend well beyond the tail-tip, a useful feature for separation from Bar-tailed Godwits in flight. The smaller birds in this photograph are Red Knots. John and Jemi Holmes.

WOODCOCKS

Three species (in one genus) in the northern hemisphere: Eurasian Woodcock *Scolopax rusticola*, Amami Woodcock *S. mira* and American Woodcock *S. minor*. There are six species of woodcocks (in a single genus) worldwide.

Woodcocks are chunky, medium-sized shorebirds, with short or very short legs that are similar in many respects to snipes. They are generally nocturnal feeders, and share with at least some species of snipe the unusual habit for shorebirds of feeding their young, though only while they are quite small. The high position of their eyes allows 360° vision (as seen in the photograph above). Most species feed and roost in woodland, as their name suggests.

Both Eurasian and American Woodcock are migratory. However some populations, particularly in the southern parts of their ranges, are sedentary, although they may move in response to cold weather. Amami Woodcock is sedentary, and lives on just a few small islands in southern Japan.

Although not generally appreciated, Eurasian Woodcock has a distinctive juvenile plumage, in which the entire underparts have dark marginal or submarginal fringes, giving a scalloped appearance, rather than the bold barring of the adult. It seems likely that these feathers are lost quite quickly, probably on the breeding grounds prior to migration. It seems possible that other species, including Amami Woodcock, may have a similar juvenile plumage.

99. EURASIAN WOODCOCK
Scolopax rusticola

A large, bulky, cryptically coloured shorebird that generally breeds and feeds in woodland areas.

Identification L 34cm (13.5"); **WS** 58cm (23"). Forms a species pair with the very similar Amami Woodcock (100). Large, short-necked and bulky, with rufous-brown, cryptically patterned plumage; long, straight, dark-tipped, pinkish-based bill and short, flesh-coloured legs. Steep, plain greyish forehead, dark eye set well back and high on head, enabling 360° vision; rear-crown and nape brown, with transverse pale bars, dark loral-line from bill rises to eye, not parallel to cheek-line. Upperparts are rusty-brown and black with some buffish fringes and tips; the wing-coverts and tertials have large, dark-brown, oval bars; underparts are buff with greyish-brown barring.

In flight the bill is held pointing down; uniformly dark above and below; broad, rounded wings; the legs do not extend beyond the tail. Sexes are of similar size, though the bill of the male averages shorter than female. The adults have similar plumage year-round; juvenile is separable. Occasionally some individuals have a bill as short as half the average length; the sex of such birds is unknown. *Feeds* largely at night, in damp grassland or damp deciduous woodland, both by picking and vigorous probing for invertebrate prey; the birds roost in densely forested woodlands in the day, where their camouflage makes them very difficult to see.

▲ **99a. Adult**. Central England, early February. Once the scalloped juvenile underparts have been lost ageing is no longer possible in the field. RJC.

Juvenile Neck, breast and belly feathers are narrowly fringed dark brown, appearing scaly, and thus differing from the more strongly barred underparts of the adult. Otherwise very similar to the adult.

First non-breeding Once juvenile neck, breast and belly feathers are moulted (probably by September, and before migration) inseparable in the field from adult. In the hand, may be aged by grey-white tips to the underside of the tail feathers retained from juvenile plumage, which are smaller than the pure-white (almost iridescent) under-feather tips of adult. Also, spots at the tips of the primary coverts are concolorous with the feather-side spots; adult has paler fringe at covert tips.

Adult Females breed in their second calendar year, most males in their third. Adult is described under Identification.

Call Male in display flight ('roding') has song composed of croaking notes; generally silent outside breeding season but may give Common Snipe like *scaap* when disturbed from fields at night.

Status, habitat and distribution Quite common, breeding in dampish woodland throughout Europe and Asia, east to eastern Siberia along the taiga belt, between about 45–50°N and 65–70°N. The more southerly of western populations are sedentary, but the birds will move south in hard weather; northern and eastern populations are migratory, spending non-breeding periods in western Europe from Ireland south to North Africa (where uncommon) and the Mediterranean, eastwards to south and south-east Asia. Vagrant to Iceland, Greenland and

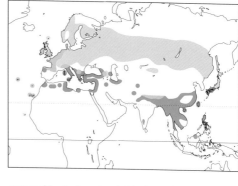

eastern North America (with the majority recorded in the 19th century).

Racial variation No races recognised.

Similar species Wood Snipe (90), which see. When non-breeding, Eurasian Woodcock occurs with Amami Woodcock (100) in the Japanese Nansei Shoto islands. Amami Woodcock has a dark fore-crown bar half the width of second bar (both bars same width in Eurasian), less steep forehead, loral line parallel to cheek-line, and a deeper-based more tapering, uniformly coloured bill; has rather plain (not barred) wing-coverts and tertials, and finer barring on the underparts. American Woodcock (101) is smaller, has more rufous unbarred underparts, and in flight its emarginated outer primaries generate a twittering noise.

References Hoodless (1995), Hoodless & Coulson (1998), Kalchreuter (2000), Mulhauser & Diaz (2003), Thorup (2006).

▲ **99b. Adult**. Central England, late January. The slightly paler fringe at the tips to the primary coverts show this to be an adult; first non-breeding has darker spots at the covert tips similar to those at the side of the feathers. RJC.

▲ **99c. Adult in flight**. Finland, early June. Appears dark in flight, with broad rounded wings. This bird is 'roding', its breeding display flight. Through much of the breeding range this occurs at dusk, but at northerly latitudes roding may take place before sunset. Matti Suopajärvi.

100. AMAMI WOODCOCK
Scolopax mira

A species endemic to Amami-Oshima and nearby islands of southern Japan. Vulnerable.

Identification L 35cm (14"); **WS** 66cm (26"). Very similar to Eurasian Woodcock (99), with which it forms a species pair. Large, short-necked and bulky, with dark olive-brown, cryptically patterned plumage; long, straight, deep-based, pinkish-brown bill (lacking Eurasian's dark tip) and medium-length flesh-coloured legs. Plain brown forehead, flatter than Eurasian Woodcock owing to deeper-based bill; dark eye set well back and high on head, enabling 360° vision. Many (but not all) have small areas of bare skin before and behind eye, which are difficult to see in the field. Rear-crown and nape brown, with transverse pale bars, first dark forehead bar is half width of second, not same width as in Eurasian Woodcock. Dark loral-line from bill to eye continues line of bill, parallel with dark cheek-line (unlike Eurasian where the lines are divergent). Upperparts olive-brown and black, with some buffish feather fringes and tips; wing-coverts and tertials largely dark-brown, not strongly barred as Eurasian; underparts buff, with fine greyish-brown barring. *In flight* bill held pointing down, uniformly dark above and below; wings rounded, broader than Eurasian Woodcock, toes do not extend beyond tail. Sexes of similar size, though bill of male may average shorter than female, as with Eurasian. Adults have similar plumage year-round, juvenile may be separable. *Feeds* largely at night, both by picking and probing, roosting in woodland by day. If disturbed as likely to run for cover as fly; flies only a short distance before dropping down, sometimes landing in trees.

Juvenile Not known, but may perhaps have lightly scalloped underparts, not barred as adult, similar to Eurasian Woodcock.

Adult Age at first breeding not known but probably similar to Eurasian Woodcock. Adult plumage described under Identification.

Calls Occasionally gives snipe-like *jeh* or *jee* calls when flushed.

Status, habitat and distribution Fairly common in damp, shady, evergreen hill forest in the subtropical Japanese Nansei Shoto islands. Sedentary.

Racial variation None.

Similar species Eurasian Woodcock (99), which see.

References Brazil & Ikenaga (1987).

▼ **100a. Adult**. Amami-Oshima, Japan, late February. Differences from Eurasian Woodcock include forecrown bars of unequal width, an area of bare skin around the eye (but see 100b), a deeper-based bill and a different wing-covert pattern. Shigeo Ozawa.

▼ **100b. Adult**. Amami-Oshima, Japan, early April. Not all individuals have the area of bare skin around the eye; the reason for the presence or absence of this feature is not known. Kazuyasu Kisaichi.

101. AMERICAN WOODCOCK
Scolopax minor

The smallest of the woodcocks, and the most distinctive.

Identification L 28cm (11"); **WS** 42cm (16.5"). Medium-sized, short-necked and bulky, with rufous-brown, cryptically patterned plumage; very long, straight, dark-tipped, pinkish-based bill and very short, brownish-flesh legs. Plain, greyish forehead, dark eye set well back and high in head, rear-crown and nape dark brown with transverse pale bars; narrow, dark eye-stripe. Upperparts black with rufous fringes and bars; broad pale-grey mantle and scapular-Vs; underparts uniformly plain reddish buff. *In flight* from above uniformly dark, outer three prmaries are shortened and emarginated (more so in males), producing a twittering noise; bill held at 45°; legs do not extend beyond short tail. Female up to 10% larger than male, with longer bill; adults have similar plumage year-round, juvenile differs marginally. *Feeds* in or near damp woodlands by picking and probing.

Juvenile Has rather duller plumage than adult, most with dark grey band on throat and upper breast. Bill typically dark grey, legs greyish.

First non-breeding/adult Breed in second calendar year. Adult described under Identification; once juvenile plumage is lost (with the birds moulting prior to migration), there are no characters for ageing in the field. In the hand, first non-breeding has white tips to middle secondaries contrasting with dark subterminal areas (retained from juvenile),

adult has pale buff tips showing less contrast with the subterminal zone.

Call Generally silent outside the breeding season.

Status, habitat and distribution Breeds in dampish, young woodland in eastern North America, from Manitoba to Newfoundland, south to Louisiana and Florida; non-breeding usually spent in the southern part of the breeding range, to Gulf Coast and lower Rio Grande valley. Scarce during migration, and as a non-breeder west of the breeding area. Rare vagrant west of the North American continental divide.

Racial variation No races recognised.

Similar species Wilson's Snipe (94), Great Snipe (95) and Eurasian Woodcock (99), all of which see.

References Keppie & Whiting (1994), Patten *et al.* (1999).

▼ **101b. Adult**. Rhode Island, mid March. The combination of grey forehead and unbarred orange-brown underparts are diagnostic among the woodcocks. Michelle L. St.Sauveur.

▼ **101a. Adult**. Ohio, mid May. A smaller version of Eurasian Woodcock. David Fisher.

GODWITS

There are four godwit species in a single genus worldwide, all of which breed in the northern hemisphere: Black-tailed Godwit *Limosa limosa*, Hudsonian Godwit *L. haemastica*, Bar-tailed Godwit *L. lapponica* and Marbled Godwit *L. fedoa*. All four species are migratory; Bar-tailed Godwit of the race *baueri* is exceptional in having the longest-known non-stop migration flight of any bird.

The godwits are all large to very large shorebirds, with long, near-straight bills; all have distinctive juvenile, non-breeding and breeding plumages. In all species the female is marginally larger, often with a noticeably longer bill than the male. Outside the breeding season Black-tailed and Hudsonian Godwits prefer freshwater, while the other two species usually occur on coasts or estuaries.

Migration of Bar-tailed Godwits wintering in Australasia Eastern (*baueri*) Bar-tailed Godwits spend the non-breeding season in eastern Australia and New Zealand. They depart from their non-breeding grounds in March and early April, already in breeding plumage, unlike nominate *lapponica* which spend the non-breeding season in western Europe and which do not acquire breeding plumage until late April at the earliest. Satellite tagging has shown that in 2007 one bird flew non-stop from North Island, New Zealand, over the western Pacific to the Yellow Sea, a distance of at least 10,000km. It departed in mid-March and flew continuously for seven days; after spending about six weeks in the Yellow Sea, replenishing its fat reserves, in early May it flew north-east to Alaska, a leg of 7,000km in six days. After presumably breeding, it returned at the end of August directly to North Island, New Zealand, covering 11,500km in nine days. The average flight speeds on each leg were about 50–60 km/hour.

The early acquisition of breeding plumage by *baueri* Bar-tailed Godwits seems likely to be an adaptation to allow them to concentrate their energy resources solely on replenishing fat reserves once they commence migration, unlike *lapponica* Bar-tailed Godwits in Europe which do not moult into their breeding plumage until much later as they only have to travel around 3,000km to their breeding areas, and do not have an extended ocean crossing. The majority of *baueri* Bar-tailed Godwits probably use the Yellow Sea as a staging area on northbound migration. To get to the Yellow Sea they fly about 5,000 to 6,000km further than the direct route to their breeding area, which clearly illustrates the importance of the threatened Yellow Sea coasts, not only to the godwits but also to the millions of other shorebirds that stage there on migration.

102. BLACK-TAILED GODWIT
Limosa limosa

Large, elegant shorebird with brick-red upperparts in breeding plumage. Near Threatened.

Identification L 40cm (16"); **WS** 66cm (26"). Large, with long neck; very long straight or very slightly upturned bill, with dark tip and pinkish or orange base (varies seasonally); long dark-grey legs. In all plumages supercilium barely extends behind eye. *In flight* darkish above, with prominent white wing-bar, white rump and uppertail-coverts and black tail; below, wing white, outlined black, with translucent wing-bar, legs and feet extend beyond tail. Shares its striking black-and-white upperwing flight pattern with Hudsonian Godwit (103). Female marginally larger than male, occasionally obviously so, with slightly longer bill; plumage varies seasonally and with age. *Feeds* in damp grassland and by wading in shallow water, sometimes belly deep, by picking and probing. Prefers fresh water, often feeding in flocks, particularly when non-breeding.

Juvenile Head and face bright buff, rather plain, with whitish supercilium broader in front of eye. Mantle and upper scapulars have blackish-brown centres, lower scapulars and tertials blackish-brown with internal buff markings, all broadly fringed buff; feather centres and fringes fade with age. Underparts darkish buff but vent paler. Base of bill pink.

First non-breeding/adult non-breeding Uniformly greyish above, with short whitish supercilium; breast and flanks light grey, remainder of underparts white. Adult wing-coverts are grey with faint whitish fringes when fresh. First non-breeding retains some juvenile wing-coverts and tertials at least until January–February, often longer; these become bleached and worn, and contrast with fresher, greyer upperparts. Base of bill pink.

First-breeding First-breeding attains variable amounts of adult plumage; bill base pink.

Adult breeding Usually breeds for the first time in third calendar year. The extent of red both above and below depending on both race and sex. *Male* mantle and scapulars brick-red with black markings; head, neck and upper-breast brick-red, lower breast and remainder of underparts white, barred dark brown. *Female* (particularly race *limosa*) typically has less depth of colour on head and neck, more grey non-breeding type plumage feathers on upperparts, and more white on underparts with less barring on belly. Both sexes have base of bill orange-pink.

Call A clear *wikka-wikka-wikka.*

Status, habitat and distribution Breeds in grassy wetlands. Race *islandica* in Iceland (almost the entire population), northern Britain and Ireland; nominate *limosa* in eastern England, France, the Netherlands and Denmark west to central Siberia; *melanuroides* in eastern Siberia, from Lake Baikal to Kamchatka. When non-breeding visits coastal freshwater wetlands: *islandica* in Britain and Ireland, the Netherlands, Denmark, France, south to the Mediterranean; *limosa* in Spain, Portugal and Morocco (where overlaps with *islandica*), and Africa south of Sahara east to India; *melanuroides* from south-east Asia to Australia and New Zealand. Vagrant to eastern North America south to the Caribbean (presumably *islandica* or *limosa*), and Aleutian and Pribilof Islands (*melanuroides*).

Racial variation Nominate *limosa* is the largest race, with longer legs, longer, deeper-based bill and longer neck. The deep bill-base of *limosa* gives it a flatter, less rounded forehead than *islandica*, though this difference seems less obvious with juveniles, perhaps because the deep bill-base takes some time to develop. Juvenile *islandica* is a richer red-brown on neck and breast, has darker, more chestnut upperpart fringes, and averages more contrast on mantle and scapulars than *limosa*. When breeding, *islandica* has red underparts with coarser black stripes extending well on to belly, while *limosa* has less coarsely striped underparts, with lighter red extending only to the upper belly. The smallest race, *melanuroides*, is similar to *islandica* in all plumages but males in particular usually have extensive dark barring on belly, in flight has darker forewing and narrower wing-bar.

Similar species Hudsonian Godwit (103) has slightly upturned bill, shorter legs and, in flight

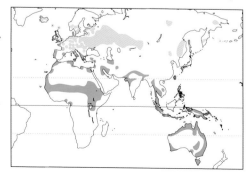

shows a less conspicuous white wing-bar and diagnostic sooty-black underwing-coverts. Bar-tailed Godwit (104) is intermediate in size between *limosa* and *melanuroides*, shorter-legged, has more upturned bill, lacks wing-bar and, in Palearctic races, has white rump extending in a V up back.

References Barter (2002), Ebels (2002), Engelmoer & Roselaar (1998), Groen *et al.* (2006), Gunnarsson *et al.* (2006), Harrison & Harrison (1965), Higgins & Davies (1996), Holling *et al.* (2007), Rasmussen & Anderton (2005), Roselaar & Gerritsen (1991), Thorup (2006), van Scheepen & Oreel (1995).

◀ **102a. Juvenile *limosa*.** The Netherlands, late July. The deep-based bill of this race seems to take some time to develop; all juvenile and non-breeding Black-tailed Godwits have a pink bill base. Astrid Kant.

◀ **102b. Juvenile *islandica*.** Southwestern England, mid September. Race *islandica* is smaller than *limosa*; in juvenile plumage it has darker markings on upperparts and averages brighter on head, neck and breast. RJC.

◀ **102c. First non-breeding *islandica*.** Eastern England, early February. The faded juvenile wing-coverts are retained for a considerable period, and contrast strongly with the plain non-breeding upperparts. RJC.

▲ **102d. Adult non-breeding** *limosa*. India, late December. Long bill suggests a female; note the deep bill-base and rather flat forehead of this race. RJC.

◄ **102e. Adult non-breeding** *islandica*. Eastern England, late November. Shallower bill-base and steeper forehead than in *limosa*. RJC.

◄ **102f. Adult male breeding** *limosa*. Poland, early June. Note orange base to bill and near-black legs of breeding individuals; the bright, well marked plumage indicates a male. RJC.

◀ **102g. Adult female breeding** *limosa*. Poland, late April. Race *limosa* is larger overall than the other two races, and has proportionally longer neck and legs. Females of this race may have very little breeding-type plumage, but still acquire an orange base to the bill and dark legs. RJC.

◀ **102h. Adult male breeding** *islandica*. Iceland, late June. A variable number of wing-coverts are replaced by breeding-type feathers. RJC.

▼ **102i. Adult female breeding** *islandica*. Iceland, late June. Less extensive colour than on the male, but more than on typical *limosa* females. RJC.

◀ **102j. Adult male breeding** *melanuroides*. South Korea, mid May. The smallest race, though very similar to *islandica*; some males have very strongly barred bellies. RJC.

◀ **102k. Non-breeding *islandica* (from location) in flight.** Eastern England, early November. From below the white wing-bars appear almost transparent. RJC.

▼ **102l. Adult breeding** *melanuroides* **in flight.** South Korea, early May. Similar pattern in all races; from above the white wing-bar and rump are striking. RJC.

103. HUDSONIAN GODWIT
Limosa haemastica

A large, elegant North American godwit with brick-red breeding plumage and a slightly upturned bill.

Identification L 39cm (15.5"); **WS** 66cm (26"). Large, with a long neck and long, very slightly upturned bill with dark tip and pinkish base; medium-length blackish legs. White supercilium, strongest in front of eye; dark line from bill to eye. *In flight* from above dark, with short, narrow, white wing-bar, white rump and uppertail-coverts, blackish tail; below, axillaries and wing-coverts sooty black with translucent white wing-bar; feet extend beyond tail. Female slightly larger, with longer bill; plumage varies with age and season. *Feeds* often in flocks, by deep vertical probing, frequently wading into rather deep waters.

Juvenile Mantle dark brown with narrow buff fringes; scapulars dark brown, notched buff, or buff with narrow, dark-brown bars; tertials dark brown, notched buff. Wing-coverts grey with dark shafts. Underparts pale, breast mottled greyish-buff, vent white.

First non-breeding/adult non-breeding Uniformly greyish above; breast light grey, remainder of underparts white. Adult wing-coverts grey, with faint off-white fringes; first non-breeding retains some worn juvenile wing-coverts.

First-breeding/adult breeding Usually breeds for the first time in third calendar year. Upperparts black with silvery edge-spots and fringes; underparts are deep brick-red, with dark-brown and white barring. Female is paler, with more white barring on underparts. Bill is more extensively black than non-breeding, with a restricted deep-pink base; legs are black. First-breeding attains a variable amount of adult plumage and retains very worn juvenile primaries.

Call *Ta-wit*, but often silent.

Status, habitat and distribution Breeds in wet meadows near tree-line, in southern Alaska, northwest MacKenzie, and locally west of Hudson Bay. Moves south via Hudson and James Bays, then probably flies direct to northern South America; non-breeding period spent in southern Argentina. Uncommon away from breeding and non-breeding

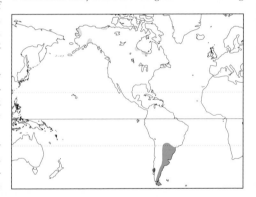

▼ **103a. Juvenile**. California, September. Aged by small, dark-centred upperparts and wing-coverts. Peter LaTourrette.

areas; rare (possibly migrant) to Pacific islands and New Zealand. Rare vagrant to western Europe, South Africa and Australia.

Racial variation No races recognised.

Similar species Black-tailed Godwit (102), which see. Bar-tailed Godwit (104) is shorter-legged, has a more upturned bill and lacks the wing-bar.

References Elphick & Klima (2002), Grieve (1987), Harrington *et al.* (1993), McCaffery & Harwood (2000), Wright (1987).

▲ **103b. Non-breeding.** Buenos Aires, Argentina, mid April. Individuals intending to breed are largely in breeding plumage by this date, so the lack of breeding plumage apart from some reddish beneath suggest this bird is unlikely to breed and will probably not return to the breeding grounds. There are no obvious remaining juvenile feathers, so precise ageing is not possible. Alec Earnshaw.

▲ **103c. Adult breeding male.** Manitoba, Canada, June. Note the deep pink base to the bill, and dark legs. Richard Sale.

▼ **103d. Adult breeding female.** Kansas, late April. Long bill and incomplete breeding plumage suggest a female. RJC.

▼ **103e. Adult breeding male lifting wing to show flight pattern.** Manitoba, Canada, mid June. The flight pattern above is similar to that of Black-tailed Godwit, but the black underwing-coverts are diagnostic at all ages. Bob Gress.

104. BAR-TAILED GODWIT
Limosa lapponica

A compact, relatively short-legged godwit with a slightly upturned bill and chestnut underparts in breeding plumage.

Identification L 39cm (15"); **WS** 65cm (25.5"). Large, fairly long-necked, with long (male) or very long (female) slightly upturned bill, with dark tip and pinkish-base (but varies seasonally); medium-length, dark greenish-grey legs (but see adult breeding below). *In flight* from above shows blackish primary coverts but is otherwise uniform, with white rump and V up back; underwing and axillaries white, latter with limited brown barring (race *lapponica*; see below for other races), toes just extend beyond tail. Female is larger, with longer bill and legs; plumage (particularly male) varies seasonally, and with age. *Feeds*, often in flocks, by deep vertical probing, both while wading and on drier ground.

Juvenile Crown dark; upperparts and wing-coverts brown, notched buff and broadly fringed off-white; upper breast streaked or washed buff; remainder of underparts white.

First non-breeding/adult non-breeding Rather pale buffish-grey upperparts with darker feather centres and broad paler fringes, less patterned and less contrasting than in juvenile. Breast washed buff, remainder of underparts white. First non-breeding retains many worn, darkish, juvenile wing-coverts and tertials, which contrast with paler adult-type upperparts.

First-breeding Attains a variable amount of adult breeding plumage but retains many worn juvenile wing-coverts. At least some remain either in the non-breeding areas or south of the breeding grounds.

Adult breeding Usually breeds for the first time in third calendar year. *Male* mantle and scapulars dark brown with red and whitish notches and fringes; entire underparts uniform deep chestnut-red. *Female* much duller than male but has similar, though typically fewer, breeding-type mantle and scapular feathers; face to upper breast and flanks variable, from pale buffy-red to off-white, similar to non-breeding. Both sexes have most (often all) of bill dark, with a very restricted pink or red area at base, but the extensive pink base of non-breeding is quickly regained; legs brown-black. Most of the red of both upperparts and underparts is the supplemental breeding plumage. In the case of birds staging on northbound migration in western Europe, this plumage is gained in a surprisingly short 12-day period in late May.

Call Frequently silent, but groups in flight give barking *kirruc* or *kak*; very noisy when breeding.

Status, habitat and distribution Breeds generally on low-lying tundra, from northern Scandinavia discontinuously through northern Russia to Alaska. Five races, each a distinct geographical population, can be distinguished on a combination of plumage and measurements: nominate *lapponica* breeds northern Norway to Kanin peninsula; *taymyrensis* from Yamal peninsula to Anabar river; *menzbieri* from the Lana delta to Chaunsk Bay; *anadyrensis* from Anadyr to southern Chukotski Peninsula); and *baueri* in Alaska. All races when non-breeding use coastal and estuarine environments: *lapponica* largely in Britain, Ireland and Denmark; *taymyrensis* south down Atlantic seaboard to west and southern Africa; *menzbieri* and *anadyrensis* locally in East Africa and Madagascar, Arabia, east to India, south-east Asia and north-west Australia; *baueri* eastern Australia and New Zealand. Vagrant (presumably *lapponica*) to Iceland, eastern North America and Caribbean, and (presumably *baueri*) Aleutians and Pacific coast North America; very rare inland in North America.

Racial variation The five races cannot all be identified in the field but three groups exist that can be separated: the '*lapponica* group' (*lapponica* and *taymyrensis*), *menzbieri*, and the '*baueri* group' (*anadyrensis* and *baueri*). The latter in flight are largely uniform brown above, lacking the white rump and V up the back of the *lapponica* group (described in Identification), and have heavily brown-barred underwing-coverts and axillaries, and thus are separable in flight from *lapponica* and many *menzbieri*; *menzbieri* is intermediate between the other

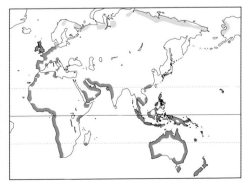

two groups and, with brown-barred white rump and back, can often be separated from individuals of the other two groups.

Similar species Black-tailed (102) and Hudsonian Godwits (103), which see.

References Atkinson (1996), Barter (2002), Ebels (2002), Engelmoer & Roselaar (1998), Higgins & Davies (1996), McCaffery & Gill (2001), Miskelly *et al.* (2001), Nieboer *et al.* (1985), Piersma & Jukema (1993), Thorup (2006), Yésou *et al.* (1992).

▲ **104a. Juvenile *lapponica*.** South-western England, mid September. Slightly upturned bill is a helpful identification feature, pink base except in breeding adults; shortish bill suggests a male. RJC.

▲ **104b. First non-breeding *lapponica*.** Eastern England, early March. Retains faded juvenile wing-coverts for an extended period, as does Black-tailed Godwit. RJC.

▲ **104c. Adult non-breeding *lapponica*.** Eastern England, late January. Shows the relatively sparse underwing barring of this race. RJC.

◄ **104d. Adult breeding male baueri** or **menzbieri.** South Korea, early May. In fresh breeding plumage; the bill is all-dark, or nearly so, when breeding. Note that these races are only separable in flight. RJC.

◄ **104e. Adult breeding female baueri** or **menzbieri.** South Korea, late April. Females acquire varying amounts of breeding-type feathers. RJC.

◄ **104f. Adult breeding male lapponica.** Eastern England, early August. Has returned to its non-breeding area; its bill has reverted to non-breeding colours and the bird is already in primary moult, though this cannot be seen in the photograph. RJC.

▲ **104g. Adult breeding female *lapponica*.** Eastern England, early August. In an equivalent state to 104f, though females, particularly those of *lapponica*, acquire much less breeding-type plumage than males. RJC.

▼ **104h. Non-breeding *lapponica* in flight.** Eastern England, late January. The tails are barred, but this race has a clear white V up the back. RJC.

▲ **104i. Adult breeding Bar-tailed Godwits in flight**. South Korea, mid May. A mixed flock of races *baueri* (uniformly dark above) and *menzbieri* (rump barred, but largely white V up back), and intergrades between the two. RJC.

▼ **104j. Adult breeding female *menzbieri* in flight**. South Korea, late April. Back and rump barred, but white V up back is readily apparent. RJC.

▼ **104k. Adult breeding female *baueri* or *menzbieri* in flight**. South Korea, late April. Showing strongly barred underwing shared by both races. The bill will be all black, or nearly so, by the time this bird reaches the breeding grounds in late May. RJC.

105. MARBLED GODWIT
Limosa fedoa

The largest godwit, cinnamon in all plumages with a very long, slightly upturned bill.

Identification L 44cm (17.5"); **WS** 71cm (28"). Large and long-necked with a very long, slightly upturned bill with a dark tip and (generally) pinkish base; medium-length dark-grey legs. Barred cinnamon-buff in all plumages. *In flight* from above uniformly cinnamon-buff with blackish primary-coverts and outer primaries; uniformly cinnamon-buff beneath; feet extend beyond tail. Sexes of similar size, but female typically has slightly longer bill; adults vary seasonally, juvenile separable. *Feeds* often in flocks when non-breeding, by vertical probing while wading, but also on drier ground, by picking and probing.

Juvenile/non-breeding Creamy-buff supercilium; upperparts uniform cinnamon-buff, with blackish bars; barring broader on mantle and scapulars, tertials black with bold buff notches. Underparts bright buff. Juveniles best distinguished by neat unworn feathering at a time when adults generally show at least some very worn (larger) scapulars or tertials. Some juveniles are apparently much paler, with narrower, brownish upperpart markings. Similarity of juvenile and adult plumages soon make ageing difficult.

First-breeding Most individuals probably remain in non-breeding area and retain a variable proportion of their juvenile plumage, but difficult to separate from adult non-breeding.

Adult breeding Probably breeds in third calendar year. Freshly acquired breeding plumage is slightly darker, more chestnut above than adult non-breeding; belly and flanks barred dark brown. At height of breeding season bill is more extensively dark, and bill-base is orange, not pink.

Call A loud *ker-reck* or *wik-wik.*

Status, habitat and distribution Fairly common, breeding in prairie wetlands in central Canada south to South Dakota, non-breeding on Pacific coast from Washington (in small numbers), south to California, Mexico and central America, locally to Peru; on Atlantic coast from Massachusetts, where scarce, becoming more common south from the

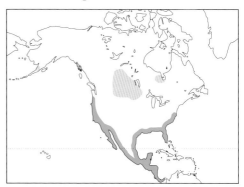

◀ **105a. Juvenile**. Florida, mid September. Aged by neat, paler wing-coverts, deeply notched, not barred, tertials, and total lack of barring on flanks, unlike adult behind. RJC.

Carolinas, in Florida and the Gulf coast. Vagrant to Nova Scotia, central Chile and Hawaii. Additionally, small populations breed in James Bay, Canada, and Ugashik Bay, Alaska Peninsula, the latter (assigned to race *beringia*) occurring non-breeding along Pacific coast, from Washington to northern California; vagrant on St Lawrence (Alaska) perhaps this race.

Racial variation The small Alaskan race *beringia*

cannot be separated on plumage characters, but it has shorter bill and legs and a more bulky body than nominate race *fedoa*.

Similar species Long-billed Curlew (112) shows similar cinnamon coloration both on ground and in flight, but is larger and has strongly decurved bill.

References Gibson & Kessel (1989), Gratto-Trevor (2000), Howell & Webb (1995).

◄ **105b. First breeding**. Florida, late June. Has remained in non-breeding area; as 105a, but has replaced some upperparts and wing-coverts with larger adult-type feathers, but has retained some very worn juvenile greater coverts and tertials. RJC.

◄ **105c. Adult non-breeding**. Florida, mid October. Totally unbarred underparts. RJC.

◄ **105d. Adult breeding**. Florida, late March. Still acquiring underpart barring of breeding plumage. Bill base turning orange; shorter bill suggests a male. RJC.

▲ **105e. Adult breeding**. Florida, late March. Unlike 105d, has not yet acquired any orange at bill base; long bill suggests a female. RJC.

◄ **105f. Adult breeding**. Florida, late August. In worn plumage. RJC.

▼ **105g. Adult breeding in flight**. Florida, early April. Entirely cinnamon above, a flight pattern shared by Long-billed Curlew, and, although less brightly cinnamon, *hudsonicus* Whimbrel. RJC.

CURLEWS

There are nine species of curlew in two genera worldwide, all of which breed in the northern hemisphere: Little Curlew *Numenius minutus*, Eskimo Curlew *N. borealis*, Whimbrel *N. phaeopus*, Bristle-thighed Curlew *N. tahitiensis*, Slender-billed Curlew *N. tenuirostris*, Eurasian Curlew *N. arquata*, Far Eastern Curlew *N. madagascariensis*, Long-billed Curlew *N. americanus* and Upland Sandpiper *Bartramia longicauda*. Eskimo Curlew, which has not been reliably reported for many years and seems likely to be extinct, is not treated here. Most curlews are long-distance migrants.

The curlews generally have brown, barred plumage that is very similar at all ages, though the juveniles can, with care, be separated. In all curlew species the female is slightly larger (though only very marginally so in Little Curlew), and often has a noticeably longer bill than the male. All curlews, but particularly Whimbrel and Bristle-thighed Curlew, are much brighter and have more contrasting plumage as juveniles than as adults. Six of the nine species have a pre-breeding moult, and gain a breeding plumage that differs slightly from the non-breeding one. The three exceptions are Little Curlew, Whimbrel and Upland Sandpiper, which have restricted pre-breeding moults (practically non-existent in the case of Whimbrel), and as a consequence have similar plumage year-round.

Whimbrel races Up to six races of Whimbrel are recognised, though only four can generally be identified in the field; see the relevant species account (p. 343). The other two races are only separable on measurements. Genetic studies show that there are significant differences between Siberian whimbrels of the race *variegatus* and the Nearctic races, perhaps sufficient to merit specific status for the two groups. Within North America, the Alaskan (*rufiventris*) and Hudson Bay (*hudsonicus*) races differ significantly on measurements; similarly, the race breeding from Scandinavia east to western Siberia (*phaeopus*) differs from the Icelandic, Faeroese and Scottish race (*islandicus*). A further distinctive race, *alboaxillaris*, breeds in the steppes of western Kazakhstan.

Bristle-thighed Curlew migration and moult The Bristle-thighed Curlew has a migration strategy that parallels that of the Australasia-wintering Bar-tailed Godwits. Bristle-thighed Curlews breed in western Alaska and spend the non-breeding season – more than nine months of the year – on islands in the central Pacific. Some birds make a single non-stop flight from Alaska to Laysan in the Hawaiian islands, a distance of about 4,000km. Birds wintering on islands further to the south seem likely to make a non-stop flight of at least 6,000km. The extended stay in the Pacific seems to be the result of the absence of predators on these islands (at least until the arrival of people), to such an extent that many adults simultaneously moult many of their flight feathers; this renders them flightless for perhaps a couple of weeks. The Bristle-thighed Curlew is the only shorebird to have evolved this moult strategy.

106. LITTLE CURLEW
Numenius minutus

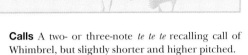

The smallest of the curlews.
Alternative name: Little Whimbrel.

Identification L 31cm (12"); **WS** 57cm (22.5"). The smallest curlew, about the same size as Common Greenshank (117) or Greater Yellowlegs (119). Medium-sized, long-necked, with a narrow buff crown-stripe and dark-brown stripes to the sides of the crown; broad buff supercilium. Lores pale, darker just before the eye, large dark eye with narrow pale eye-ring, medium-length, slender, dark bill with pink base to lower mandible, slightly decurved at tip. Medium-length greyish, dull yellow or flesh legs. Upperparts and wing-coverts brown, with buff fringes; neck and upper breast buff with dark brown streaks; remainder of underparts pale buff. *In flight* from above brown, paler on inner-wing-coverts, rump and tail; underwing uniformly medium-brown, finely barred; toes extend beyond tail. Adults similar year round, juveniles separable. *Feeds* mainly by picking and shallow probing, largely on insects and seeds, often in large flocks when non-breeding.

Juvenile Upperparts and tertials dark brown with prominent buff notches, wing-coverts brown with broad buff fringes; darker, showing more contrast than adult. In flight can show darker secondaries than adult.

First non-breeding Buff notches rapidly fade and wear; darker, more contrasting tertials are perhaps the best distinction from adult.

First-breeding Most individuals probably breed in their second calendar year. Some moult all the primaries, others none, or just outer ones; worn primaries, or contrast between new outer and worn inner primaries, may allow ageing. Extent of body-moult is variable.

Adult Described in Identification; a variable number of body feathers are replaced pre-breeding.

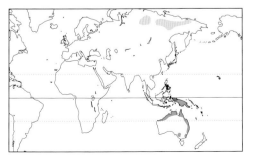

Calls A two- or three-note *te te te* recalling call of Whimbrel, but slightly shorter and higher pitched.

Status, habitat and distribution Uncommon to locally common, breeding in Arctic areas of mixed grassland and small trees, discontinuously in northern-central and north-east Siberia; on migration in Mongolia, China, Japan and Indonesia. Non-breeding on short dry grassland near fresh water in New Guinea, northern Australia, less commonly eastern Australia, rare in New Zealand. Vagrant Thailand, Britain, Norway and Pacific coast of North America from Alaska to California.

Racial variation None.

Similar species Whimbrel (107) is larger, has a different face pattern (with dark eye-stripe), and

▼ **106a. Probable juvenile or first non-breeding.**
Northern Territory, Australia, early October. This is a difficult species to age, but the neat pale-fringed wing-coverts and the uniform pattern on the upperparts suggest a juvenile. Robert Lewis.

has proportionally longer, more decurved bill (bill longer than tarsus; shorter than tarsus in Little Curlew). Slender-billed Curlew (109) is larger and has a long decurved bill, streaked and spotted whitish underparts and a different flight pattern.

Upland Sandpiper (113) is slightly smaller, has a shorter straight bill and a longer tail. All three species have different calls.

References Barter (2002), Barter *et al.* (1999), Collins & Jessop (2001), Moon (1983).

▲ **106b. Adult**. China, late April. The smallest curlew; rather pale, with a relatively short bill and pale legs. Ming-Li Pan.

▲ **106c. Little Curlew in flight**. Western Australia, mid October. Lacks any distinctive flight pattern, though the darker primaries contrast with the paler inner-wing. Rohan Clarke.

107. WHIMBREL
Numenius phaeopus

A medium-sized curlew with a circumpolar breeding range, readily identified by its striped head pattern.

Identification L 43cm (17"); **WS** 76cm (30"). Forms a species pair with Bristle-thighed Curlew (108). Large, with fairly long neck; long, strongly decurved, greyish bill, usually with some pink at base of lower mandible (but see below); medium-length, bluish-grey legs. Dark crown; narrow, buffish crown-stripe and prominent whitish supercilium. *In flight* from above brown with blackish primary-coverts and outer primaries and, variably (depending on race, see below), white rump with V up back, uniformly coloured rump and back, or intermediate pattern; below, white lower belly and sparsely streaked undertail-coverts, but underwing-coverts and axillaries are variably patterned, toes do not extend beyond tip of tail. Female marginally larger, with longer bill; adult plumages similar year-round, juvenile separable. *Feeds* intertidally by probing and picking (particularly on crabs), and on coastal grassland when non-breeding; often feeds on berries on the breeding grounds.

Juvenile Crown dark, with rather indistinct crown-stripe; scapulars, wing-coverts and tertials blackish with prominent buff lateral notches, overall darker and showing more contrast than adult.

First non-breeding/first-breeding/adult Probably does not breed in second calendar year. Adult has head pattern as juvenile but crown-stripe more prominent; upperparts are slightly paler and less contrasting than juvenile, broadly but diffusely fringed and notched. There is little or no replacement of non-breeding upperpart feathers prior to breeding, so these feathers are worn. Breeding adult has all-dark bill. First non-breeding is difficult to distinguish from adult, as wear of retained juvenile wing-coverts reduces size of pale spots and fading reduces contrast. Some first-breeding birds commence primary moult as early as January, others not until May; since adults do not moult primaries until on the non-breeding grounds from about September, those in primary moult prior to September are immatures.

Call A fluty, rapidly repeated *pu*, typically given about seven times.

Status, habitat and distribution Fairly common, breeding in Arctic tundra near tree-line: *islandicus* in Iceland and Scotland, nominate *phaeopus* from Scandinavia to western Siberia, with both these races non-breeding coastally from Portugal and Spain to west and south Africa, Middle East to western India and Sri Lanka (*islandica* probably to west of range, *phaeopus* to the east); vagrant to eastern North America and Caribbean. Race *variegatus* breeds in eastern Siberia, non-breeding in eastern India to Taiwan, Indonesia, Philippines, Australia and New Zealand; vagrant to west coast North America. Race *hudsonicus* breeds west Hudson Bay, presumed non-breeding from North Carolina south to southern South America; vagrants to Europe lacking white above in flight are presumably *hudsonicus*. Race *rufiventris* breeds Alaska and Yukon, presumably non-breeding coastally from California south to southern South America. Race *alboaxillaris* breeds in the steppes of western Kazakhstan, non-breeding in eastern Africa and adjacent islands of Indian Ocean south to Mozambique.

Racial variation The six races can be separated into four groups that can be recognised in the field, particularly when in flight: 'Western Whimbrel' (the '*phaeopus* group', races *phaeopus* and *islandicus*), 'Steppe Whimbrel' (*alboaxillaris*), 'Eastern Whimbrel' (*variegatus*), and 'Hudsonian Whimbrel' (the '*hudsonicus* group', *hudsonicus* and *rufiventris*). The *hudsonicus* group are darker above and buffish beneath, with a more contrasting head-pattern and a slightly longer bill than *phaeopus*. The two races in each of the *phaeopus* and *hudsonicus* groups can only be separated on measurements, *phaeopus* and *hudsonicus* each being the smaller race in their group. In flight the *phaeopus* group and *alboaxillaris* have white rump with V up back, the *hudsonicus* group has back and rump coloured as wings and tail, while *variegatus* is intermediate; below, underwing-coverts and axillaries are whitish in the *phaeopus* group, largely white in *alboaxillaris*, strongly barred brown and buff in *hudsonicus*, and greyish in *variegatus*.

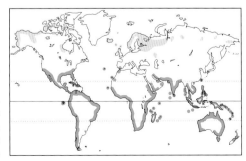

Similar species Bristle-thighed Curlew (108) is most similar to juvenile *hudsonicus*, lacking streaking and barring beneath (as do a few Whimbrels), has rump and tail-coverts bright buff, and tail with more widely spaced dark-brown and cinnamon bars; underwing has cinnamon base colour as has *hudsonicus*, though *variegatus* has greyish underwing. Greater primary underwing-coverts of Bristle-thighed Curlew are spotted or notched (heavily barred in *hudsonicus*), and this species has a diagnostic 'wolf-whistle' call. Slender-billed Curlew (109) is similar in size to Whimbrel (race *phaeopus* is most likely in Slender-billed's range), but is generally paler, has more slender, tapered bill, lacks Whimbrel's 'stripy' head and usually has flanks spotted, not diffusely streaked and barred; in flight Slender-billed shows considerable contrast between near-black outer wing and much paler inner wing, and has all-white underwing-coverts. Compared to both Eurasian (110) and Long-billed Curlews (112), Whimbrel is smaller, has shorter bill more abruptly decurved towards tip, a striped head and more strongly barred underwing.

References Barter (2002), Bosanquet (2000), Ebels (2002), Engelmoer & Roselaar (1998), Gerasimov & Gerasimov (2002), Hobson *et al.* (1984), Morozov (1999, 2000), Skeel & Mallory (1996), Thorup (2006), Zöckler (1998).

▲ **107a. Juvenile, *phaeopus* group**. Canary Islands, mid September. Note buff notches on entire upperparts and wing-coverts. RJC.

▼ **107b. Juvenile, *hudsonicus* group**. Florida, late September. Except when in flight very similar in all respects to 107a. RJC.

◄ **107c. First non-breeding, *hudsonicus* group**. California, early November. Has replaced upperparts and body with adult-type feathers, the neck and upper breast being more coarsely streaked and the flanks more barred than juvenile. The retained juvenile coverts and tertials are faded, and the buff notches towards the feather tips are largely worn away. RJC.

▲ **107d. Adult, *hudsonicus* group**. California, mid November. In fresh plumage; the adult has less contrasting, more diffuse feather patterns than the juvenile. RJC.

◄ **107e. Adult, *variegatus* group**. South Korea, late April. Adults of *phaeopus* and *variegatus* groups have less contrasting head patterns than *hudsonicus*. Note the all-black bill acquired during the breeding season. RJC.

▲ **107f. Adult, *phaeopus* group**. Iceland, late June. Whimbrels have a very restricted pre-breeding moult, and are quite worn at this stage. RJC.

▲ **107g. Adult, *phaeopus* group, in flight**. The Gambia, early January. Barred tail and rump, and a clear white V up the back. RJC.

▲ **107h. Adult, *phaeopus* group, raising wings**. Wales, mid August. Underwing is white, with narrow dark bars. RJC.

◄ **107i. Adult,** *variegatus* **group, in flight**. South Korea, early May. Some barring on back, and heavy dark grey barring on white underwing. RJC.

▲ **107j. Adult,** *hudsonicus* **group, in flight**. Florida, mid June. No white on rump or back; well south of the breeding grounds in June and in moult, so probably in second calendar year. RJC.

◄ **107k. Adult,** *hudsonicus* **group**. Florida, late August. Strong brown and buff underwing barring; in heavy wing moult. RJC.

108. BRISTLE-THIGHED CURLEW
Numenius tahitiensis

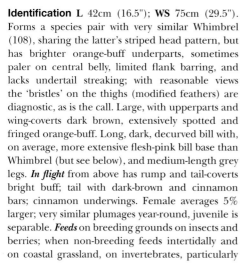

A scarce Alaskan species with bristle-like thigh feathers and an extraordinary migration strategy. Vulnerable.

Identification L 42cm (16.5"); **WS** 75cm (29.5"). Forms a species pair with very similar Whimbrel (108), sharing the latter's striped head pattern, but has brighter orange-buff underparts, sometimes paler on central belly, limited flank barring, and lacks undertail streaking; with reasonable views the 'bristles' on the thighs (modified feathers) are diagnostic, as is the call. Large, with upperparts and wing-coverts dark brown, extensively spotted and fringed orange-buff. Long, dark, decurved bill with, on average, more extensive flesh-pink bill base than Whimbrel (but see below), and medium-length grey legs. *In flight* from above has rump and tail-coverts bright buff; tail with dark-brown and cinnamon bars; cinnamon underwings. Female averages 5% larger; very similar plumages year-round, juvenile is separable. *Feeds* on breeding grounds on insects and berries; when non-breeding feeds intertidally and on coastal grassland, on invertebrates, particularly crabs: these are dismembered in the manner of Whimbrels and other curlews (p. 28) or (unlike in Whimbrel) may be beaten on hard surface to smash the carapace before eating the flesh.

Juvenile Compared to adult quite bright and contrasting, with large, clear, bright buff notches on scapulars and tertials. Migrates south to non-breeding area in juvenile plumage.

First non-breeding Very similar to adult but retains juvenile coverts.

First/second-br/second n-br/adult n-br Very similar to adult breeding; probably do not breed until their fourth calendar year, and until then remain in non-breeding areas.

Adult Has incomplete upperpart moult January–April, and gains more heavily streaked underparts. Described in Identification but the bright, orange-buff upperpart fringes characteristic of spring

▲ **108a. Sub-adult.** French Polynesia, late September. Bristle-thighed Curlews do not breed until their fourth calendar year. Apart from one or two dark breeding-type upperpart feathers, this bird is rather faded, suggesting that it is in its second or third calendar year. Pete Morris.

become much duller later in the year, though some birds may be dull in spring. When breeding, bill all dark or with restricted pink at base of lower mandible, males generally having less (duller) pink than females. Some adults become flightless when moulting on the non-breeding grounds.

Calls 'Wolf-whistle' call, *whee-whe-oo* or *chiu-eet*, the latter recalling Grey Plover, is diagnostic.

Status, habitat and distribution Scarce, breeding in Arctic tundra with dwarf shrubs in western Alaska, staging on southbound migration in the Yukon Delta and, following a non-stop flight of at least 4,000km, non-breeding in a variety of open habitats in the central Pacific, from the Hawaiian islands west to Guam, south to Tonga and the Pitcairn Islands. Vagrant to Chukotka, Pribilofs and Aleutians, New Guinea, Japan and the North American west coast from British Columbia south to California; also Equador.

Racial variation None.

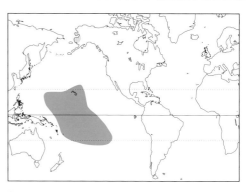

Similar species Whimbrel (107), which see. In North America, comparison when breeding is with race *hudsonicus*, when non-breeding with race *variegatus*.

References Gill *et al.* (1991), Konyukhov & McCafferty (1993), Marks *et al.* (2002), Mlodinow *et al.* (1999), Pyle (1999).

▼ **108b. Adult breeding**. Alaska, June. Note species-diagnostic 'bristles' between legs. When breeding, acquires a variable number of dark upperpart feathers, and the throat and breast become more streaked; the bill is all-dark. René Pop.

▲ **108c. Adult breeding**. French Polynesia, late September. Again shows the diagnostic 'bristles'. Similar to 108b, but more worn and faded, and bill is pale at the base. Pete Morris

▲ **108d. Adult in flight.** French Polynesia, mid September. Note the diagnostic buff rump and narrowly barred tail. Pete Morris.

▲ **108e. Bristle-thighed Curlew in flight**. French Polynesia, mid September. Underwing has cinnamon ground-colour, and underparts are pale cinnamon. Pete Morris.

109. SLENDER-BILLED CURLEW
Numenius tenuirostris

A small, pale, and extremely rare curlew with
a slender decurved bill. Critically Endangered.

Identification L 39cm (15"); **WS** 79cm (31"). Large,
with long, dark, slender decurved bill, tapered
over distal third (and typically shorter than that
of Whimbrel); medium-length dark grey legs, with
tibia relatively shorter than Eurasian Curlew. Some
have dark, heavily streaked rusty crown (occasionally with an indistinct pale crown-stripe) and pale
streaked supercilium; others more 'plain faced'.
Upperparts have dark, elongate, irregular centres;
extensive streaks and spots on buff neck and breast,
and prominent spots on lower breast and flanks
(sometimes difficult to see). Underparts otherwise
white (not buff as Eurasian Curlew), particularly
on flanks (but juvenile is streaked, not spotted). *In
flight* rump white, white V up back, near-black outer
primaries and primary coverts contrast with paler
inner wing; prominent white shaft on outer primary
(perhaps outer two); tail white with incomplete dark
bars; entirely white underwing-coverts; toes just
extend beyond tail. Female marginally larger, with
longer bill; plumage similar year-round, juvenile
separable. Small webs between outer toes, slight
between inners. *Feeds* on invertebrates by picking
and probing.

Juvenile Streaking on head, mantle, scapulars,
neck and breast finer than in adult; upperpart and
wing-covert fringes uniformly buff; tertials and some
scapulars with buff notches; flanks streaked, not
spotted.

First non-breeding Retains some juvenile scapulars
and tertials, and most wing-coverts. Body moult
results in the acquisition of some adult-type flank
spotting.

First-breeding/adult non-breeding Probably does
not begin to breed until at least third calendar year,
so slowly acquires adult non-breeding during second
calendar year, gradually replacing the body and
flight feathers (the latter from April). Wing-coverts
become particularly worn and faded, contrasting
with the replaced upperparts. Adult non-breeding
plumage as adult breeding, but all feathers are
fresh.

Adult breeding Acquires many new body feathers
but perhaps not many upperpart ones, with some
contrast between worn wing-coverts and replaced
upperparts. Upperpart feathers have irregular,
broad, dark, elongate, central streak and wide

bright buff fringes (new feathers) or off-white
fringes (faded, retained from non-breeding); dark
spots on lower breast, broader ('heart-shaped')
on flanks. Bill base probably becomes largely or
completely dark.

Calls *Cour-ee*, briefer and higher-pitched than
the similar call of Eurasian Curlew, and a rapidly
repeated *bi-bi-bi*.

Status, habitat and distribution Extremely rare,
and possibly extinct (no definitive records since the
late 1990s). Breeding grounds remain unknown,
thought to be south-west Siberia; non-breeding
in Morocco, also Iran and Iraq. Vagrant in Japan,
Canada, Yemen, Eritrea and western Europe.

Racial variation No races recognised.

Similar species Little Curlew (106) and Whimbrel
(107), which see. Eurasian Curlew (110) is larger,
has a proportionally longer and different bill shape,
adult has heavily marked flanks (streaked with chevrons) on a buff ground, not discreet round spots
on a white ground. Also, Eurasian Curlew in flight
shows less contrast between dark primaries and
inner wing than Slender-billed. Western race *arquata*
of Eurasian Curlew generally has barred (not white)
underwing-coverts; juveniles of both species have
streaked flanks and are best separated on size,
though Eurasian Curlew may have a confusingly
short bill.

References Beardslee & Mitchell (1965), Boere
& Yurlov (1998), Bojko & Nowak (1996), Cleeves
(2002), Corso (1996), de Smet (1997), Gretton
(1991), Gretton *et al.* (2002), Heard (1998a, 1998b),
Porter (2004), Serra *et al.* (1995), Steele & Vangeluwe
(2002), van den Berg (1988).

▲ **109a. Presumed adult (left) with Eurasian Curlew (right)**. Yemen, January. Note the size difference with the Eurasian Curlew, which is of the long-billed race *orientalis*. The Slender-billed Curlew has a more prominent supercilium, a largely (or completely) dark bill, darker grey legs and a more upright stance. R. F. Porter.

▲ **109b. Adult non-breeding Slender-billed Curlew**. Morocco, January. The 'slender', all-dark bill, with relatively limited curvature is well shown. The face of the bird is stained with mud. Arnoud B. van den Berg.

▲ **109c. Adult**. Morocco, early February. 'Slender' all-dark bill with relatively limited curvature; heavily spotted underparts are characteristic of adults, particularly when breeding. Chris Gommersall.

▲ **109d. Presumed adult in flight**. Yemen, January. Flight pattern similar to Eurasian Curlew, but there is considerable contrast between near-black primaries and the inner-wing, and the tail has less barring so appears near-white. Same bird as 109a. R. F. Porter.

▲ **109e. Adult Slender-billed Curlew, flight**. Morocco, January. The underwing-coverts are entirely white, similar to many Eurasian Curlews of race *orientalis*. Those of nominate Eurasian Curlew *arquata* are usually barred, but variably so. Arnoud B. van den Berg

110. EURASIAN CURLEW
Numenius arquata

The common large curlew of Europe and Asia.

Identification L 55cm (21.5"); **WS** 92cm (36"). Forms a species pair with Far Eastern Curlew (111). Very large, long-necked, with a long (male) or very long (female) strongly decurved, dark-brown bill, with pinkish base to lower mandible; medium-length grey legs. *In flight* from above has blackish primary-coverts and outer primaries, white rump and V up back, tail barred brown. Underwing-coverts and (particularly) axillaries barred brown and white, but occasionally these feathers are largely white; toes usually reach just beyond tail. Female slightly larger but with a considerably longer bill; plumages vary somewhat seasonally and with age. Small web between outer toes, slight web between inner toes. *Feeds* both in dampish grassland and in estuarine environments, probing deeply; sometimes wades.

Juvenile Head finely streaked dark brown; mantle and scapulars dark brown, broadly and neatly fringed buff, rear scapulars with prominent buff notches; tertials blackish to dull brown with large rounded (but variable) buff notches. Wing-coverts rather uniform brown, fringed pale-buff; larger inner coverts similar to tertials. Neck, breast and flanks washed warm buff, neatly streaked dark brown; remainder of underparts white. Bill often shorter than adult.

First non-breeding Acquires mantle and scapulars as adult non-breeding but retains a variable number of juvenile scapulars, wing-coverts and, particularly, tertials, which fade and wear with age, appearing browner and less contrasting than in juvenile. Underparts as adults.

First-breeding First-breeding gradually acquires adult-type tertials and a variable number (sometimes none) of adult breeding-type upperpart feathers.

Second non-breeding/adult non-breeding Head greyish, finely streaked dark grey-brown; mantle and scapulars have blackish-brown centres and the scapulars fine lateral bars, with greyish-brown fringes. Tertials greyish-brown, dark brown along shafts, with dark brown oblique bars. Wing-coverts dull brown with blackish bars and extensive pale notches and fringes. Neck, breast and flanks have buffish wash with bold, dark-brown streaks and spots; remainder of underparts white.

Adult breeding Probably breeds in third calendar year. Head, neck and breast warmer buff than adult non-breeding. A variable number (a few to nearly all) of mantle feathers, scapulars and (less frequently) wing-coverts are replaced, the new feathers having irregular, elongate, blackish centres and pale rufous fringes, which fade with age; when fresh, rufous upperparts contrast with pale buff wing-coverts.

Call Fluty *cur-lee* or *cour-loo*, with rising inflection.

Status, habitat and distribution Common, breeding in temperate to subarctic dampish open areas, often moorland, from Britain and Ireland, France, Scandinavia east to Urals (race *arquata*), through Asia to about 125°E (*orientalis*), and in the southern Urals and Kazakhstan (*suschkini*). Non-breeding birds occur both coastally and inland: *arquata* from Iceland (few), western Europe (where some are sedentary) south to West Africa; *orientalis* in Mediterranean (where overlaps with *arquata*), Africa (not north-west), Middle East, south and south-east Asia, Japan, south-east China south to Philippines; *suschkini* reportedly in West Africa, but more data needed. Vagrant to eastern North America (probably *arquata*) and Australia (probably *orientalis*).

Racial variation Compared with *arquata* (described in Identification), *orientalis* is larger with a longer bill, some brown barring on rump, streaked rather than heavily spotted, underparts, and unbarred white underwing and axillaries, but differences are clinal and not all *arquata* have barred axillaries. Race *suschkini* is similar in size and bill length to *arquata*, but shares most plumage features with *orientalis*, from which it probably cannot be separated in the field, though tail apparently has contrasting black-and-white (not brown-and-white) barring.

Similar species Whimbrel (107) and Slender-billed Curlew (109), which see. Long-billed Curlew (112) is typically longer-billed, has overall cinnamon

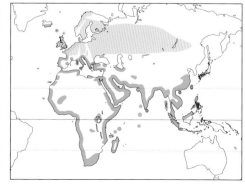

coloration and in flight shows no white above and has cinnamon underwings. Far Eastern Curlew (111) appears similar in size to eastern race *orientalis* Eurasian Curlew (with which it occurs in far east), but has a deeper bill-base, marginally steeper forehead, buff, not white, underparts (particularly undertail-coverts), and in flight shows no white on back or rump; also has somewhat different call.

References Barter (2002), Engelmoer & Roselaar (1998), Odin (1985), Thorup (2006).

◀ **110a. Juvenile *arquarta*.** Wales, late August. Aged by short bill (which will grow considerably), dark-centred, pale-fringed upperparts and wing-coverts, and boldly pale-notched rear scapulars and tertials (which are not always as contrasting as this). The neatly streaked breast and flanks lack the broad spots and chevrons shown by this race in all other plumages. RJC.

◀ **110b. First non-breeding *arquarta*.** Eastern England, late January. Has non-breeding dark-centred upperparts (larger feathers with pale lateral bars), and strongly marked flanks, but has worn, pale-notched juvenile tertials. Bill is almost certainly fully grown at this age, its length suggesting a female. RJC.

◀ **110c. Adult non-breeding *arquarta*.** Eastern England, mid December. Much as 110b, but tertials are more diffusely patterned with narrow dark bars. Shortish bill suggests a male. RJC.

◀ **110d. Adult breeding *arquarta*.**
Eastern England, mid April. Newly
acquired upperparts (with bold
black centres and deep, pale-buff,
notches) contrast with (mainly)
worn and faded wing-coverts;
bill and face are covered with
mud. RJC.

◀ **110e. Adult breeding/non-
breeding *arquarta*.** Eastern
England, early August. Moulting
some upperparts and tertials. This
bird has chequered, breeding-type
wing-coverts; usually, breeding
wing-coverts are as non-breeding,
with only a few breeding-type
feathers, as seen on 110d. RJC.

▼ **110f. Adult breeding
orientalis.** South Korea, late
April. Longer-billed than *arquata*
and has streaked rather than
strongly barred or spotted flanks.
Race *orientalis* seems to lack
the transverse barring on the
wing-coverts shown by breeding
arquarta.The particularly long bill of
this bird suggests a female; this is
another with mud on its face! RJC.

◄ **110g. Adult *arquarta* in flight**. Eastern England, early February. Race *orientalis* has a similar flight pattern above though typically more barred at rump. RJC.

▼ **110h. Adult non-breeding *arquata* in flight**. South-west England, mid September. The top left bird is a male just completing primary moult; the barring on the underwing can be quite variable in this race. RJC.

◄ **110i. Adult in flight (race uncertain)**. The Gambia, early January. This individual has largely white underwings as in *orientalis*, but the extensive flank markings suggest this may be an intergrade between *arquarta* and *orientalis*, or even an example of *suschkini*. RJC.

111. FAR EASTERN CURLEW
Numenius madagascariensis

The largest and longest-billed shorebird.
Alternative name: Eastern Curlew.

Identification L 64cm (25"); **WS** 97cm (38"). A very large curlew, though in the field usually appears similar in size to eastern race of Eurasian Curlew (110), with which it forms a species pair. Long-necked, with plain, uniformly dark-streaked head and neck, obscure pale supercilium; very long, deep-based, decurved dark bill with pink base to lower mandible; medium-length grey legs. *In flight* from above uniformly dark, lacking pale rump and white back of Eurasian Curlew, with heavily barred under-wing, and buff to cinnamon underbody; feet extend beyond tail. Female slightly larger and longer-billed; plumage largely similar year-round; juvenile is separable. Small webs between all three toes. *Feeds* in typical curlew manner in muddy coastal and estuarine areas, by picking and deep probing for invertebrate prey, particularly crabs.

Juvenile Best aged by smaller, neater upper-part feathers with buff fringes and broad notches, particularly on the tertials, though some birds only show notching near the feather tips. Underparts buff, with less and finer streaking than adults. Typically has much shorter bill than adult, which is probably not fully grown until end of first calendar year.

First and second non-breeding/breeding Some may retain juvenile coverts and/or tertials until the moult to second non-breeding. Otherwise as adult non-breeding. Second non-breeding only recognisable for as long as the very worn juvenile primaries are retained.

Adult non-breeding Acquired in third calendar year. Mantle feathers, scapulars and coverts have dark-brown centres and pale grey-brown fringes, larger scapulars and tertials diffusely barred brown; heavily streaked neck, breast and upper belly; barred on flanks, underparts otherwise off-white to buff.

Adult breeding Most individuals probably do not breed until fourth calendar year. Some uncertainty about breeding plumage, perhaps only shown by males. Head, neck and breast with heavy dark streaks on a bright buff wash; gains some breeding-type upperpart feathers, tertials and wing-coverts which are dark-centred with bars and bright buff fringes; remainder of underparts bright buff with little streaking. Pink at bill-base duller, more restricted than non-breeding.

Call A Eurasian Curlew-like *quee-quee* or *coor-ee*, but flatter, lacking much of Eurasian's rising inflection.

Status, habitat and distribution Uncommon, breeding in marshes and wet meadows in eastern Siberia, non-breeding on coastal and estuarine mud-flats and salt-marshes from South Korea south to Australia (majority of population) and New Zealand; non-breeding immatures in Australia tend to move northward during the boreal summer but not known if they leave Australia. Vagrant to north-west North America, Aleutians and Pribilofs.

Racial variation None.

Similar species Eurasian Curlew (110), which see.

References Barter (2002), Higgins & Davies (1996), Wilson (2000).

▼ **111a. Juvenile**. Japan, mid September. As in Eurasian Curlew, juvenile Far Eastern Curlews take some time to grow their bills to full length. The tertial pattern recalls that of juvenile Whimbrel and juvenile Eurasian Curlew. The rich buff underparts will fade somewhat, and are not as bright as this in any other plumage. Nobuhiro Hashimoto.

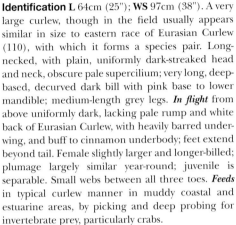

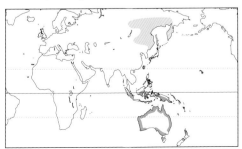

◀ **111b. Adult**. Western Australia, early April. Once adult, the plumage appears largely similar year-round. RJC.

◀ **111c. Adult breeding**. Kamchatka, Russia, June. This bird has acquired a few bright buff breeding feathers on the upperparts and flanks. Yuri Artukhin.

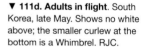

▼ **111d. Adults in flight**. South Korea, late May. Shows no white above; the smaller curlew at the bottom is a Whimbrel. RJC.

112. LONG-BILLED CURLEW
Numenius americanus

The only large North American curlew.
Near Threatened.

Identification L 57cm (22.5"); **WS** 87cm (34"). Very large, long-necked, with very long, strongly decurved bill with pink base to lower mandible, medium-length grey legs. Cinnamon body with darker upperparts. *In flight* from above fairly uniform, though with darker forewing and blackish primary-coverts and outer primaries; below, uniformly cinnamon, feet extend to just beyond tail. Sexes are of similar size, but female typically has a considerably longer bill. Plumages are generally similar year-round; juvenile is separable. Has partial webbing between all three toes. *Feeds* both intertidally and on coastal grassland, both by picking and probing deeply; also wades.

Juvenile Mantle and scapulars blackish-brown, extensively notched buffish-cinnamon; neck, upper breast and flanks lightly streaked. Newly fledged birds have relatively short bills. Very similar to adult; best distinguished by neat, unworn plumage.

First non-breeding/adult Females breed in third or fourth calendar year, males not until fourth or fifth. Mantle and scapulars black, extensively notched cinnamon; neck, breast and flanks have dark brown streaks. Has pre-breeding moult, but breeding plumage very similar to non-breeding. Some non-breeders remain in non-breeding areas; others may return to breeding area.

Call *Cur-lee* with upward inflection.

Status, habitat and distribution Breeds in grassland from British Columbia and Saskatchewan south to Texas. Race *parvus* from Oregon to South Dakota northwards, nominate *americana* to the south; non-breeding birds occur from California east to Texas, Louisiana, and south to Mexico and Costa Rica, and South Carolina to Florida, where uncommon. Vagrant north to Alaska in west and James Bay in east, south to Panama, Colombia and Venezuela.

Racial variation Mainly clinal, *parvus* being the smaller.

Similar species Marbled Godwit (105), and Eurasian Curlew (110), both of which see.

References de Bruin (2006), Dugger & Dugger (2002).

▼ **112a. Juvenile**. Texas, mid August. Aged by neat, bright, uniform plumage, and by narrow black triangles (not barred as adult) on wing-coverts. Rather early for the bill to be well-grown, but Long-billed Curlews fledge from mid June. RJC.

▲ **112b. Adult male**. California, early November. Still moulting to non-breeding; most of the plumage is rather bleached. Bill length suggests a male. RJC.

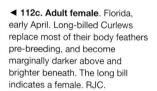

◄ **112c. Adult female**. Florida, early April. Long-billed Curlews replace most of their body feathers pre-breeding, and become marginally darker above and brighter beneath. The long bill indicates a female. RJC.

◄ **112d. Adult in flight**. Florida, mid September. Very similar to Far Eastern Curlew, but darker and more rufous; this bird is in heavy wing-moult. RJC.

113. UPLAND SANDPIPER
Bartramia longicauda

A small, very short-billed, atypical
North American curlew.

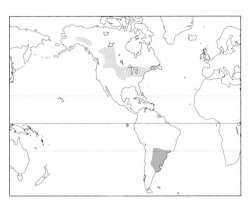

Identification L 30cm (12"); **WS** 53cm (21").
Medium-sized, with small head, long neck, short
straight bill with yellow base to lower mandible,
medium-length yellow legs. Tail extends well beyond
folded wings. Dark crown contrasts with plain face
and prominent dark eye; brownish above, whitish
underparts. *In flight* from above dark, with blackish
primary coverts, primaries and secondaries, rump
and tail dark centrally with narrow white sides; feet
do not reach the tip of the long tail. Sexes are of
similar size; adults have similar plumage year-round,
juvenile is separable. Small web between outer and
middle toes, slight web between inner toes. Feeds in
grassland by picking, in a plover-like manner, with
alternating short runs and sudden stops.

Juvenile Mantle and scapulars dull brown, with
darker markings and narrow buff fringes; tertials
dull brown, notched pale buff with dark brown spots
between notches. Wing-coverts fringed buff, with
all-dark centres or U-shaped dark markings with buff
centres. Throat and upper breast streaked brown,
flanks with brown chevrons and bars; remainder of
underparts pale buff.

First non-breeding Acquires adult-type mantle and
scapulars but retains most of its juvenile wing-coverts
and tertials.

Adult Age at first breeding not known. Mantle and
scapulars blackish with buffish fringes, the longer
scapulars and tertials having evenly spaced black
chevron-bars across the feathers; median coverts
have black subterminal anchor-marks. Underparts
are as juvenile. Has a very restricted pre-breeding
moult and consequently retains essentially the same
plumage year-round.

▼ **113a. Juvenile**. Southwest England, mid October. Neat, scalloped, dark-centred pale-fringed plumage. Gordon
Langsbury.

Call A mellow, liquid *ch-wut*, and a whistled *quip-ip-ip-ip*.

Status, habitat and distribution Breeds in inland North America in temperate and subarctic open grassland, in Alaska and Yukon, and from British Columbia south to Oregon in the west, eastwards through the Plains and Great Lakes to western Virginia and Maryland. Uncommon in eastern North America as most migrate overland through the interior. Non-breeding birds occur in grasslands in Surinam, Paraguay, Uruguay, Brazil and Argentina. Vagrant to Greenland, Iceland, the Azores, western Europe, most usually between September and October; also West Africa and Australasia (December–February).

Racial variation No races recognised.

Similar species Little Curlew (106), which see.

References Alexander-Marrack (1992), Houston & Bowen (2001).

▲ **113b Adult**. Kansas, late April. Once adult plumage is acquired retains similar plumage year-round. RJC.

▲ **113c. Adult in flight**. Alberta, Canada, late June. Lacks any particular flight pattern on upperwing, though primaries are notably dark and the tail is long. Gerald Romanchuk.

TRINGA AND RELATED SANDPIPERS

Three genera containing sixteen species, all of which breed in the northern hemisphere: Spotted Redshank *Tringa erythropus*, Common Redshank *T. totanus*, Marsh Sandpiper *T. stagnatilis*, Common Greenshank *T. nebularia*, Nordmann's Greenshank *T. guttifer*, Greater Yellowlegs *T. melanoleuca*, Lesser Yellowlegs *T. flavipes*, Solitary Sandpiper *T. solitaria*, Green Sandpiper *T. ochropus*, Wood Sandpiper *T. glareola*, Grey-tailed Tattler *T. brevipes*, Wandering Tattler *T. incana*, Willet *T. semipalmata*, Terek Sandpiper *Xenus cinereus*, Common Sandpiper *Actitis hypoleucos* and Spotted Sandpiper *A. macularius*. A recent review of the *Tringa* group of species (Pereira & Baker 2005) has resulted in the conclusion that the Willet (formerly *Catoptrophorus semipalmatus*) and the two tattler species (formerly *Heteroscelus brevipes* and *H. incanus*) should be placed in *Tringa*, and these revisions are followed here.

The *Tringa* are elegant, delicately built, small or medium-sized shorebirds with medium to long generally straight bills, long necks, and medium-length to long legs. All have hind toes and a limited amount of webbing between the toes, features shared by Terek Sandpiper and the two small *Actitis* sandpipers, Common and Spotted, of Eurasia and America respectively.

All members of the *Tringa* group are migratory, many strongly so, breeding in the more northerly temperate regions or the low Arctic. The more southerly breeding populations of Common Redshank are sedentary, or are only short-distance migrants. All species have distinct juvenile, non-breeding and breeding plumages. Some species, notably Common Redshank, Common Greenshank and the western race of Willet *inornata*, acquire a mix of non-breeding and breeding-type upperparts in breeding plumage.

The *Tringa* group are both freshwater and coastal species (on coasts particularly outside the breeding season); the larger species may wade quite deeply, and all can occasionally be seen swimming. A characteristic feature of many *Tringa*, but especially the smaller species (and also the *Actitis* sandpipers), is their habit of bobbing, or 'teetering', wagging their bodies with an up-and-down motion.

The two species of *Actitis* sandpipers generally have a complete post-juvenile moult, in both cases in the non-breeding areas, though the primary moult may be incomplete or delayed well into the second calendar year.

114. SPOTTED REDSHANK
Tringa erythropus

A particularly elegant *Tringa* with a unique black breeding plumage.

Identification L 30cm (12"); **WS** 52cm (20.5"). Medium-sized with a long neck; long, straight, fine bill with red base to lower mandible in all seasons, and long red legs (but see Adult Breeding); narrow white eye-ring. *In flight* from above uniformly darkish grey, white V up back and barred (appearing greyish) tail; feet extend beyond tail. Sexes are of similar size; plumage varies seasonally and with age. Small web between outer toes, slight between inner toes. *Feeds* usually in fresh water, wading fairly deeply (sometimes swimming), probing and scything; often in small flocks which sometimes chase prey cooperatively.

Juvenile Greyish crown, prominent off-white supercilium in front of eye only; dark eye-stripe. Upperparts and wing-coverts greyish-brown, all strongly spotted white; underparts rather paler; throat, foreneck and upper breast are mottled grey-brown, remainder strongly barred grey-brown. Legs orange-red.

First non-breeding/adult non-breeding Crown brownish-grey; white supercilium in front of eye only; dark eye-stripe. Mantle, scapulars and wing-coverts grey, with narrow white fringes; tertials grey, edge-spotted dark and white. Underparts white with pale grey wash across upper breast. First non-breeding retains many worn juvenile wing-coverts. Legs red.

First-breeding Age when first breeds not known but first-breeding only partially attains adult breeding plumage, so perhaps in third calendar year.

Adult breeding Head, neck and entire underbody sooty-black; mantle, scapulars and most tertials, together with a variable number of wing-coverts, black with prominent white spots. Breast feathers, and underparts from flanks to undertail-coverts, have white fringes of variable extent, females showing more white than males. Legs darkish red, sometimes near-black.

Call A distinctive, clear *chu-wit* with rising inflection.

Status, habitat and distribution Fairly common, breeding in wooded to open Arctic tundra from northern Scandinavia through northern Russia to eastern Siberia. Non-breeding in western Europe in small numbers on fresh water or sheltered estuaries; many more occur in Africa just south of Sahara; also throughout southern and south-east Asia. Vagrant throughout North America and Caribbean.

Racial variation No races recognised.

Similar species From Common Redshank (115) by slightly larger size, longer legs, barred underparts of juvenile, unique black adult breeding plumage and, when non-breeding, by more elegant appearance together with prominent pale fore-supercilium and longer, finer bill. In flight lacks the white secondaries of Common Redshank.

References Anderson & Baldock (2001), Barter (2002), Ebels (2002), Mlodinow (1999), Taverner (1982), Thorup (2006).

▼ **114a. Juvenile.** Western Germany, late August. Juveniles are surprisingly dark, both above and below, and are much darker than juvenile Common Redshank. The long, fine bill with red base to lower mandible and prominent supercilium restricted to the front of eye are helpful for separation from Common Redshank. Mathias Schäf.

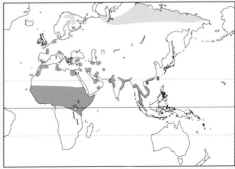

◄ **114b. First non-breeding**. India, late December. Has non-breeding upper- and underparts, but still retains a few white-notched dark juvenile wing-coverts; now much paler than juvenile. RJC.

◄ **114c. Adult non-breeding**. Eastern England, late February. Greyer, less brown, than Common Redshank when non-breeding. RJC.

▼ **114d. Adult non-breeding/ breeding**. Eastern England, early April. Commonly seen on migration moulting both in and out of the distinctive black breeding plumage; the first upperpart breeding feathers to be grown are not so dark as those acquired later. RJC.

◀ **114e. Adult breeding.** South Korea, early May. Striking and unmistakable in this plumage; females have more white beneath. The legs are near-black when breeding. RJC.

◀ **114f. Adult breeding/non-breeding.** Eastern England, late July. Moulting out of breeding plumage. RJC.

◀ **114g. Adult non-breeding/breeding in flight.** Eastern England, early April. In similar plumage to 114d. RJC.

115. COMMON REDSHANK
Tringa totanus

The most common larger *Tringa* throughout Europe and much of Asia.

Identification L 28cm (11"); **WS** 51cm (20"). Medium-sized with fairly long neck; medium-length, straight, horn-coloured red-based bill, and medium-length orange or red legs. Brown or greyish-brown above, with paler underparts; white eye-ring. *In flight* from above has a unique pattern of a broad, white trailing edge to the inner wing, formed by the white-tipped inner primaries and white secondaries (see Adult Breeding below); white uppertail-coverts, rump and V up back; feet project beyond tail. Sexes of similar size; plumages vary seasonally and with age. Small web between outer toes, slight between inner toes. *Feeds* on freshwater margins and coastal mudflats by picking and probing; often wades.

Juvenile Dark crown, sparsely streaked buff; indistinct buff supercilium; upperparts and wing-coverts dark greyish-brown, with feather edges extensively spotted buff; foreneck and upper breast white, with dark brown streaks. Colour of bill-base varies with increasing age from brownish to dull red; legs orange.

First non-breeding/adult non-breeding Crown brownish-grey; upperparts and wing-coverts uniformly brownish-grey, with relatively inconspicuous small dark spots at feather edges and, especially in fresh plumage, with narrow white fringes. Foreneck and upper breast washed grey with dark brown streaks. First non-breeding retains some worn juvenile wing-coverts and tertials. Base of bill and legs orange-red.

First-breeding/adult breeding Many breed in second calendar year. Acquires a variable proportion of plain, brownish, non-breeding-type upperparts and wing-coverts, mixed with strongly brown-barred buffish feathers. White underparts are heavily streaked brown, sometimes with brown barring on flanks. Legs and base of bill bright orange-red. First-breeding occasionally identifiable but only if some very worn juvenile wing-coverts are retained. During moult to adult non-breeding (July–September) the diagnostic white secondaries are often exposed in the folded wing.

Call Most frequent is a fluty *teu-tu-tu*, often given in flight.

Status, habitat and distribution Common; nests in temperate, open, wet grasslands in Iceland (race *robusta*); Britain and Ireland (where many are sedentary), Scandinavia and locally around the Mediterranean (nominate race *totanus*); and east through Russia, Asia to northern China (other races). When non-breeding generally coastal, occurring from Britain and Ireland (both *totanus* and *robusta*), continental Europe south to Africa and Arabia and throughout south and south-east Asia, rarely Australia. Vagrant to Newfoundland.

Racial variation Race *robusta* is slightly larger

◄ **115a. Juvenile.** Wales, early August. The buff fringes fade quite quickly. RJC.

(though with a shorter bill), in breeding plumage has smaller proportion of barred feathers on upperparts than *totanus* and is more heavily marked below. Other races have been proposed, mainly on the basis of measurements, but there appears to be insufficient consistent difference to allow separation in the field away from their breeding areas.

Similar species Spotted Redshank (114), which see.

References Barter (2002), Hale (1980), Mactavish (1996), Thorup (2006), Valle & Scarton (1996).

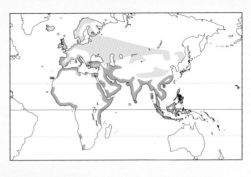

◀ **115b. First non-breeding**. Wales, mid October. Retains juvenile coverts and tertials, but buff fringes and spots have faded to white. RJC.

◀ **115c. Adult non-breeding**. Eastern England, early February. Upperparts, coverts and tertials have rather inconspicuous dark edge-spotting, and also have a narrow white fringe when fresh; underparts largely unmarked. RJC.

◀ **115d. Adult breeding *totanus***. Eastern England, early April. Usually acquire only a few dark upperpart breeding-type feathers, the remainder being either retained from non-breeding or newly grown non-breeding type feathers. The underparts are quite strongly marked with dark streaks, spots and chevrons. RJC.

◄ 115e. Adult breeding *totanus*. Eastern England, late July. Wing-stretching, showing extensive white on secondaries. This bird is in wing moult, and has lost a few inner primaries. RJC.

▼ 115f. Adult breeding *robusta*. Iceland, late June. Very similar to *totanus*. Generally has fewer upperpart breeding feathers, but can be more heavily marked below; *robusta* also has a shorter bill, on average. RJC.

◄ 115g. Adult in flight. Eastern England, mid June. The combination of broad white trailing edge to the wing and white V on back is unique. RJC.

116. MARSH SANDPIPER
Tringa stagnatilis

A small, delicate *Tringa* with a long, fine bill and very long greenish legs.

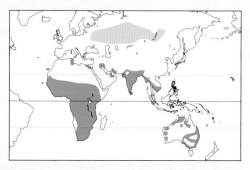

Identification L 23cm (9"); **WS** 42cm (16.5"). Medium-sized with a long neck; long, fine, straight bill, and very long greenish or yellowish legs. Generally grey above and white below; dark area near bend of folded wing. *In flight* from above wings uniformly sooty-grey, white V up back, white rump and tail, the latter lightly barred in the centre; legs and feet extend beyond tail. Sexes of similar size; plumages vary seasonally and with age. Web between outer toes, slight between inner toes. *Feeds* in freshwater areas by picking, often wading, occasionally running fast through water after prey.

Juvenile Crown and hindneck brown-grey with darker streaks, supercilium white, upperparts and wing-coverts greyish-brown with darker markings and narrow, white fringes; underparts white. Legs greenish.

First non-breeding/adult non-breeding Very similar to juvenile but upperparts and wing-coverts are a cleaner grey, with narrower dark subterminal lines and fringes. First non-breeding retains some juvenile wing-coverts and tertials. Legs greenish.

First-breeding Some breed in second calendar year and largely attain adult breeding plumage. Many replace outer primaries. Non-breeders that do not return to breeding area often retain much of their first non-breeding plumage. Legs greenish.

Adult breeding Crown and neck brownish, with dark streaks; whitish supercilium. A proportion of the upperparts, wing-coverts and tertials are brown with dark bars. Underparts white with dark spots on upper breast; flanks spotted and barred. Legs generally yellow.

Calls *Tew,* or *chip* in alarm.

Status, habitat and distribution Breeds in open wetlands in a band between about 48°N and 58°N,

◀ **116a. Juvenile**. Northeast China, early September. The rather brown juvenile upperparts with white fringes and the notched and spotted rear scapulars and tertials are very different from the grey non-breeding feathers that replace them. Alister Benn.

from Finland and the Baltic States (where scarce) eastwards across Russia to central Siberia. Non-breeding period spent in freshwater areas very locally in southwest Europe and more widely throughout Africa south of the Sahara, in India, south-east Asia and Australasia. On migration occurs Japan, Korea, China. Widespread but scarce on passage in western Europe, north to Sweden, mainly during May and July–September. Vagrant Alaska and Maldives.

Racial variation No races recognised.

Similar species Separated from Common Greenshank (117) by smaller size and more delicate proportions, proportionally longer legs, and significantly finer and straighter bill that is not slightly upturned.

References Anderson & Baldock (2001), Barter (2002), Pettet (1980), Thorup (2006).

▲ **116b. First non-breeding**. Taiwan, early September. Has adult-type upperparts, but retains juvenile coverts and tertials. Ming-Li Pan.

▲ **116c. Adult non-breeding**. Thailand, early December. RJC.

▲ **116d. Adult breeding**. Cyprus, mid April. Greyish when fresh, becoming browner later in breeding season. RJC.

▼ **116e. Adult breeding in flight**. Eastern China, mid May. Note browner tinge to upperparts than in 116d. Yuan Xiao.

117. COMMON GREENSHANK
Tringa nebularia

One of the larger *Tringa* species, with a relatively stout, slightly upturned bill.

Identification L 32cm (12.5"); **WS** 58cm (23"). Forms a species pair with Nordmann's Greenshank (118), though also very similar to Greater Yellowlegs. Medium-sized, elegant, with a fairly long neck; medium-length, slightly upturned bill (bill shorter than tarsus) with dark tip and grey base; long yellowish-green (rarely pale yellow) legs. Greyish above and white beneath; narrow white eye-ring. *In flight* from above wings uniformly grey, rump white with white V up back, and white tail with dark barring; feet extend beyond tail. Sexes of similar size; plumages vary seasonally and with age. Small web between outer toes. *Feeds* generally by wading, usually in freshwater habitats; periodically runs after prey, frequently changing direction.

Juvenile Crown dark grey, streaked white; indistinct white supercilium. Upperparts, wing-coverts and tertials are rather variable, some individuals being dark grey with white fringes and dark edge-spotting, while others are browner-grey with neat pale fringes, lacking spotting. Grey streaking on neck, remainder of underparts white.

First non-breeding/adult non-breeding Crown and hindneck streaked dark grey, otherwise head and neck white. Upperparts grey, with neat, narrow, white fringes and dark subterminal lines; some dark edge-spotting, especially on tertials and greater coverts. Underparts white. First non-breeding retains many worn juvenile wing-coverts and tertials.

First-breeding Breeds in third calendar year, though may return to breeding area in second calendar year. Extent of moult to first-breeding is highly variable; some retain worn first non-breeding plumage, others attain near-adult breeding-type plumage.

Adult breeding Whole of head and neck heavily streaked blackish. Mantle feathers black, with narrow pale-brown fringes; scapulars and tertials a mixture of pale-fringed black and uniform brownish-grey feathers, the latter sometimes having dark edge-spots. Upper breast and flanks heavily streaked blackish, remainder of underparts white.

Call A mellow *teu*, repeated two to four times, very similar to Greater Yellowlegs.

Status, habitat and distribution Fairly common; breeds in open moorland or freshwater marshes, from Scotland eastward throughout Scandinavia, eastern Europe to far-eastern Siberia, in a band between about 53°N and 70°N. Migrates south on a broad front; non-breeding occurs in small numbers in western Europe (usually in estuarine habitats) from Scotland to the Mediterranean, and more commonly at freshwater sites throughout Africa south of Sahara; also in south and south-east Asia to Australasia. Vagrant to Iceland, the Aleutians, eastern Canada and the Caribbean.

Racial variation No races recognised.

▼ **117a. Juvenile**. Scotland, late July. Juvenile Common Greenshanks have variable amounts of edge-spotting on their upperparts, coverts and tertials; this is a strongly spotted individual. RJC.

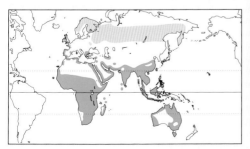

Similar species Marsh Sandpiper (116), which see. Greater Yellowlegs (119) is very similar to Common Greenshank but has bright yellow legs, a slightly less abruptly upturned bill, browner (less grey) general coloration, and folded wings extend beyond tail (just reach the tail-tip in Common Greenshank); also has white rump but lacks white V up back in flight. Nordmann's Greenshank (118) has a more compact build, a strongly two-toned, slightly upturned bill, and shorter, more yellow legs (particularly when breeding); Nordmann's bill is longer than the tarsus, it has a white underwing (grey in Common Greenshank), and also has a different call.

References Barter (2002), Chandler (1990b), Ebels (2002), Miskelly *et al.* (2001), Mlodinow (1999), Nethersole-Thompson & Nethersole-Thompson (1979), Ruttledge (1978), Thorup (2006).

◀ **117b. Juvenile.** Southwest England, early September. A much less strongly marked bird than 117a; some have no edge-spotting whatsoever. RJC.

◀ **117c. First breeding.** Eastern England, mid July. Has very worn wing-coverts, a scattering of breeding feathers, and has just started to acquire adult-type non-breeding upperparts; Common Greenshanks do not breed in their second calendar year. RJC.

◀ **117d. Non-breeding.** Spain, early September. An early date for an adult to be in non-breeding plumage, so probably a second calendar year bird (as 117c) that has moulted early. RJC.

▲ **117e. Adult breeding**. South Korea, late April. Adults acquire a variable proportion of breeding feathers on their upperparts, and have extensive streaking and spotting on the neck, upper breast and flanks. RJC.

▲ **117f. Adult breeding**. Spain, early September. A worn individual just commencing moult to non-breeding; this is typical timing for an adult, compared to which 117d has moulted very early to non-breeding plumage. RJC.

▼ **117g. Adult breeding Common Greenshanks in flight**. South Korea, late April. No wing-bar, a whitish tail and a white V on the back. RJC.

▼ **117h. Non-breeding in flight**. The Gambia, mid January. The grey barring on the underwing-coverts and underside of the primaries is quite dark, unlike Nordmann's (see 118d). RJC.

118. NORDMANN'S GREENSHANK
Tringa guttifer

A little-known migratory species that breeds only on Sakhalin and in adjacent eastern Siberia. Alternative name: Spotted Greenshank. Endangered.

Identification L 31cm (12"); **WS** 55cm (21.5"). Very similar to Common Greenshank (117), with which it forms a species pair, and with which it sometimes associates, but paler (except when breeding), more bulky, with heavier two-toned bill (but see below) and shorter yellow legs, particularly tibia (bill is longer than tarsus, *vice versa* in Common Greenshank); also has different call. Nordmann's is medium-sized, pale above and white below (except when breeding), with indistinct off-white supercilium that is more prominent in front of eye; medium-length, slightly upturned bill with outer half dark, inner-half grey; medium-length, pale yellow or greyish-flesh legs. Darker than Common Greenshank above, with heavy spotting on the breast and flanks (recalling Great Knot) when breeding. *In flight* from above wings uniformly grey, rump and V up back white; white tail with pale grey barring often appears completely white; below, underwing-coverts completely white (Common Greenshank has grey underwing); toes extend slightly beyond tail. Sexes similar; plumages vary seasonally and with age. Slight webbing between all three toes, more extensive between outer ones. *Feeds* on intertidal mudflats, much as Common Greenshank, picking, sometimes probing, often on small crabs; horizontal carriage recalls a leisurely Terek Sandpiper (127), though sometimes runs fast.

Juvenile Crown diffusely streaked brown. Otherwise head white, with fewer brown streaks, upperparts grey-brown with broad, scalloped pale fringes (unlike alternating dark and white edge-spotting of many – but not all – juvenile Common Greenshanks); coverts brown-grey with narrow pale fringes, underparts almost unmarked, with limited fine streaks on sides of breast and flanks, legs pale yellowish grey.

First non-breeding/adult non-breeding Head much less streaked than in juvenile, so head whiter; upperparts and coverts pale grey, with narrow, dark shaft-streaks and narrow white fringes when fresh (again lacking Common Greenshank's edge-spotting); underparts completely white. First non-breeding identifiable while juvenile brown-grey coverts and tertials are retained. Legs pale yellow.

First-breeding Age when first breeds not known but probably not until third calendar year; plumage probably similar to adult non-breeding but some (perhaps first-breeding) similar to adult breeding but with grey (not black) feather centres to upperparts.

Adult breeding Crown with heavy dark streaking; mantle, scapulars and tertials blackish with prominent white edge-spotting; throat and upper breast densely spotted black; flanks, lower breast and belly with heavy black spots. Bill often all-dark; legs yellow. This plumage is gained at migration staging areas, particularly the Yellow Sea.

Calls A nasal, slightly rasping *greeek* or *gwaak* and a high-pitched *keyew*; calls less frequently than Common Greenshank.

▼ **118a. First non-breeding**. Malaysia, November. The strongly bi-tonal yellowish-based bill is a feature of all non-breeding plumages, with the bill base becoming quite dark when breeding. The grey first non-breeding upperparts contrast with the worn, brownish juvenile wing-coverts. Ang Teck Hin.

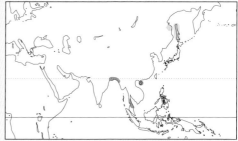

Status, habitat and distribution Very rare, breeding on coastal marshes with scattered larch woodland, on Sakhalin Island and adjacent mainland Siberia; on migration, coastal China and Korea, rare Japan, particularly in spring. Non-breeding occurs very locally on extensive mudflats from Bangladesh east to the Malay peninsula.

Racial variation None.

Similar species Common Greenshank (117), which see.

References Barter (2002), Bijlsma & de Roder (1986), Howes & Lambert (1987), Kennerley & Bakewell (1991).

▲ **118b. Adult non-breeding.** South Korea, early May. Compared to Common Greenshank is bulkier, larger headed, and has shorter, yellow legs; the tibia, particularly, are noticeably shorter. This bird is moulting to breeding plumage; the upperparts eventually become even blacker. RJC.

▲ **118c. Adult breeding.** South Korea, mid May. The breeding plumage is gained quite quickly at migration stop-over sites, as here on the Yellow Sea. The leg flags were put on near Shanghai, China, and show that this individual was in its fourth calendar year at least when photographed. The 'yellow' flag is stained, and was originally white. RJC.

▼ **118d. Adult breeding in flight.** South Korea, mid May. Nordmann's Greenshank has a very similar flight pattern above to Common Greenshank, but below has a white, not grey, underwing. The upper bird is a Great Knot. RJC.

119. GREATER YELLOWLEGS
Tringa melanoleuca

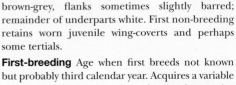

A large North American *Tringa* with a long, slightly upturned bill and long, bright yellow legs.

Identification L 31cm (12"); **WS** 60cm (23.5"). Medium-sized with a longish neck; long, slightly upturned bill with dark tip and grey, greenish or yellowish-base; long, bright yellow legs. Grey-brown above and white beneath; narrow white eye-ring. *In flight* from above uniformly darkish (but secondaries show marginal buff spotting), with white rump and barred whitish tail; feet extend well beyond tail. Sexes similar; plumage varies seasonally and with age. Small web between outer toes, slight between inner toes. *Feeds* most frequently while wading in shallow water, usually by picking, often chasing small fish.

Juvenile Streaked crown and hindneck. Upperparts, wing-coverts and tertials dark grey, with extensive and fairly large white marginal spots; lower neck and upper breast white, streaked dark brown; remainder of underparts white.

First non-breeding/adult non-breeding Head and upperparts much as juvenile but mantle feathers grey-brown, with fine, alternately dark and white edge-spots; wing-coverts and tertials dark grey, with white edge-spots. Foreneck streaked

▼ **119a. Juvenile.** Ontario, mid September. Upperpart feathers neat and small (though some may already have been replaced). Wing-coverts give impression of regular pattern of white spots across the wing, and tertials have relatively large white lateral notches. The tertials in particular are similar to those of breeding adult, so care is needed when ageing moulting adults at this time. Steve Pike.

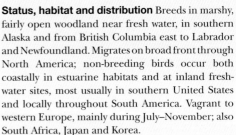

brown-grey, flanks sometimes slightly barred; remainder of underparts white. First non-breeding retains worn juvenile wing-coverts and perhaps some tertials.

First-breeding Age when first breeds not known but probably third calendar year. Acquires a variable proportion of adult breeding plumage but retains some juvenile feathers, most usually wing-coverts.

Adult breeding Similar to juvenile but mantle and scapulars black, fringed and edge-spotted white. Foreneck and upper breast heavily streaked black, flanks have black bars.

Call *Teu*, typically repeated two or three times, very similar to Common Greenshank.

Status, habitat and distribution Breeds in marshy, fairly open woodland near fresh water, in southern Alaska and from British Columbia east to Labrador and Newfoundland. Migrates on broad front through North America; non-breeding birds occur both coastally in estuarine habitats and at inland freshwater sites, most usually in southern United States and locally throughout South America. Vagrant to western Europe, mainly during July–November; also South Africa, Japan and Korea.

Racial variation No races recognised.

Similar species Common Greenshank (117), which see. Lesser Yellowlegs (120) is smaller, has a proportionally shorter, straighter, finer, virtually all-dark bill and proportionally longer legs; feathering at base of bill reaches nostril (separated from nostril in Greater). Lesser's bill is marginally longer than head, 1.5x longer in Greater. In flight, given good views, secondaries are plain (Greater Yellowlegs shows pale marginal spotting); adult Greater tends to show wing-moult before Lesser.

References Elphick & Tibbitts (1998), Park & Kim (1994).

◄ **119b. First non-breeding.**
Florida, late December. Still retains most of its juvenile wing-coverts and tertials, but has adult-type upperparts. An adult would have lost its similarly patterned breeding coverts and tertials by late December. Useful distinctions from Lesser Yellowlegs are that the bill is about 50% longer than the head, and the feathering on the lower mandible does not reach the nostril. Compared to Common Greenshank (which can occasionally have yellowish legs) the wing-tips typically extend beyond the tail; in Common Greenshank they usually just reach the tail-tip. RJC.

◄ **119c. Adult non-breeding.**
Florida, late December. Upperparts, wing-coverts and tertials have much smaller white edge-spots than juvenile. RJC.

◄ **119d. Adult non-breeding.**
Florida, late March. Starting to gain a few breeding coverts and tertials (dark, with contrasting large white lateral notches). RJC.

◀ **119e. First-breeding.** Texas, mid April. As in Common Greenshank, Greater Yellowlegs acquire only partial breeding plumage in their second calendar year, and probably do not breed; the worn wing-coverts suggest this is a second calendar year bird. RJC.

◀ **119f. Adult breeding.** California, early April. Contrasting black-and-white breeding plumage feathers; strong flank-barring is also characteristic of this plumage, a feature not seen in either Common Greenshank or Lesser Yellowlegs when breeding. RJC.

◀ **119g. Adult in flight.** Florida, late October. Lacks the white V on back of the otherwise rather similar Common Greenshank. This bird is still growing outer primaries; the edge-spotted secondaries differ from those of Lesser Yellowlegs, which are plain. RJC.

120. LESSER YELLOWLEGS
Tringa flavipes

A North American *Tringa* with yellow legs, smaller and more delicate than Greater Yellowlegs.

Identification L 24cm (9.5"); **WS** 49cm (19"). Medium-sized with indistinct supercilium (more prominent in front of eye), longish neck; medium-length, fine, straight, dark bill with dull yellow at very base; tips of folded wings extend beyond tail; long bright yellow legs. Brownish-grey above, white beneath; narrow white eye-ring. *In flight* from above uniformly darkish (has plain secondaries), with white rump and barred whitish tail; feet extend well beyond tail. Sexes similar; plumage varies seasonally and with age. Small web between outer toes, slight between inner toes. *Feeds* by picking; often wades.

Juvenile Brown-grey crown, streaked white; diffuse whitish supercilium. Upperparts, wing-coverts and tertials brownish-grey, extensively edge-spotted pale-buff, darker between spots. Foreneck streaked brownish-grey, remainder of underparts white.

First non-breeding/adult non-breeding Crown, hindneck and upperparts uniform grey; wing-coverts and tertials have off-white fringes and dark side-spots, the latter particularly on greater coverts and tertials. Foreneck and upper breast finely streaked brownish-grey, remainder of underparts white. First non-breeding retains some worn juvenile wing-coverts and tertials.

First-breeding/adult breeding Age when first breeds uncertain, either second or third calendar year, but not uncommon in non-breeding areas during breeding season. At both ages replaces a variable number of upperparts, tertials and coverts, sometimes all; first-breeding presumably replaces fewer than adult. Head and neck heavily streaked brown-grey; mantle and most scapulars dark brown-black with off-white fringes and edge-spots, often with a few non-breeding-type scapulars. Upper breast heavily streaked and spotted dark brown-black, flanks sparsely barred dark brown-black; remainder of underparts white. Bill may be completely dark when breeding.

Call A quiet *tu*, usually given only once or twice.

Status, habitat and distribution Breeds in similar habitat to Greater Yellowlegs, from central Alaska across Canada to southern Hudson Bay and James Bay. Migrates southward on broad front, less coastal on northward passage. Non-breeding birds occur both coastally and on inland wetlands, south from New York in the east (uncommon) and California in the west, more frequently in far south of United States, and Central and South America south to Argentina and Chile. A widespread vagrant to

▼ **120a. Juvenile**. Florida, mid September. In virtually complete juvenile plumage. The bill-length is similar to that of the head, and feathering on the lower mandible reaches the nostril, both useful distinctions from Greater Yellowlegs. RJC.

western Europe (mostly September–October), Africa, the Far East and Australasia.

Racial variation No races recognised.

Similar species Greater Yellowlegs (119), which see. Wood Sandpiper (123) is marginally smaller, has a strong supercilium before and behind the eye, has a slightly shorter bill and shorter dull-yellow legs, and its wing-tips do not quite reach the end of the tail.

Reference Tibbitts & Moskoff (1999).

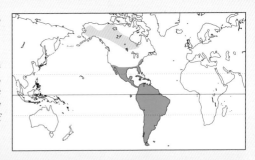

▲ **120b. First non-breeding**. Eastern England, early January. Adult-type upperparts, but retains worn and faded juvenile coverts and tertials. RJC.

▼ **120c. Adult non-breeding**. Florida, late December. Upperparts much as 120b, but adult coverts are less contrasting with narrower, more diffuse fringes, and tertials show some barring. RJC.

▲ **120d. Adult breeding**. Texas, mid April. Breeding plumage recalls Greater Yellowlegs, but the flank-barring is rarely so extensive as in that species. A number of species of shorebirds will perch in trees, particularly on the breeding grounds; this behaviour is more unusual when non-breeding or on migration. RJC.

▲ **120e. Non-breeding Lesser Yellowlegs in flight**. Florida, late December. Like Greater Yellowlegs has white rump, but no white V on back. RJC.

121. SOLITARY SANDPIPER
Tringa solitaria

A dark North American *Tringa*, lacking the white rump of its Old World counterpart, Green Sandpiper.

Identification L 20cm (8"); **WS** 42cm (16.5"). Very similar to Green Sandpiper (122), with which it forms a species pair. Medium-sized; small, straight, medium-length bill with dark tip and greenish-base, and medium-length, dull greenish or yellowish legs. In all plumages dark greenish-brown above, spotted whitish; prominent short supercilium in front of eye only, white eye-ring (slightly wider than Green Sandpiper). Throat white, neck and upper breast brown, remainder of underparts white. Following post-juvenile moult, undertail-coverts have dark bars, apparently not acquired by Green Sandpiper. *In flight* from above uniformly dark, with dark rump and dark-centred tail with white sides and dark bars extending to outer feather (outer feather white or with much less dark in Green Sandpiper); under-wing dark, toes just extend beyond tail. Sexes similar; plumage varies seasonally and with age. Small web between outer and middle toes. *Feeds* in fresh or brackish water, often by wading, when thrusts head well under the surface; frequently bobs tail.

Juvenile Crown brown; upperparts, wing-coverts and tertials greenish-brown, with small pale-buff edge-spots. Foreneck and upper breast mottled pale brown. Yellow or greenish-yellow legs.

First non-breeding/adult non-breeding Head, neck and upper breast uniformly grey-brown, breast darker than juvenile; upperparts and tertials rather dull, with sparse, tiny, pale edge-spots. First non-breeding retains most of its (smaller) juvenile wing-coverts and some juvenile tertials, but these worn feathers are difficult to see.

Adult breeding Age at first breeding not known. Dark crown finely streaked white; those upper-parts, wing-coverts and tertials which are replaced are bronze-brown with small white edge-spots, similar to but whiter than those of juvenile. Throat and upper breast streaked brown, flanks sparsely barred brown; remainder of underparts white. Legs greenish.

Calls *Peet-weet* or *peet-weet-weet*, less penetrating than Green Sandpiper.

Status, habitat and distribution Breeds in marshy areas with scattered trees, from central Alaska and throughout southern Canada to Labrador and Quebec: nominate race *solitaria* in east as far as eastern British Columbia; race *cinnamomea* in the rest of British Columbia, and further north and west. Race *cinnamomea* migrates through western United States, *solitaria* on a broad front from mid-western states eastward, some perhaps by a direct flight across the western Atlantic. Non-breeding birds occur in southern United States (rarely), south to northern Argentina and Uruguay; distribution of the two races when non-breeding unknown. Vagrant to Greenland, Iceland and western Europe (typically during July–October); also Africa, South Georgia.

Racial variation The two races are very similar and show overlapping characters that are further confused by fading and wear. Upperpart spotting in *cinnamomea*, particularly the juvenile, is more cinnamon, but effect of bleaching is uncertain.

◄ **121a. Juvenile.** Southwest England, early October. Best aged by neat, small, uniformly edge-spotted and rather dull upperparts and wing-coverts, and uniformly mottled upper breast. Has moulted one or two scapulars. The edge-spots are smaller than when breeding; the relatively pale spots and the dark loral area suggest this may be of the nominate race *solitaria*. Steve Young.

Similar species Green Sandpiper (122) is slightly larger but generally very similar; best distinguished by its white rump, which is conspicuous in flight. Spotted Sandpiper (129) is smaller, less dark above, has a less prominent eye-ring and shows a white wing-bar in flight; adult breeding is spotted beneath.

References Dowsett *et al.* (1999), Lethaby (1995), McGowan & Weir (2002), Moskoff (1995), Prince & Croxall (1996).

◄ **121b. Adult non-breeding**. Venezuela, mid November. Upperparts dull, with very small, sparse, pale edge-spots. RJC.

◄ **121c. Adult breeding**. Texas, late April. Upperparts darker than non-breeding, with white edge-spots. RJC.

▲ **121d. Adult breeding**. Venezuela, mid November. Still largely in breeding plumage but beginning moult to non-breeding. Note the bars on the undertail, which are apparently not shown by (non-juvenile) Green Sandpipers. RJC.

▼ **121e. Juvenile in flight**. New Jersey, early September. Note the dark rump; the otherwise very similar Green Sandpiper has a white rump and less barring on outer tail. Richard Crossley.

122. GREEN SANDPIPER
Tringa ochropus

A widespread Eurasian *Tringa* with a distinctive bright white rump in flight.

Identification L 22cm (9"); **WS** 44cm (17"). Very similar to Solitary Sandpiper (121), with which it forms a species pair. Medium-sized; straight, medium-length bill with dark tip and greenish-base, and dull yellowish-green, medium-length legs. In all plumages is dark greenish-brown, spotted whitish, above; pale supercilium in front of eye only, bold white eye-ring, brownish upper breast; otherwise white beneath. *In flight* from above is uniformly dark with a bright white rump, blackish tail with white sides and dark bars, though with outer feather entirely white or only restricted dark; underwing all dark; feet extend just beyond tail. When in flight the contrast of dark upperparts and dark underwing with white rump and underbody gives a very black-and-white appearance. Sexes similar; plumage varies seasonally and with age. Small web between outer and middle toes, slight between inner toes. *Feeds* generally at muddy freshwater margins, often wading, thrusting head well under surface.

Juvenile Crown brownish-grey; greenish-brown upperparts, wing-coverts and tertials (duller than in the adult) have small, pale-buff edge-spots. Head is brownish-grey, foreneck and upper breast brown-grey, streaked white at centre; some barring on upper flanks; remainder of underparts white.

First non-breeding/adult non-breeding Very similar to juvenile but with smaller buff spots at feather edges; foreneck and upper breast rather more streaked. First non-breeding retains most of its worn juvenile wing-coverts; some replace outer primaries.

Adult breeding Age at first breeding not known. Crown dark greenish-brown, streaked white; upperparts dark greenish-brown with bright-white marginal spots, though not all feathers are replaced; tertials are similar but any retained worn non-breeding feathers may lack spots. Foreneck and upper breast strongly streaked dark brown.

Call Loud, distinctive *weet, tweet, wit, wit* on taking flight.

Status, habitat and distribution Breeds in marshy areas with scattered trees, from Scotland (rare), Scandinavia, continuously through southern Siberia to the Kolyma, in a band between 50° and 68°N. Migrates overland; non-breeding birds occur by fresh water, often inland, from central England and the Netherlands south to central Africa, Arabia, south and south-east Asia to China and Japan. Vagrant to Aleutians, New Guinea.

Racial variation No races recognised.

Similar species Solitary Sandpiper (121), which

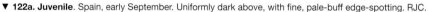

▼ **122a. Juvenile**. Spain, early September. Uniformly dark above, with fine, pale-buff edge-spotting. RJC.

see. Wood Sandpiper (123) is smaller, less bulky, has a different face pattern with a long pale supercilium, lacking Green's prominent white eye-ring; Wood is paler above with more prominent pale spotting; in flight much less 'black-and-white', with feet extending further beyond tail. Common Sandpiper (128) is smaller and paler above, has a less prominent eye-ring and shows a white wing-bar in flight.

References Higgins & Davies (1996), Holling *et al.* (2007), Thorup (2006), van den Berg (1992).

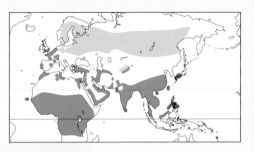

◀ **122b. Non-breeding. Eastern England, mid August.** Shows the green tinge to the feathers that gives the species its name. The small buff edge-spots on the upperparts are very similar to those of the juveniles. Some birds are already largely in non-breeding plumage by the beginning of August. RJC.

◀ **122c. Non-breeding. Eastern England, mid August.** Wing-lifting, showing white rump and white-striped axillaries. The same individual as 122b, with (on the furthest wing) newly-grown inner primaries, a gap, then old unmoulted outer primaries. RJC.

◀ **122d. Non-breeding. India, early January.** The tertials on this bird are breeding-type feathers, with large edge-spots; such feathers may be acquired from about January by both first non-breeding and adult birds. RJC.

◄ **122e. Adult breeding**. Israel, early April. Extensive bold white spotting above, strongly streaked breast. RJC.

◄ **122f. Adult breeding**. South Korea, early May. Upperpart spots are rather small, so may be a first-breeding bird. RJC.

◄ **122g. Adult in flight**. Eastern England, mid September. RJC.

123. WOOD SANDPIPER
Tringa glareola

An elegant *Tringa* with yellowish or greenish legs, readily identified by its spangled upperparts.

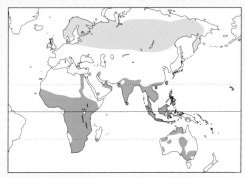

Identification L 20cm (8"); **WS** 39cm (15.5"). Small; medium-length straight bill with dark tip and dull greenish base; long, dull-yellowish or greenish legs. Prominent creamy supercilium, off-white eye-ring, brown upperparts with copious, fairly large, pale spots; underparts white. *In flight* from above uniform brown (outer primary shaft white), with white rump and brown barred tail; underwing off-white, feet project beyond tail. Sexes similar; plumage varies seasonally and with age. Small web between outer and middle toes, slight between inner toes. *Feeds* by picking, at muddy freshwater margins.

Juvenile Crown brownish-black with creamy streaks; upperparts brown with cream marginal spots and fringes, wing-coverts and tertials have bold cream notches. Upper breast lightly mottled brown; remainder of underparts white, with sparse brown streaking on undertail-coverts.

First non-breeding/adult non-breeding Much as juvenile but pale spots and fringes on upperparts duller and slightly less conspicuous. Foreneck and breast streaked and washed brownish-grey. First non-breeding retains some worn juvenile wing-coverts and tertials but is difficult to distinguish from adult.

First-breeding/adult breeding Age at first breeding not known but some perhaps in second calendar year. Adult breeding similar to juvenile, but upperparts, replaced tertials and wing-coverts are darker brown, with prominent white spots. Throat white, foreneck and breast strongly streaked dark brown, flanks barred brown. Some first-breeding remain in non-breeding area and acquire only partial breeding plumage; others moult outer primaries and perhaps return to breed.

Call A sharp, high-pitched *chiff-iff-iff.*

Status, habitat and distribution Breeds in marshland, usually in areas with some trees, from Scotland (rare), Denmark and Fennoscandia eastwards through Russia to Kamchatka, in a band between 50° and 70°N. Migrates on a broad front; non-breeding birds occur inland in sub-Saharan Africa and throughout south and south-east Asia south to Australia. Uncommon on passage in Britain and Ireland; rare but regular on the Aleutians. Vagrant to Iceland, Greenland, eastern North America and the Caribbean.

Racial variation None.

Similar species Lesser Yellowlegs (120) and Green Sandpiper (122), which see.

References Chisholm (2007), Ebels (2002), Thorup (2006).

◀ **123a. Juvenile.** Eastern England, early August. Has large pale buff edge-spots in juvenile plumage. RJC.

▲ **123b. First non-breeding**. Namibia, mid December. The upperparts have been replaced, but there are still a few very worn juvenile wing-coverts (partly hidden under the plain grey upper scapulars). RJC.

▼ **123d. Adult breeding**. South Korea, early May. Bright upperparts and strongly streaked breast. RJC.

▲ **123c. Non-breeding**. India, late December. A difficult species to age when non-breeding; this individual has juvenile-like tertials, but the edge-spotting on these is smaller and more closely spaced than on the juvenile (compare with 123a), so these represent non-breeding feathers. Starting to acquire breeding-type dark upperparts with white edge-spots. The yellow legs of this individual invite comparison with Lesser Yellowlegs, but note their slight brownish tone. Compared to Lesser Yellowlegs, Wood Sandpiper has a shorter bill and a stronger supercilium, and the wing-tips only just reach the tail, not beyond. RJC.

▼ **123e. Wood Sandpiper wing-lifting**. South Australia, mid January. Showing the flight pattern with white rump. David Harper.

124. GREY-TAILED TATTLER
Tringa brevipes

A rather short-legged grey shorebird with a stout bill, common on coastal mudflats outside the breeding season.

Identification L 25cm (10"); **WS** 51cm (20"). Very similar to Wandering Tattler (125), with which it forms a species pair. Outside breeding season uses mudflats, unlike Wandering Tattler, which is a bird of rocky coasts. Medium-sized, pale brown-grey above, tail paler than upperparts (but variable), with medium-length, straight, dark bill, typically with paler base. White supercilia sometimes meet on forehead and usually extend behind eye (but variable); white eye-ring; folded primaries do not project beyond tail or do so only slightly; short yellow legs. *In flight* from above uniformly grey with no obvious features; toes do not extend beyond tail. Sexes similar; plumage varies seasonally and with age. Small web between outer and middle toe only. *Feeds* on muddy areas, occasionally on rocks, by picking, sometimes moving stones or weed like a turnstone, running frequently; bobs like a Common Sandpiper. See Table 13 (p. 393) for a detailed comparison of Grey-tailed and Wandering Tattlers.

Juvenile Upperparts pale to medium grey with scattered small white edge-spots; coverts grey with white edge-spots and dark but obscure subterminal lines, and narrow white terminal fringes; tertials have white edge-spots. Grey wash across the breast, slightly mottled when fresh; flanks largely white with a pale grey wash. Bill yellowish-grey with a dark tip.

First non-breeding/adult non-breeding Once juvenile pale fringes and edge-spotting of upperparts have worn off it is difficult to distinguish these two age classes, both appearing uniformly medium grey.

First-breeding Probably does not breed until third calendar year, perhaps later. Typically acquires scattered narrowly white-fringed upperparts and a few indistinct barred underpart feathers; many remain in non-breeding area. Birds of this age seem to have particularly pale tails, perhaps the result of bleaching.

Adult breeding Upperparts and wing-coverts have neat pale fringes when fresh (sometimes with dark submarginal line); widely spaced medium-grey barring on breast and flanks only, and restricted dark markings on undertail, not barred as Wandering Tattler. Bill darker but still two-toned, with small orange area at base of lower mandible.

Calls A whistled *wee-hoo*, a rising *hoo-weet* and, less commonly, *too-doo-weet*, the latter two calls recalling Grey Plover.

Status, habitat and distribution Fairly common, breeding by mountain streams and lakes in north-central and north-east Siberia to Kamchatka, non-breeding on intertidal reefs and mudflats in Taiwan, Thailand, Sumatra, east to New Guinea and

▼ **124a. Juvenile**. Japan, late September. Compared to juvenile Wandering Tattler, has more white fringes and spotting on the upperparts, particularly on tertials. Nobuhiro Hashimoto.

south to east Indian Ocean and Australia; also New Zealand, where scarce. Regular but scarce migrant in western Alaska. Vagrant western North America from Washington to California; also Britain.

Racial variation None.

Similar species Wandering Tattler (125); for details see Table 13 below.

References Hirst & Proctor (1995), Paulson (1986), Stenning & Hirst (1994), Thorpe (1995).

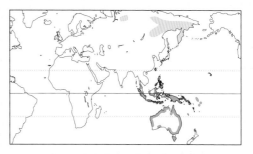

Table 13. The main distinguishing features of Grey-tailed and Wandering Tattlers. Note that the tarsal scales are only visible in the hand.

		Grey-tailed Tattler	Wandering Tattler
Structure	Nasal groove	Half bill length or less	> Half bill length
	Primary projection	None or short beyond tail; four primary tips beyond tertials	Variable beyond tail, typically long; five primary tips beyond tertials
	Tarsal scales *	Overlapping on front and rear	Usually overlapping on front only
	Bill	Deeper base; usually more extensive pale area at base than Wandering	Whole bill more slender, less deep at base; usually smaller pale area at base than Grey-tailed
Plumage	Upperparts and tail: all plumages	Upperparts pale/medium buff-grey, tail paler	Upperparts and tail both medium-dark grey
	Supercilia: all plumages	Sometimes meet on forehead, extending behind eye, but variable	Usually only front of eye, but variable, often extends behind when breeding
	Upperparts: juvenile	Extensive pale fringes and edge-spotting, especially coverts and tertials	Narrow pale upperpart fringes, minor edge-spotting on larger rear coverts
	Flanks: juvenile & non-breeding	Whitish or pale grey, contrasting with folded wing	Grey, similar to folded wing
	Underparts: breeding	Widely spaced grey barring on breast and flanks only	Closely spaced dark grey barring on entire underparts, including under tail
Call		Diagnostic - see text	Diagnostic - see text

◀ **124b. Non-breeding**. Western Australia, mid March. The lack of any underpart barring, at a time when adults are largely in breeding plumage, suggests that this is a second calendar year bird that will remain in the non-breeding area. RJC.

◀ **124c. Non-breeding**. Northern Territory, Australia, mid September. A second calendar year individual that probably did not return to the breeding grounds; a worn version of 124b. Non-breeding individuals often have a paler tail than other plumages, presumably as a consequence of bleaching. The absence of grey on the flanks and the lack of primary projection beyond the tail both help to separate from Wandering Tattler. RJC.

◀ **124d. Adult breeding**. South Korea, early May. Underpart barring when breeding is much less than in Wandering Tattler, being restricted to the breast and flanks. RJC.

◀ **124e. Adult breeding wing-stretching**. Japan, late May. In flight both species of tattler are a uniform plain grey above, lacking any obvious features. Nobuhiro Hashimoto.

125. WANDERING TATTLER
Tringa incana

Very similar to Grey-tailed Tattler but with a preference for rocky coasts outside the breeding season.

Identification L 27cm (11"); **WS** 54cm (21"). Very similar to Grey-tailed Tattler (124), with which it forms a species pair. Medium-sized, mid-grey above, tail of similar colour to upperparts (not paler), with a medium-length, straight, dark bill, typically with a small pale area at base of lower mandible. White supercilium usually restricted to front of the eye (but variable), white eye-ring, folded primaries typically project beyond the tail (often more obvious on older birds), short yellow legs. *In flight* from above uniformly grey with no obvious features; toes do not extend beyond the tail. Sexes similar; plumage varies seasonally, and with age. Small web between outer and middle toes only. *Feeds* by picking and probing, usually on intertidal rocks, much less frequently on sand or mudflats; bobs like a Common Sandpiper. See Table 13 (p. 393) for a detailed comparison of Grey-tailed and Wandering Tattlers.

Juvenile Very narrow pale fringes to upperparts and coverts when fresh, with fine, pale edge-spotting to the rear wing-coverts and some tertials, but variable, some showing very little edge-spotting. Breast is mottled grey, flanks are grey with diffuse white barring. Bill usually has a dark tip and a grey base.

First non-breeding/adult non-breeding Once juvenile pale fringes and edge-spotting have worn off it is difficult to distinguish these two age classes, both appearing uniformly grey.

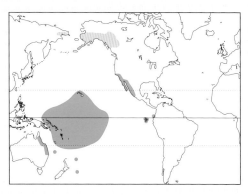

First-breeding Probably does not breed until at least third calendar year. Many birds remain south of the breeding grounds in non-breeding-type plumage; others acquire a variable amounts of breeding plumage; this can sometimes be near-complete, though retaining worn and faded primaries.

Adult breeding Supercilium streaked dark, often extending behind the eye; upperparts largely replaced, new feathers having very narrow pale fringes, less obvious than in Grey-tailed Tattler. The whole of the underparts have closely spaced, dark-grey barring, including the undertail. Bill largely dark, with a small orange area at the base of the lower mandible.

Call A Whimbrel-like *peet-peet-peet-…* rapidly repeated

◀ **125a. Juvenile**. California, early September. White edge-spots and fringes on upperparts are very restricted compared to juvenile Grey-tailed Tattler (see 124a). RJC.

at the same pitch about ten times; very different from the call of Grey-tailed Tattler.

Status, habitat and distribution Uncommon, breeding by mountain streams and lakes in north-east Siberia (rare), and western Alaska to British Columbia. Non-breeding birds occur on rocky coasts from California to western Mexico, and in the central and south-western Pacific from Hawaii to eastern New Guinea and north-east Australia. Vagrant to eastern North America.

Racial variation None.

Similar species Grey-tailed Tattler (124); for details see Table 13 (p. 393).

References Gill *et al.* (2002), Hirst & Proctor (1995), Paulson (1986), Schenk & Ebels (2004).

▲ **125b. Adult non-breeding.** California, mid November. When non-breeding the best distinctions from Grey-tailed Tattler are the extensive grey on flanks and the primary extension beyond the tail (though this can be variable). In flight plain grey above, almost identical to Grey-tailed Tattler (124e). RJC.

▲ **125c. Adult breeding.** California, mid April. Extensive barring beneath, including undertail, with the markings darker than on Grey-tailed Tattler. It is unusual to see a Wandering Tattler away from rocks. RJC.

126. WILLET
Tringa semipalmata

The largest *Tringa*, with a very stout bill and striking black-and-white wings in flight.

Identification L 37cm (14.5"); **WS** 63cm (25"). Large, long-necked, with medium-length, straight, grey-based and black-tipped bill and long legs. Short supercilium in front of eye only; narrow white eye-ring. *In flight* from above has striking broad white wing-bar, black primary coverts, grey innerwing-coverts; black trailing edge to wings, broad on outer primaries but extending only to outer secondaries; white inner secondaries; white rump and greyish tail, feet extend beyond tail. Females typically average slightly larger, but size difference can sometimes be substantial, particularly in the western race (*inornata*). Plumages vary seasonally and with age. Web between outer toes, restricted web between inner toes. *Feeds* by picking and probing, often wading.

The two races, 'Eastern Willet' *semipalmata* and 'Western Willet' *inornata*, can often be separated on both plumage and structure, and are described separately below; see also 'Racial variation'.

Juvenile *Eastern* Upperparts darkish grey-brown, fringed buff, with bold, dark subterminal lines; wing-coverts with paler fringes; tertials plainer, notched buff, often with pale margins and some dark edge-spotting towards tips. Upper breast washed brownish grey, remainder of underparts white. Primary tips fall close to tail-tip. *Western* Similar, but upperparts greyer and paler owing to almost complete absence of dark markings, these often restricted to dark shaft-streaks. Primary tips extend slightly beyond tail-tip.

First non-breeding/adult non-breeding *Eastern* Pale grey-brown upperparts show slight contrast with greyer wing-coverts, all of which have narrow paler fringes when fresh; breast washed pale grey, remainder of underparts white. First non-breeding retains many worn juvenile wing-coverts. *Western* As Eastern Willet but pale grey, lacking brown tones on upperparts.

First/second breeding Age at first breeding uncertain but some probably breed in fourth calendar year. Many probably do not do not visit breeding grounds and remain in non-breeding area either with non-breeding-type plumage or acquiring restricted breeding-type plumage.

Adult breeding *Eastern* Upperparts light brown, most feathers strongly barred dark brown; neck, breast and flanks strongly streaked and barred brown, remainder of the underparts white. There is some dark barring on the pale grey uppertail, particularly on the central feathers. Bill is typically all-dark, with a slightly paler pinkish area at base. *Western* Upperparts pale grey, with less barring, both above and below; can have barred centre tail. Males are often more strongly marked. Bill has a grey base.

Calls *Kip*, and *pill-will-willet*.

Status, habitat and distribution *Eastern* Common, breeding in coastal salt marsh in Nova Scotia, from New Jersey south to Florida and Texas, and in northern South America; leaves North America around mid August. Non-breeding birds occur in coastal South America south to Chile and Brazil, perhaps also in the West Indies (unknown whether the two races overlap when non-breeding); returns from early March, already largely in breeding plumage. *Western* Breeds in prairie marshes in western Canada and western United States. Non-breeding birds occur coastally from North Carolina to Florida and Texas, the Caribbean to Uruguay, and from Oregon (small numbers) and California south to Chile and the Galápagos. Eastern adults leave their breeding areas from mid July, as non-breeding Western adults arrive. As returning Eastern Willets are in breeding plumage they are rarely if ever seen in North America in non-breeding plumage. Rare vagrant to western Europe, where both races have occurred.

Racial variation Typical plumages of the two races are described above. The races differ in average size (Western averages about 15% larger on most

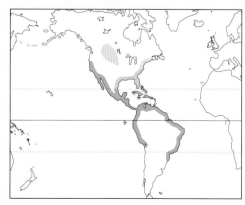

dimensions) and in structure. Eastern has shorter, deeper, more parallel bill except towards tip, and marginally shorter legs; Western has longer, finer, more tapered bill and a more attenuated rear-end. In all plumages Western is usually slightly paler and greyer; when breeding has fewer dark feathers on upperparts, breast and flanks, with less streaking and barring on white ground (Eastern has denser markings on buff ground). Juvenile Eastern tends to have more strongly marked dark marginal lines on darker upperpart feathers, while non-breeding Western has paler breast and throat. There is variation in all plumages but the combination of size, structure and plumage features may often allow racial identification.

Similar species None.

References Antonucci & Corso (2008), Howell & Webb (1995), O'Brien (2006), O'Brien *et al.* (2006), Siblet & Spanneut (1998), Sondebø (1992).

◄ **126a. Juvenile *semipalmata*.** Texas, early August. Juvenile *semipalmata* Willets have dark brown upperparts (though this bird is acquiring greyish non-breeding scapulars) and well marked wing-coverts. The bill shape, rather short with deep base and blunt tip, is typical of *semipalmata*. RJC.

◄ **126b. Juvenile *semipalmata*.** Texas, early August. The breeding range of *semipalmata* Willet includes Texas, but some Texan birds have characters intermediate between the two races. This is an example, with the plumage and structure of juvenile *semipalmata*, but a slender bill more typical of *inornata*. RJC.

◄ **126c. Juvenile *inornata*.**
Florida, early September. Western
Willets of race *inornata* can be big
and bulky, as this individual, which
is presumably a female; the bill
is typically tapered over most of
its length. In juveniles (not adults)
the primary projection beyond the
tertials averages longer than in
semipalmata. Upperparts are lighter
than juvenile *semipalmata*, and the
markings on the wing-coverts are
less contrasting, but there is much
variation. RJC.

◄ **126d. First non-breeding
inornata.** Florida, mid October.
Adult-type non-breeding
upperparts, and worn juvenile
wing-coverts and tertials. RJC.

▼ **126e. Adult non-breeding
inornata with Marbled Godwits.**
Florida, mid December. Western
Willets are very variable in size,
as seen here! The centre bird is a
female, the right-hand birds are
presumably males. RJC.

▲ **126f. First breeding *inornata*.** Florida, late June. Willets do not breed until their fourth calendar year, and remain in the non-breeding areas. This individual has not acquired any breeding plumage, and still has one or two very worn juvenile coverts. The long tapered bill and bulky shape show this to be *inornata*. RJC.

▼ **126g. Adult breeding *semipalmata*.** Florida, mid May. Upperparts are brownish with extensive dark markings, flanks strongly barred on a buff ground. Pink base to bill is also a character of *semipalmata*, though perhaps only when breeding. RJC.

◀ **126h. Adult breeding** *inornata*. Kansas, late April. Limited dark markings on greyish upperparts, flanks have narrow barring on a white ground. RJC.

◀ **126i. Adult breeding** *semipalmata* **in flight**. Florida, early April. Dark central barring on the tail can be shown by both races when breeding. RJC.

▼ **126j. Adult non-breeding** *inornata* **in flight**. California, early November. Plain grey or faintly barred tail when non-breeding. RJC.

127. TEREK SANDPIPER
Xenus cinereus

A distinctive sandpiper with an upturned bill and orange-yellow legs.

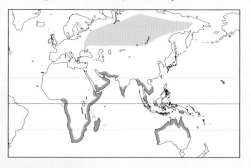

Identification L 24cm (9"); **WS** 42cm (16.5"). Has an unmistakable combination of compact shape with horizontal carriage, steep forehead, slightly upturned bill and yellow or orange-yellow legs. Medium-sized, largely grey above, with poorly developed supercilium, dark lores and eye-stripe; long, slightly upturned dark bill with dull-yellowish base; upper breast finely streaked or washed grey-brown; remainder of underparts white; short yellow legs. *In flight* recalls Common or Spotted Sandpiper with flicked downbeats; has unique pattern of black primaries and primary coverts, narrow black bar across secondary coverts, white secondaries and grey tail; toes do not quite reach tail-tip. Sexes similar; plumage varies seasonally and with age. Moderate webbing between all three toes. *Feeds* energetically in a variety of open habitats, with frequent short runs, picking, probing and sweeping bill avocet-like when wading, largely on insects and other invertebrates, particularly crabs; frequently bobs tail.

Juvenile Very similar to adult non-breeding but upperparts, coverts and tertials have short, dark, subterminal bars and narrow white tips.

First non-breeding Retains some juvenile coverts and tertials.

Adult non-breeding Upperparts and coverts plain grey with narrow black shaft streaks; sometimes with restricted black on upper scapulars, less than adult breeding; white supercilia more prominent than in other plumages, particularly in front of eyes, and may meet above bill.

First-breeding A very few may breed in second calendar year. Plumage of some as adult breeding; many moult some outer primaries. Birds that remain in the non-breeding area are much as adult non-breeding.

Adult breeding Most probably breed in third calendar year. Head streaked brown, darker on crown, supercilium restricted to small white area in front of eye; upper scapulars with broad, black, central streak, sometimes largely black, merging to give elongate dark scapular line; many lower scapulars have a bold black centre-streak. Probably has all-dark bill when breeding.

Calls Rather variable, a fluty *wee-we* with falling inflection, a rapid *wit-wit-wit-wit* or *witty-wit-wit*, depending on degree of alarm; twittering call given when in flock.

Status, habitat and distribution Fairly common, breeding in lowland valleys, either open or forested

◀ **127a. Juvenile**. Japan, late August. Terek Sandpipers are very similar in all plumages; juvenile has restricted black on upper scapulars and short dark subterminal bars on the upperparts, particularly tertials. Nobuhiro Hashimoto.

with open areas, in subarctic from north-east Europe east to central and eastern Siberia. Coastal when non-breeding (on largely tropical and subtropical open mudflats and estuaries, often near mangroves, which are used for roosting) in southern and eastern Africa, Arabia, south India and south-east Asia to Australia; annual New Zealand; vagrant to both North American seaboards, Mexico, Caribbean, the Canary Islands, western Africa and western Europe.

Racial variation None.

Similar species None.

References Barter (2002), Bijlsma & de Roder (1991), Carmona *et al.* (2003), Ebels (2002), Schrijvershof & Steedman (1986), Stemple *et al.* (1991), Thorup (2006), Wilson (1989).

◀ **127b. Non-breeding**. Western Australia, mid March. Restricted black on upper scapulars. RJC.

◀ **127c. Adult breeding**. South Korea, early May. Extensive black on upper scapulars forms a near-continuous band; some lower scapulars have black central streak. RJC.

▼ **127d. Adult breeding in flight**. South Korea, early May. Flight pattern, with black outerwing and white secondaries but otherwise grey above, is unique. RJC.

128. COMMON SANDPIPER
Actitis hypoleucos

A widespread small Eurasian sandpiper that bobs its tail constantly as it forages.

Identification L 20cm (8"); **WS** 34.5cm (13.5"). Very similar to Spotted Sandpiper (129), with which it forms a species pair. Small; straight, medium-length, brownish bill with dark tip, and short, dull-yellow (sometimes greyish) legs. Brown above, white beneath, with sharply demarcated brown area on upper breast. Tail extends well beyond tips of folded wings (only marginally so in Spotted Sandpiper). Shares with Spotted Sandpiper characteristic flicking flight on downward-bowed wings, usually low over water. *In flight* from above brown, with narrow white wing-bar across inner primaries continuing on secondaries, reaching to body (shorter, not quite reaching body on Spotted Sandpiper), white trailing edge to secondaries, and white outer tail; feet do not extend beyond tail. Sexes of similar size; plumage varies seasonally and with age. Web between outer toes only. *Feeds* generally by picking at edge of water, occasionally wading; constantly bobs tail.

Juvenile Crown brown, indistinct supercilium, white eye-ring. Upperparts brown with narrow pale-brown fringes and narrow, dark, subterminal lines; wing-coverts brown with black-brown subterminal bars bordered buff on either side, so folded wing strongly barred. Tertials brown, with small dark-and-light edge-spots. Brown upper breast finely streaked. Legs dull greyish or yellowish.

First non-breeding/adult non-breeding Upperparts fairly uniform brown; wing-coverts with darker subterminal bars bordered buff on either side, similar to juvenile but less contrasting. Brown on upper breast less extensive and more uniformly coloured than juvenile, sometimes not meeting at

centre; remainder of underparts white. Moult to first non-breeding is generally complete, including primaries, so difficult to age once moult is finished. Legs dull greyish or yellowish.

First-breeding/adult breeding Some breed for first time in second calendar year; most probably do so in third. Upperparts and tertials bronze-brown with broad, dark-brown shaft streaks and irregular narrow barring, wing-coverts with black subterminal bars and narrow pale fringes when fresh; brown upper breast is streaked dark brown. Some first-breeding may replace only outer primaries and/or retain a few non-breeding plumage wing-coverts. Legs dull pale yellow.

Call Typically a clear high-pitched, three-note *swee-wee-wee*.

Status, habitat and distribution A widespread species, breeding near clear, fresh water, especially upland streams, throughout most of Europe and northern Asia to about 70°N. Migrates on a broad front. Non-breeding birds occur usually by fresh water, but also coastally; locally from southern England (uncommon) south to sub-Saharan Africa, throughout south and south-east Asia to Australia, rarely New Zealand. Vagrant on the Aleutians and in Alaska.

Racial variation and hybridisation No races recognised. Mating has been observed between Common and Spotted Sandpiper.

Similar species Spotted Sandpiper (129) is very similar but has very short extension of tail beyond tips of folded wings (long in Common), generally

◄ **128a. Juvenile**. Eastern England, late August. The white 'peak' between folded wing and dark breast (shared by Spotted Sandpiper) is useful for separation from other small sandpipers, particularly those that also 'bob' or 'teeter', such as Solitary, Green and Wood Sandpipers. Aged by edge-spotted upperparts and tertials. In juvenile plumage distinguished from Spotted Sandpiper by longer tail extension beyond wing-tips, and by extensive edge-spotting on tertials (only at tip of the tertials on Spotted Sandpiper). RJC.

has brighter yellow legs and in flight shows less extensive wing-bar and less white at sides of tail. Juvenile Spotted Sandpiper is greyer, has more boldly barred wing-coverts, and spotting on tertials and greater coverts is confined to the tips (not extending along the feather edges as on Common); adult breeding Spotted Sandpiper has a boldly spotted breast. Typical call of Spotted Sandpiper is softer and usually disyllabic.

References Holland & Yalden (1994), Lawrence (1993), Thorup (2006).

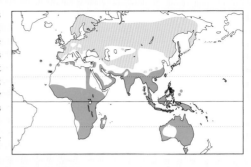

◄ **128b. Non-breeding**. Brunei, early January. Upperparts plain brown, lacking any edge-spotting. RJC.

◄ **128c. Adult breeding**. Poland, late April. When breeding upperparts have dark shaft streaks and prominent dark bars. RJC.

▲ **128d. Adult breeding.** Eastern England, early May. Some have plainer upperparts, with fewer dark markings; perhaps these are second calendar year individuals. RJC.

◄ **128e. Juvenile wing-stretching**. Eastern England, late July. The wing-bar reaches the body, which it does not do with Spotted Sandpiper (see 129f). RJC.

129. SPOTTED SANDPIPER
Actitis macularius

A widespread North American sandpiper with heavily spotted underparts in breeding plumage.

Identification L 19cm (7.5"); **WS** 32.5cm (13"). Very similar to Common Sandpiper (128), with which it forms a species pair. Small, with medium-length brown bill and short, yellowish or flesh-coloured legs. Brown above, white beneath (heavily spotted in adult breeding), with sharply demarcated brown area on upper breast. Tail extends only a short distance beyond tips of folded wings. Has characteristic low flicking flight on downward-bowed wings, as Common Sandpiper. *In flight* from above brown, with restricted white wing-bar across inner primaries and outer secondaries, not reaching body, white trailing edge to secondaries, and white outer tail; toes just reach tail-tip. Sexes of similar size; plumage varies seasonally and with age. Web between outer toes only. *Feeds* generally by picking at edge of water, occasionally wading; constantly bobs tail.

Juvenile Crown brown; indistinct supercilium; white eye-ring. Upperparts greyish-brown with buff fringes and narrow, dark, subterminal lines; wing-coverts brown with brown-black subterminal bars bordered buff on either side, so folded wing strongly barred; greater coverts and tertials uniform brown, barred only at tip and lacking edge-spotting of juvenile Common Sandpiper. Brown upper breast almost unstreaked, paler at centre. Bill pinkish with a dark tip; legs yellowish flesh to bright yellow.

First non-breeding/adult non-breeding As juvenile but with fairly uniform brown upperparts; wing-coverts with darker subterminal buff on either side, less contrasting than in juvenile. Some adults (probably also second calendar year birds) retain some spots on underparts well into November and usually have obvious dark flecks on the undertail. First non-breeding has complete moult, including primaries, but may retain some worn wing-coverts. Bill brown with pinkish lower mandible; legs variably pale grey, yellowish or flesh.

Adult breeding Many breed in second calendar year. Upperparts, tertials and wing-coverts bronze-brown with dark-brown irregular bars and shaft streaks; upper breast washed brown; irregular round, dark-brown spots on underparts from throat to vent. Females on average have larger, darker spots. Bill pinkish-orange with dark tip; legs pink.

Call Typically a disyllabic *peet-weet*, less clear than Common Sandpiper.

Status, habitat and distribution Breeds in open areas near fresh water, in Alaska, throughout Canada except far north, and United States south to

▼ **129a. Juvenile**. Florida, mid September. Aged by upperparts, and tertials with dark subterminal bars and white tips. Note lack of edge-spotting, particularly on the tertials, compared with juvenile Common Sandpiper (128a). RJC.

California and South Carolina. Non-breeding birds occur from southern United States south through Central America to Chile and Argentina. Vagrant to Iceland, the Azores, western Europe and eastern Siberia. Has bred once in Scotland.

Racial variation and hybridisation No races recognised. For Common Sandpiper x Spotted Sandpiper, see Common Sandpiper (128).

Similar species Common Sandpiper (128), which see.

References Oring *et al.* (1997), Wallace (1970), Wilson (1976).

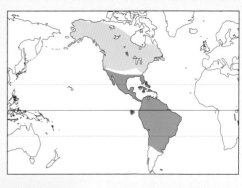

◀ **129b. Adult non-breeding**. California, early November. Upperparts largely fresh, plain non-breeding, but retains a trace of breeding plumage on underparts. RJC.

◀ **129c. Adult non-breeding**. California, early November. Showing partial webbing between outer toes, a feature shared with Common Sandpiper. RJC.

◀ **129d. Adult breeding**. Florida, late August. Unmistakable when breeding; females (which this probably is) on average have larger, darker spots, which extend to lower belly. RJC.

◀ **129e. Adult breeding/ non-breeding**. California, early November. Has replaced upperparts, but still has either extremely worn or breeding tertials and significant spotting on underparts. RJC.

◀ **129f. Non-breeding in flight**. Venezuela, mid November. Flight pattern differs from Common Sandpiper in that wing-bar does not extend to body. RJC.

TURNSTONES

There are two species of turnstones in a single genus, both of which occur in the northern hemisphere: Ruddy Turnstone *Arenaria interpres* and Black Turnstone *A. melanocephala*. Both species are migratory, Ruddy Turnstone particularly so; although a high Arctic breeding species, when non-breeding Ruddy Turnstone is one of the most widespread of all shorebirds, occurring coastally virtually worldwide. In contrast, Black Turnstone has a restricted distribution, and is limited to rocky shores of the Pacific coast of North America, from Alaska south to Mexico. As the name turnstone suggests, both species frequently feed by turning stones, seaweed and other debris with their bills as they search for food, and they share a short wedge-shaped bill that reflects this distinctive method of feeding.

Both species of turnstone have distinctive juvenile, non-breeding and breeding plumages, and are of similar size, with similar flight patterns.

Some of the first non-breeding Ruddy Turnstones that spend the non-breeding season in parts of southern Africa and Australia are unusual for shorebirds in that they have a primary moult that commences with the innermost feathers; most other shorebirds that moult primaries at this age replace a few of their outer primaries. A small proportion of first non-breeding Ruddy Turnstones in Australia have a complete primary moult, others (both in Australia and in southern Africa) have an arrested moult, the birds retaining their outermost worn and faded juvenile feathers, and still others retaining all their juvenile primaries. This latter strategy is shared by all European birds of similar age, which also do not have a primary moult; it is not known which primary moult strategy is used by South American first non-breeding birds.

▲ Adult breeding Ruddy Turnstone in a feeding territory dispute with another turnstone. South Korea, early May. RJC.

130. RUDDY TURNSTONE
Arenaria interpres

A circumpolar breeding bird, and perhaps the most widespread shorebird outside the breeding season.

Identification L 24cm (9"); **WS** 48cm (19"). Perhaps the most widespread shorebird along with Grey Plover. Medium-sized, with short neck; short, black-tipped, wedge-shaped, grey bill; short orange legs. Variable plumage, but always has distinctive black or brown breast containing pale or white areas; remainder of underparts white. *In flight* from above appears almost black-and-white, dark with white up back, white area parallel to body at base of wing, white wing-bar and white uppertail-coverts, and white outline to black tail; toes do not reach tail-tip. Sexes similar; plumage varies seasonally and with age. *Feeds* in a variety of coastal habitats, particularly on rocky shores, using bill to turn stones and weed, seeking invertebrates; has been recorded eating an extraordinary variety of food, including such unlikely items as soap, potato peel and the flesh of dead vertebrates.

Juvenile Crown brown, finely streaked darker; upperparts, tertials and wing-coverts blackish-brown, with neat, off-white upperpart fringes and buff wing-covert fringes. Face and throat white, with brown ear-coverts. Breast-patches are generally brownish.

First non-breeding/adult non-breeding Head and face greyish, chin and throat white; upperparts, wing-coverts and tertials dark greyish brown, shading paler at margins. Pale brownish breast patches. First non-breeding retains most of its worn juvenile wing-coverts and some juvenile tertials.

First-breeding Usually breeds first in third calendar year. First-breeding usually remains in non-breeding area and has mixture of adult breeding and worn first non-breeding plumage.

Adult breeding Crown white, streaked black; rest of head and neck white, with striking black markings continuous with black breast; white breast patches. Mantle and scapulars largely black with a broad line of rusty chestnut running across scapulars and wing-coverts. Extent of black streaking on crown varies with race (see below) and sex; in breeding pairs male has whiter crown and is generally brighter.

Call A rapid, low pitched and musical rattle: *tuk-a-tuk-tuk.*

Status, habitat and distribution Common, breeding coastally near rocky or pebbly shores, with a circumpolar distribution. Nominate *interpres* breeds on north-eastern Canadian islands, north and north-east Greenland (but not Iceland), and from Denmark and Norway across northern Russia to Alaska; race *morinella* breeds from north-east Alaska through northern Canada, perhaps western Greenland. Coastal when non-breeding, with *morinella* in southern United States, the Caribbean, Central America south to Tierra del Fuego; *interpres* worldwide elsewhere beyond the range of *morinella*, in and south of the temperate zone. Ringing recoveries show *morinella* to be a vagrant to western Europe.

Racial variation Race *interpres* described in Identification; *morinella* slightly smaller but with

◀ **130a. Juvenile *interpres*.** Southwest England, late September. Juvenile is dark sooty brown, with neat pale upperpart fringes; primary tips also have narrow white fringes. Upper breast has paler mottling than adult. RJC.

longer bill and tarsus (can be distinguished on measurements), and generally brighter, more rufous, in all plumages. Juvenile *morinella* has pale chestnut fringes to upperparts and wing-coverts, shading to rufous on feather margins; non-breeding with rufous in mantle and scapulars. Adult breeding has whiter crown and more extensive, brighter, chestnut area on upperparts. Breeding birds in Alaska appear to be intermediate in coloration between *interpres* and *morinella* but are regarded as *interpres* on measurements; east Siberian birds are darker (dark rusty chestnut) than other *interpres* populations.

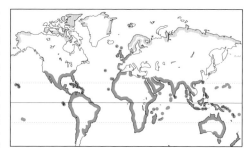

Similar species Black Turnstone (131), whose range is restricted to the North American Pacific coast, is entirely black above and on the breast.

References Dennis (1994), Engelmoer & Roselaar (1998), Nettleship (2000), Skewes *et al.* 2004, Thorup (2006).

▲ **130b. Juvenile *morinella***. Florida, early September. On average *morinella* is more rufous in all plumages – compare the wing-coverts here with 130a – and is longer-billed. RJC.

▼ **130c. First non-breeding *interpres***. Eastern England, mid December. Adult-type upperparts, but retains juvenile coverts. RJC.

▲ **130d. Adult non-breeding**
interpres. Southwest England,
late September. Very similar to first
non-breeding (130c), but larger
adult wing-coverts give a more
untidy appearance. RJC.

◄ **130e. Adult non-breeding**
morinella. Florida, mid December.
Subtly brighter and more rufous
than *interpres*, particularly on the
scapulars; compare with 130d.
RJC.

▼ **130f. Adult breeding *interpres*.**
Eastern England, early May. Fresh
plumage, still with white fringes on
black of breast. Probably a male,
with much white on head and
breast-side patches, and most of
wing-coverts replaced with bright
breeding-type feathers. RJC.

◄ **130g. Adult breeding** *interpres*.
Eastern England, early August.
Probably a male; by the end of the
breeding season the wing-coverts
have faded significantly compared
to 130f. RJC.

◄ **130h. Adult breeding** *morinella*.
Florida, late May. This individual
with an almost completely
white head can only be a male!
Upperparts and coverts bright
chestnut in *morinella*. RJC.

◄ **130i. Adult breeding** *interpres*.
South Korea, early May. Probably
a male. East Siberian breeders,
which this is presumed to be from
location, are darker rusty-chestnut
than other *interpres* populations.
RJC.

▲ **130j. Adult breeding *interpres*.** Spain, early June. A rather late date for migrants, so these may not breed. Individuals can only be sexed with confidence in breeding pairs, but these two show the typical difference between a possible male (left) and possible female (right). Though the 'male' has more streaking on crown than many, it has white breast-side patches, and has replaced most coverts with breeding-type feathers. The 'female' has a dark crown lacking clear white streaks, the breast-side patches are suffused with buff, and the wing-coverts are non-breeding type, all features typical of breeding females. RJC.

◄ **130k. Non-breeding *morinella* in flight.** Florida, mid December. The striking flight pattern is shared only by Black Turnstone. RJC.

▼ **130l. Ruddy Turnstones in flight.** North Wales, mid April. The almost black-and-white flight pattern is unmistakable, and diagnostic of Ruddy Turnstone away from the Pacific coast of the Americas. RJC.

131. BLACK TURNSTONE
Arenaria melanocephala

A nearly all-black turnstone, restricted as a breeding bird to to the Pacific north-west of North America.

Identification L 24cm (9"); **WS** 48cm (19"). Medium-sized, with uniformly black head and upperparts, a faint white supercilium behind eye only (but see adult breeding below) and a short neck; short, wedge-shaped, near-black bill, and short orange-brown or black legs. Head pattern variable; upperparts, breast and upper belly black, apart from white fringes; remainder of underparts white. *In flight* from above similar to Ruddy Turnstone but black where Ruddy is merely dark; white up back, white area parallel to body at base of wing, white wing-bar and uppertail-coverts, and white outline to black tail; feet do not reach the tail-tip. Sexes are similar; plumages vary seasonally; juvenile is separable. *Feeds*, when not on breeding grounds, on rocky shores in the intertidal zone and much less frequently (particularly at high tide) on sandy and pebbly beaches; feeds by picking, using bill to turn stones and seaweed to reveal hidden invertebrates beneath; on breeding grounds mainly eats insects.

Juvenile Dark sooty brown rather than black. Scapulars, tertials and coverts have narrow, pale-buff fringes that fade with age, broader and paler laterally; similar to adult but smaller and neater. Primary tips rather pointed, narrowly outlined white when fresh. Legs dull orange-brown; bill dark, sometimes with paler base.

First non-breeding The majority probably moult on non-breeding grounds, replacing juvenile body feathers but retaining some juvenile wing-coverts, scapulars and tertials, and all flight feathers; replaced feathers are near-black. Legs dull orange or brown. Quickly becomes difficult to separate from adult non-breeding.

Adult non-breeding As with first non-breeding, most moult on non-breeding grounds. Head, mantle and breast uniform, sooty, brown-black; breast speckled black; remainder of upperparts brown-black, with prominent white edges to scapulars and greater coverts. Legs usually dark.

First-breeding Many, but not all, breed in their second calendar year; those that do not breed apparently remain in non-breeding area. As adult breeding, except that (juvenile) primaries are worn and faded and legs are dull brown.

Adult breeding Black crown with extensive white streaking, particularly on forehead and above eye; oval white patch on lores; white spots on upper breast, legs black. In pairs on breeding grounds, male has slightly more white on head and breast and larger white, oval loral spot.

Call A dry, piping chatter or trill, recalling Ruddy Turnstone but more rapid and higher pitched.

Status, habitat and distribution Fairly common,

▼ **131a. Juvenile**. California, mid September. Easily identified by black plumage; in September best aged by entirely neat plumage at a time when adults are moulting and untidy; see 131c. RJC.

breeding on near-coastal salt-grass meadows and dwarf-shrub tundra in western Alaska; non-breeding birds on rocky coasts from southern Alaska south to central west coast of Mexico. Vagrant to Bering Sea islands and coastal north-east Siberia; also in inland North America as far east as Wisconsin.

Racial variation None.

Similar species Surfbird (63) and Ruddy Turnstone (130), which see.

References Handel & Gill (1992, 2001).

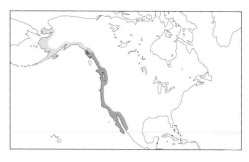

◄ **131b. Juvenile**. California, early November. Quite a worn individual. Juvenile is slightly browner than adult, with black upperparts; has similarly white-edged scapulars and greater coverts to adult, but these feathers are smaller and neater. Primary tips more pointed, narrowly fringed white, though the white wears quite quickly. Breast more uniformly coloured, with less black spotting than adult non-breeding; slightly paler legs than adult. RJC.

◄ **131c. Adult non-breeding**. California, mid September. Still moulting to non-breeding plumage, with missing scapulars and other upperparts giving an untidy look, and very faded outer primaries still not replaced. Compare with much neater juvenile at same date (131a). RJC.

◄ **131d. Adult non-breeding**. California, early November. Blacker than juvenile; the white streaking on scapulars and coverts is more prominent owing to greater colour contrast and larger adult feathers. Breast has dark spotting and legs are dark. RJC.

◄ **131e Adult breeding**. California, mid April. Black, with white patch on lores, extensive white spotting on sides of breast and black legs. RJC.

▼ **131f. Non-breeding Black Turnstones in flight**. California, early November. Has the same flight pattern as Ruddy Turnstone, but is black where Ruddy Turnstone is brown. RJC

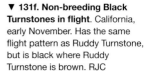

PHALAROPES

There are three species of phalarope in a single genus, all of which breed in the northern hemisphere: Wilson's Phalarope *Phalaropus tricolor*; Red-necked Phalarope *P. lobatus*; and Grey (or Red) Phalarope *P. fulicaria*. All are migratory, and all have distinctive juvenile, non-breeding and breeding plumages.

All three phalaropes are small or medium-sized and relatively long-necked, with medium-length or short straight bills and shortish legs. The Grey Phalarope's bill is flattened and broad, but the two other species have needle-fine bills. The phalaropes are well known for their fast 'spinning' as they swim in relatively shallow water; this stirs up aquatic insects and other invertebrates that can then be caught and eaten. Red-necked and Grey Phalaropes have partially webbed feet and lobed toes, Wilson's only has lobed toes and no webbing. All three species have reversed sexual roles, with the more brightly coloured breeding females being both strongly territorial and polyandrous, laying clutches for more than one male. The male is left by the female to incubate the eggs and care for the chicks and thus needs its duller and more cryptic plumage so as to be less conspicuous to potential predators.

Wilson's Phalarope breeds only in North America, on temperate inland wetlands, non-breeding on similar inland wetlands in South America. Red-necked and Grey Phalaropes are circumpolar Arctic or high Arctic breeders that spend the non-breeding season at sea; they are the only pelagic shorebirds. Both species winter in the south Pacific, with the Red-necked also in the Arabian Sea, and Grey in the Atlantic off Africa. Red-necked Phalaropes in Europe and Asia migrate overland, and are sometimes encountered in some numbers on inland lakes, but Grey Phalaropes generally occur inland only when storm-driven. Wilson's Phalarope, which is occasionally blown west across the Atlantic while migrating south, is a regular vagrant to western Europe.

▲ **Adult breeding male Red-necked Phalarope.** Insects are the main item on the menu for many shorebirds on the breeding grounds. Iceland, late June. RJC.

132. WILSON'S PHALAROPE
Phalaropus tricolor

A colourful North American phalarope
with a long, needle-fine black bill.

Identification L 23cm (9"); **WS** 40cm (16").
Medium-sized, rather small-headed, pot-bellied when
standing, with a long neck; needle-fine, straight,
black bill with greenish base, and medium-length
yellow or dark green-grey legs. **In flight** from
above fairly uniformly dark, with white rump and
uppertail-coverts, and dark tail; feet extend beyond
tail. Female slightly larger, sexes separable when
breeding. Plumage varies seasonally and with age.
Toes with narrow lateral lobes. **Feeds** by swimming in
typical phalarope manner but, more often than the
other two phalaropes, also by picking while walking
on mud or floating vegetation, or when wading.

Juvenile Crown dark brown; white supercilium.
Upperparts, wing-coverts and tertials black or
blackish, with neat buff or whitish fringes. Buffish
wash on neck, breast and flanks soon lost; rest of the
underparts white. Legs bright yellow.

First non-breeding Post-juvenile moult usually
complete by end August, so many migrants are in
first non-breeding plumage. Crown grey; white
supercilium, sometimes in front of eye only;
upperparts grey with narrow white fringes. Entire
underparts white. Legs initially bright yellow, gradually darkening. Retains dark-grey, juvenile wing-coverts, and perhaps some tertials.

Adult non-breeding Moults while staging on
southward migration, though late migrants may
delay until they reach non-breeding area. As first
non-breeding but with fresh pale-grey upperparts,
tertials and wing-coverts; legs dull yellow, becoming
darker.

Adult breeding Breeds in second calendar year.
Female whitish forehead, light-grey crown, hindneck, nape, and mantle; black running from lores,
ear-coverts and side of neck to edge of mantle,
forming indistinct mantle-V; chestnut area across
lower scapulars. Chin and throat white, foreneck
pinkish-buff; remainder of underparts white. Legs
black. *Male* duller than female; head, neck and
upperparts brown, small white area in front of eye;
otherwise as female.

Call Generally silent except when breeding, when
most frequently gives a short, nasal *ernt.*

Status, habitat and distribution Fairly common,
breeding in wetlands in interior south-west Canada,
north-west United States, and around the Great
Lakes, migrating south overland; non-breeding
birds in wetlands in southern South America.
Vagrant to Iceland and western Europe, most
frequently from late August to mid October; also,
more rarely, in Africa, Australia, New Zealand and
Antarctic islands.

Racial variation No races recognised.

▼ **132a. Juvenile**. Alberta, Canada, August. Though this individual still retains the buff wash on fore-neck and breast
typical of the juvenile, it already has a number of grey non-breeding scapulars. Alister Benn.

Similar species From the other two phalarope species (133, 134) by larger size, proportionally longer needle-fine bill, yellow or black legs, absence of dark eye-patch when non-breeding, and flight pattern with white rump but lacking any significant wing-bar. From Lesser Yellowlegs (120) by lack of spotting on upperparts in any plumage.

References Colwell & Jehl (1994), Fishpool & Demey (1991), Higgins & Davies (1996), Prince & Croxall (1996).

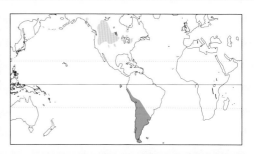

▲ **132b. First non-breeding**. Southwest England, mid September. Larger size, small head compared to bulky body, and relatively long bill help separate from the other two phalaropes. Migrant first calendar year birds (and vagrants, as this individual) are usually seen in this plumage. Has replaced virtually all upperparts, but retains juvenile coverts and tertials. RJC.

▼ **132c. Adult breeding**. California, mid April. In very fresh breeding plumage. RJC.

◄ **132d. Adult male breeding.** Kansas, late April. Has darker crown and is generally duller and less contrasting than breeding female. RJC.

▼ **132e. Adult female breeding.** California, mid April. As with all phalaropes, the female is brighter than the male. RJC.

◄ **132f. Non-breeding in flight.** California, August. Largely uniform above, but with white rump and darker tail. Peter LaTourette.

133. RED-NECKED PHALAROPE
Phalaropus lobatus

A common and colourful cicumpolar phalarope, pelagic and sociable outside the breeding season. Alternative name: Northern Phalarope.

Identification L 21cm (7"); **WS** 34cm (13.5"). Small, with a longish neck; short, straight, needle-fine, black bill, and short, yellowish or grey legs. *In flight* from above has white wing-bar formed by tips to greater coverts, extending narrowly onto tips of primary coverts; white sides to rump and uppertail-coverts; toes do not extend beyond dark tail. Below, white underwing contrasts with grey primary coverts and flight feathers, not all pale as in Grey Phalarope. Female slightly larger and brighter than male when breeding. Plumages vary seasonally and with age. Toes lobed, with basal webs. *Feeds* by picking from the water surface while swimming; on breeding grounds may also feed by walking at water's edge, usually at fresh water.

Juvenile Forehead, crown, nape and hindneck dark grey; dark patch through eye onto ear-coverts. Upperparts, wing-coverts and tertials dark grey, with neat buff fringes. Chin, throat and underparts white, foreneck sometimes pinkish. Legs initially yellow.

First non-breeding/adult non-breeding Whitish head with greyish crown and black patch through eye; upperparts, wing-coverts and tertials grey with narrow white fringes, underparts white. Legs dark grey. First non-breeding completes post-juvenile moult in non-breeding area, retaining a variable amount of worn juvenile plumage; legs become grey.

First-breeding Some remain in non-breeding area and do not acquire breeding plumage.

Adult breeding Most breed in second calendar year. *Female* dark grey head and upper breast with tiny white spot over eye, white chin and throat; red foreneck and side of neck extends around rear of ear-coverts. Upperparts, wing-coverts and tertials black or grey, with warm buff fringes and broad

▼ **133a. Juvenile**. British Colombia, Canada, August. The buff-fringed upperparts of the juvenile are quickly lost. Alistair Benn.

warm-buff mantle-V. Breast and flanks grey; remainder of underparts white. Legs dark grey. *Male* as female but duller.

Calls A sharp *hit* and *kerrek, kerrek.*

Status, habitat and distribution Common, breeding in wet Arctic marshes, with a circumpolar distribution, from Iceland, Ireland, Scotland and Norway, east throughout northern Siberia, Alaska and northern Canada to Labrador and southern Greenland. Pelagic when non-breeding. Scandinavian individuals migrate overland to the south-east, using freshwater staging posts, to Arabian Sea. Occurs in numbers offshore in eastern North America south to New York during migration, but scarcer on the coast; the majority of North American birds probably non-breeding in the Pacific off Peru. Siberian and Alaskan populations spend non-breeding period in south-west Pacific, from Philippines to New Guinea. Scattered records near-worldwide away from main non-breeding areas.

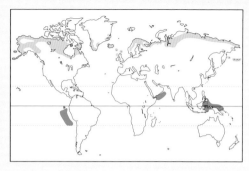

Racial variation No races recognised.

Similar species Wilson's Phalarope (132), which see. From Grey Phalarope (134) by slightly smaller size, longer and finer bill. In flight more erratic that Grey Phalarope and underwing has grey flight feathers contrasting with white coverts; underwing entirely pale in Grey Phalarope.

References Barter (2002), Thorup (2006).

▲ **133b. Juvenile/first non-breeding**. California, mid September. Has some non-breeding scapulars. RJC.

▲ **133c. Adult male breeding**. Iceland, late June. Males are much duller than females. Red-necked (and Grey) Phalaropes are much more usually seen feeding while swimming, rather than walking at the water's edge as here. RJC.

▲ **133d. Adult female breeding**. Iceland, late June. RJC.

▲ **133e. Adult female (front) and male (rear) breeding**. Iceland, late June. Note the lobed feet. RJC.

▼ **133f. Juvenile/first non-breeding in flight**. China, mid September. Note the narrow white wing-bar and white sides to tail. Yuan Xiao.

▼ **133g. Juvenile/first non-breeding wing-lifting**. China, mid September. Underwing has grey primary coverts, unlike Grey Phalarope (which has white underwing-coverts and so a uniformly paler appearance). Yuan Xiao.

134. GREY PHALAROPE
Phalaropus fulicaria

A widespread circumpolar phalarope with one of the most colourful breeding plumages of all shorebirds. Alternative name: Red Phalarope.

Identification L 21cm (8"); **WS** 41cm (16"). Small, with longish neck, pot-bellied when standing, with short, rather broad, horizontally flattened, straight bill and short, bluish or greyish legs. *In flight* from above has white wing-bar formed by tips to greater coverts, extending narrowly to tips of primary coverts. Underwing more or less uniformly pale, not contrasting with grey flight-feathers as in Red-necked Phalarope; feet do not extend beyond tail. Female slightly larger, brighter when breeding; plumages vary seasonally and with age. Toes lobed, with basal webs. *Feeds* by picking from the water surface while swimming; on breeding grounds may also feed by walking at water's edge.

Juvenile Off-white forehead, brown crown; short white supercilium. Darkish around eye, but lacks dark eye-patch, which is quickly attained at start of post-juvenile moult. Upperparts and tertials dark brown, neatly fringed rusty-buff; wing-coverts dark

▼ **134a. Juvenile**. Northern Alaska, early August. A young bird, still with down on the forehead and neck. Note rusty-buff fringes to upperparts, pinkish-grey neck and breast and all-dark bill. Emily Weiser.

brown with whitish fringes. Throat, foreneck, breast and flanks salmon-pink; remainder of underparts white. Bill usually all black; grey legs.

First non-breeding First non-breeding initially shows pinkish suffusion on front and side of neck and upper breast retained from juvenile plumage; also retains many juvenile wing-coverts and very dark tertials, whose rusty fringes quickly fade to off-white. Bare parts as juvenile but bill often has yellowish base.

First-breeding Some remain in non-breeding area and do not acquire breeding plumage.

Adult non-breeding Forehead and forecrown are whitish; blackish patch around eye and on ear-coverts; hind-crown and hindneck grey, merging into pale grey upperparts, which have narrow white fringes when fresh; wing-coverts and tertials also pale grey with narrow white fringes. Usually shows some yellow at base of black bill.

Adult breeding Most breed in second calendar year. *Female* forehead, crown and area at base of bill black; white patch around eye and on ear-coverts. Nape and hindneck black; upperparts black with rusty and buff fringes, giving streaked appearance. Whole of underparts chestnut-red. Bill yellow with black tip; legs bluish-grey. *Male* duller and less contrasting than female, head streaked brown where female is black; bill with more extensive black on tip. Both sexes acquire a variable number of dark breeding-plumage tertials, which contrast with retained, grey, non-breeding plumage feathers.

Call A whistled *wit*.

Status, habitat and distribution Has circumpolar breeding distribution along Arctic Ocean coasts, slightly more northerly than Red-necked Phalarope.

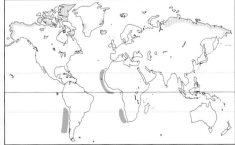

Migrates south over sea, occurring inland only when storm-driven. Thus uncommon either coastally or inland. Non-breeding occurs in flocks offshore, in the Pacific off Chile and in Atlantic off western and southern Africa.

Racial variation No races recognised.

Similar species Red-necked Phalarope (133), which see.

References Thorup (2006).

◀ **134b. First non-breeding**. Eastern England, early October. Juvenile plumage is usually lost quickly, and away from the breeding grounds first calendar year birds are most usually seen in this plumage. RJC.

◀ **134c. Adult non-breeding**. California, January. Adult non-breeding is similar to first non-breeding, except the tertials are grey, coloured as the upperparts, and there is a small yellow area at the bill base, which may also be shown by older first non-breeding individuals. Peter LaTourrette.

◀ **134d. Adult female breeding**. Iceland, late June. The bill is surprisingly broad, almost duck-like, when seen head-on (useful for separation from Red-necked Phalarope), and becomes yellow when breeding, extensively so in the brighter female. RJC.

◀ **134e. Adult female breeding in flight**. Spitsbergen, early July. Flight pattern differs from Red-necked Phalarope in the absence of white on sides of rump and uppertail, and white, not grey, underwing-coverts. Gordon Langsbury.

▼ **134f. Adult male (left) and female (right) breeding**. Iceland, late June. The female is much brighter. RJC.

▲ **Adult female *osculans* Eurasian Oystercatcher**. South Korea, early May. RJC.

REFERENCES

Alexander-Marrack, P. 1992. Nearctic vagrant waders in the Cap Lopez area, Gabon. *Malimbus* 14: 7–10.

Ali, S. & Ripley, D. S. 1983. *Handbook of the birds of India and Pakistan*, Compact Edition. Delhi, Oxford University Press.

Alström, P. 1990. Calls of American and Pacific Golden Plovers. *British Birds* 83: 70–72.

Alström, P. & Olsson, U. 1989. The identification of juvenile Red-necked and Long-toed Stints. *British Birds* 82: 360–372.

Alström, P. & Olsson, U. 1991. Spoon-billed Sandpiper. *Birding World* 4: 169–171.

Anderson, R. C. & Baldock, M. 2001. New records of birds from the Maldives, with notes on other species. *Forktail* 17: 67–73.

Andres, B. A. & Falxa, G. A. 1995. Black Oystercatcher (*Haematopus bachmani*). *In* The Birds of North America. No. 155. (A. Poole & F. Gill, eds). Academy of Natural Sciences, Philadelphia, PA, & The American Ornithologists' Union, Washington, D.C.

Anon. 1981. Bird topography. *British Birds* 74: 239–241.

Anon. 1985. Plumage, age and moult terminology. *British Birds* 78: 419–427.

Antonucci, A. & Corso, A. 2008. The Willet in Italy – a new bird for the Mediterranean basin. *Birding World* 21: 75–79.

Archer, A. L. & Iles, D. B. 1998. New bird records for Unguja (Zanzibar Island). *Bulletin of the British Ornithologists' Club* 118: 166–172.

Arkhipov, V. Y. 2005. Exceptional numbers of Oriental Plovers in southern Siberia in 2003. *British Birds* 98: 156–157.

Arvin, J. 1992. Return of the [Double-striped] Thickknee. *Birding* 24: 2–3.

Atkinson, P. W. 1996. The origins, moult, movements and changes in number of Bar-tailed Godwits *Limosa lapponica* on the Wash, England. *Bird Study* 43: 60–72.

Baicich, P. J. 2000. Attu reflections. *Birding* 32: 488–489; An Attu showcase, 545–552.

Baker, A. J., Pereira, S. L., Rogers, D. I., Elbourne, R. & Hassell, C. 2007. Mitochondrial-DNA evidence shows that the Australian Painted Snipe is a full species, *Rostratula australis*. *Emu* 107: 185–189.

Bakewell, D. N. & Kennerley, P. R. 2008. Field characteristics of an overlooked *Charadrius* plover from South-East Asia. *BirdingASIA* 9: 14–23.

Balachandran, S. 1998. Population, status, moult, measurements, and subspecies of Knot *Calidris canutus* wintering in south India. *Wader Study Group Bulletin* 86: 44–47.

Balachandran, S. & Hussain, S. A. 1998. Moult, age structure, biometrics and subspecies of Lesser Sand Plover *Charadrius mongolus* wintering along the south-east coast of India. *Stilt* 33: 3–9.

Bamford, M. J., Talbot, J., Rogers, D. I., Minton, C. D. T. & Rogers, K. G. 2005. Wader ageing series No. 1 – Red-necked Stint. *Stilt* 48: 28–33.

Barter, M. A. 2002. *Shorebirds of the Yellow Sea: Importance, threats and conservation.* International Wader Study Group International Wader Studies 12: Canberra, Australia. [CD-rom.]

Barter, M. A., Tonkinson, D., Wilson, J. R., Li, Z. W., Lu, J. Z., Shan, K. & Zhu, S. Y. 1999. The Huang He delta – An important staging site for Little Curlew *Numenius minutus* on northward migration. *Stilt* 34: 11–17.

Battley, P.F., Rogers, D.I. & Hassell, C.J. 2006. Prebreeding moult, plumage and evidence for a presupplemental moult in the Great Knot *Calidris tenuirostris*. *Ibis* 148: 27–38.

Beardslee, C.S. & Mitchell, H.D. 1965. Birds of the Niagara Frontier Region. *Bulletin of the Buffalo Society of Natural Sciences*, 22: 212–213.

Bland. B. 1998. The Wilson's Snipe on the Isles of Scilly. *Birding World* 11: 382–385.

Bland. B. 1999. The Wilson's Snipe on Scilly revisited. *Birding World* 12: 56–61.

Bent, A. C. 1929. Life histories of North American shorebirds. Order *Limicolae* (Part 2). *Smithsonian Institution Bulletin* 146: Washington.

Betts, B. J. 1973. A possible hybrid Wattled Jacana x Northern Jacana in Costa Rica. *Auk* 90, 687–689.

Bhuva, V. J. & Soni, V. C. 1998. Wintering population of four migratory species of waders in the Gulf of Kachchh and human pressures. *Wader Study Group Bulletin* 86: 48–51.

Bijlsma, R. G. & de Roder, F. 1986. Notes on Nordmann's Greenshank *Tringa guttifer* in Thailand. *Forktail* 2: 92–94.

Bijlsma, R. G. & de Roder, F.E. 1991. Foraging behaviour of Terek Sandpipers *Xenus cinereus* in Thailand. *Wader Study Group Bulletin* 61: 22–26.

Boere, G. C. & Yurlov, A. K. 1998. In search of the Slender-billed Curlew *Numenius tenuirostris*. *Wader Study Group Bulletin* 85: 35–38.

Bojko, G. W. & Nowak, E. 1996. Observations of a Slender-billed Curlew *Numenius tenuirostris* in west Siberia. *Wader Study Group Bulletin* 81: 79.

Bosanquet, S. 2000. The Hudsonian Whimbrel in Gwent. *Birding World* 13: 190–193.

Bouwman, R. G. 1987. Crab Plover in Turkey in July 1986. *Dutch Birding* 9: 65–67.

Boyd, H. & Piersma, T. 2001. Why do few Afro-Siberian Knots *Calidris canutus canutus* now visit Britain? *Bird Study* 48: 147–158.

Branson, N. J. B. A. & Minton, C. D. T. 2007. Measurements, weights and primary wing moult of Oriental Plover from North-west Australia. *Stilt* 50, 235–241.

Brazil, M. A. & Ikenaga, H. 1987. The Amami Woodcock *Scolopax mira*: its identity and identification. *Forktail* 3: 3–16.

Britton, D. 1980. Identification of Sharp-tailed Sandpipers. *British Birds* 73: 333–345.

Browning, M. R. 1991. Taxonomic comments on the Dunlin *Calidris alpina* from northern Alaska and eastern Siberia. *Bulletin of the British Ornithologists' Club* 111: 140–145.

Buchanan, J. B. 1999. Recent changes in the winter distribution and abundance of Rock Sandpipers in North America. *Western Birds* 30, 193–199.

Buckley, F. 1980. Broad-billed Sandpiper in Co. Cork. *Irish Birds* 1: 548–549.

Buden, D. W. 2006. The birds of Satawan Atoll and the Mortlock Islands, Chuuk, including the first record of Tree Martin *Hirundo nigricans* in Micronesia. *Bulletin of the British Ornithologists' Club* 126: 137–152.

Bundy, G. 1983. Call of Pintail Snipe. *British Birds* 76: 575–576.

Burns, D. W. 1993. Oriental Pratincole: new to the Western Palaearctic. *British Birds* 86: 115–120.

Byrkjedal, I. & Thompson, D. B. A. 1998. *Tundra Plovers*. T & AD Poyser, London.

Campbell, B. & Lack, E. 1985. *A Dictionary of Birds*. T & AD Poyser, Calton.

Carmona, R., Galindo, D. & Sauma, L. 2003. New and noteworthy shorebird records from south Baja California Peninsula, Mexico. *Wader Study Group Bulletin* 101/102: 62–66.

Carter, M. & Rogers, D. 1998. Little Ringed Plover *Charadrius dubius*: a Kimberley record with comments on morphological and vocal aspects of eastern populations. *Australian Bird Watcher* 17: 269–277.

Carey, G. & Olsson, U. 1995. Field identification of Common, Wilson's, Pintail and Swinhoe's Snipes. *Birding World* 8: 179–190.

Chandler, R. J. 1987a. Mystery Photographs 125: Semipalmated Plover. *British Birds* 80: 238–241.

Chandler, R. J. 1987b. Yellow orbital ring of Semipalmated and Ringed Plovers. *British Birds* 80: 241–242.

Chandler, R. J. 1990a. Mystery Photographs 151: Semipalmated Sandpiper. *British Birds* 83: 67–69.

Chandler, R. J. 1990b. Plumage variations of juvenile Ruffs and Greenshanks. *British Birds* 83: 117–121.

Chandler, R. J. 1998. Dowitcher identification and ageing: a photographic review. *British Birds* 91: 93–106.

Chandler, R. J. & Marchant, J. H. 2001. Waders with non-breeding plumage in the breeding season. *British Birds* 94: 28–34.

Chandler, R. J. & Shirihai, H. 1995. Kentish Plovers with complete breast-bands. *British Birds* 88: 136–140; see also 89: 93–94; and 94: 246–247.

Chiozzi, G. & De Marchi, G. 2003. Confirmed breeding record of Crab Plover *Dromas ardeola* in Eritrea. *Bulletin of the British Ornithologists' Club* 123: 46–47.

Chisholm, K. 2007. History of the Wood Sandpiper as a breeding bird in Britain. *British Birds* 100: 112–121.

Clark, N. A. 1987. A probable hybrid Dunlin/Sanderling. *Scottish Birds* 14: 211–213.

Cleeves, T. 2002. Slender-billed Curlew in Northumberland: new to Britain and Ireland. *British Birds* 95: 272–278, 464–465.

Collar, N. J. 1994. Bannerman and the [Canary Islands] Black Oystercatcher. *British Birds* 87: 492.

Collins, P. & Jessop, R. 2001. Arrival and departure dates and habitat of Little Curlew *Numenius minutus* at Broom, North-Western Australia. *Stilt* 39: 10–12.

Colwell, M. A. & Jehl, J. R. Jr. 1994. Wilson's Phalarope (*Phalaropus tricolor*). In The Birds of North America Online (A. Poole, ed.). Cornell Laboratory of Ornithology, Ithaca. http://bna. birds.cornell.edu/bna

Connors, P. G. 1983. Taxonomy, distribution and evolution of golden plovers (*Pluvialis dominica* and *Pluvialis fulva*). *Auk* 100: 607–620.

Connors, P. G., McCaffery, B. J. & Maron, J. L. 1993. Speciation in Golden-plovers, *Pluvialis dominica* and *P. fulva*: evidence from the breeding grounds. *Auk* 110, 9–20.

Corbat, C. A. & Bergstrom, P. W. 2000. Wilson's Plover (*Charadrius wilsonia*). In *The Birds of North America*, (A. Poole & F. Gill, eds), no. 516. The Birds of North America, Inc., Philadelphia.

Corso, A. 1996. Slender-billed Curlew in Sicily in March 1996. *Dutch Birding* 18: 302.

Corso, A. 1998. Pintail Snipe in Sicily – a new European bird. *Birding World* 11: 435–437.

Cooper, J. M. 1994. Least Sandpiper (*Calidris minutilla*). In *The Birds of North America*, (A. Poole & F. Gill, eds), no. 115. Academy of Natural Sciences, Philadelphia & The American Ornithologists' Union, Washington, D.C.

Cox, J. B. 1987. Some notes on the perplexing Cox's Sandpiper. *South Australian Ornithology* 30: 85–97.

Cox, J. B. 1988a. Some records and notes on the identification of the Oriental Plover. *South Australian Ornithologist* 30: 120–121.

Cox, J. B. 1988b. Some South Australian records of the Little Stint. *South Australian Ornithologist* 30: 113–116.

Cox, J. B. 1989. The measurements of Cooper's Sandpiper and the occurrence of a similar bird in Australia. *South Australian Ornithologist* 31: 38–43.

Cox, J. B. 1990. The enigmatic Cooper's and Cox's Sandpipers. *Dutch Birding* 12: 53–64.

Cramp, S. & Simmons, K. E. L. (Eds). 1983. *The Birds of the Western Palearctic. Volume III. Waders to Gulls.* Oxford University Press, Oxford.

Curry, R., McLaughlin, K. & Crins, B. 2003. First "Greenland" Dunlin for Ontario and Canada. *Ontario Birds* 21: 23-30.

Dare, P. J. & Mercer, A. J. 1974. The white collar of the Oystercatcher. *Bird Study* 21: 180–184.

de Bruin, A. 2006. Long-billed Curlew at Riohacha, Colombia, in March 2006. *Dutch Birding* 28: 301–302.

del Hoyo, J., Elliott, A. & Sargatal, J (eds). 1996. *Handbook of the Birds of the World, Volume 3: Hoatzin to Auks.* Lynx Edicions, Barcelona.

del Nevo, A. 1984. Non-buff Buff-breasted Sandpipers. *British Birds* 77: 358.

Dennis, J. V. 1986. European encounters of birds ringed in North America. *Dutch Birding* 8: 41–44.

Dennis, J. V. 1994. Transatlantic migration by ringed birds from North America. *Dutch Birding* 16: 235–237.

Dennis, R. H. 1983. Purple Sandpipers breeding in Scotland. *British Birds* 76: 563–566.

de Smet, G. 1997. Slender-billed Curlew at Canisvlet [The Netherlands] in September 1896. *Dutch Birding* 19: 230–232. In Dutch.

Donahue, P. K. 1996. Bill length and bill shape of Semipalmated Sandpiper. *British Birds* 89: 234–235.

Dowsett, R. J., Aspinwall, D. R. & Leonard, P. M. 1999. Further additions to the avifauna of Zambia. *Bulletin of the British Ornithologists' Club* 119: 94–103.

Driessens, G. 2005. Field characters of Afrotropical Collared Pratincole. *Dutch Birding* 27: 35–40.

Driessens, G. & Svensson, L. 2005. Identification of Collared Pratincole and Oriental Pratincole – a critical review of characters. *Dutch Birding* 27: 1–35.

Driessens, G. & Zekhuis, M. 2007. Oriental Pratincole near Doniaburen [Netherlands] in August 1997. *Dutch Birding* 29: 19–25. In Dutch.

Dring, G. 1988. Black-winged Pratincole in Avon. *Birding World* 1: 206.

Duckworth, J. W. 2006. Records of some bird species hitherto rarely found in DPR Korea. *Bulletin of the British Ornithologists' Club* 126: 252–290.

Dudley, S. P., Gee, M., Kehoe, C., Melling, T. M. and the British Ornithologists' Union Records Committee, 2006. The British List: A checklist of birds of Britain (7th Edition). *Ibis* 148: 526–563.

Dugger, B. D. & Dugger, K. M. 2002. Long-billed Curlew (*Numenius americanus*). In *The Birds of North America*, (A. Poole & F. Gill, eds), no. 628. The Birds of North America, Inc., Philadelphia.

Dukes, P. A. 1980. Semipalmated Plover: new to Britain and Ireland. *British Birds* 73: 459–464.

Dunn, J. L. 2003. Vagrant Oriental Plovers. *Birding World* 16: 305.

Dunn, J. L., Morlan, J. & Wilds, C. P. 1987. Field identification of forms of the Lesser Golden Plover. *International Field Identification – Proceedings of the 4th International Identification Meeting, Eilat,* 28–33.

Ebels, E. B. 2002. Transatlantic vagrancy of Palearctic species to the Caribbean region. *Dutch Birding* 24: 202–209.

Elner, R. W., Beninger, P. G., Jackson, D. L. & Potter, T. M. 2005. Evidence of a new feeding mode in Western Sandpiper (*Calidris mauri*) and Dunlin (*Calidris alpina*) based on bill and tongue morphology and ultrastructure. *Marine Biology* 146: 1223–1234.

Elphick, C. S. & Tibbitts, T. L. 1998. Greater Yellowlegs (*Tringa melanoleuca*). The Birds of North America Online (Poole, A. Ed.). Cornell Laboratory of Ornithology, Ithaca. http://bna.birds.cornell.edu/bna

Elphick, C. S. & Klima, J. 2002. Hudsonian Godwit (*Limosa haemastica*). In *The Birds of North America*, (A. Poole & F. Gill, eds), no. 620. The Birds of North America, Inc., Philadelphia.

Engelmoer, M. & Roselaar, C. S. 1998. *Geographical Variation in Waders.* Kluwer Academic Publishers, Dordrecht.

Ernst, S. 2004. Observations in the breeding area of Solitary Snipe *Gallinago solitaria* in the Russian Altai. *Limicola* 18: 65–100. In German.

Étchécopar, R. D. & Hüe, F. 1978. *Les Oiseaux de Chine, de Mongolie et de Corée: non passereaux.* Les Editions du Pacifique, Tahiti.

Evans, L. 1994. Killdeers in Britain and Ireland. *Birding World* 7: 57–60.

Evans, T. D. 2001. Ornithological records from Savannakhet Province, Lao PDR, January-July 1997. *Forktail* 17: 21–28.

Ferns, P. N. 1981. Identification, subspecific variation, ageing and sexing in European Dunlins. *Dutch Birding* 3: 85–98.

Ferns, P. N. & Siman, H. Y. 1994. Utility of the curved bill of the Curlew *Numenius arquata* as a foraging tool. *Bird Study* 41: 102–109.

Fishpool, L. D. C. & Demey, R. 1991. The occurrence of both species of 'Lesser Golden Plover' and of Nearctic scolopacids in Côte d'Ivoire. *Malimbus* 13: 3–10.

Gantlett, S. J. M. 1997. The Spur-winged Plover in Kent – a new British bird? *Birding World* 10, 217–219.

Gantlett, S. J. M. & Grant, P. J. 1989. The Saltholme Sandpiper. *Birding World* 2: 357–360.

Garner, M., Lewington, I. & Slack, R. 2003. Mongolian and Lesser Sand Plovers: an identification overview. *Birding World* 16: 377–385.

Gerasimov, Y. N. & Gerasimov, N. N. 2002. Whimbrel *Numenius phaeopus* on Kamchatka, Russia. *Stilt* 41: 48–54.

Gibson, D. D. & Kessel, B. 1989. Geographic variation in the Marbled Godwit and description of an Alaska subspecies. *Condor* 91: 436–443.

Gibson, D. & Kessel, B. 1992. Seventy-four new avian taxa documented in Alaska 1976–1991. *Condor* 94: 454–467.

Gill, J. A., Clark, J., Clark, N. & Sutherland, W. J. 1995. Sex differences in the migration, moult, and wintering areas of British-ringed Ruff. *Ringing and Migration* 16: 159–167.

Gill, R. E., Lanctot, R. B., Mason, J.D. & Handel, C. M. 1991. Observations on habitat use, breeding chronology and parental care in Bristle-thighed Curlews on the Seward Peninsula, Alaska. *Wader Study Group Bulletin* 61: 28–36.

Gill, R. E., McCaffery, B. J. & Tomkovich, P. S. 2002. Wandering Tattler (*Heteroscelus incanus*). In *The Birds of North America*, (A. Poole & F. Gill, eds), no. 642. The Birds of North America, Inc., Philadelphia.

Gill, R. E. & Tomkovich, P. S. 2004. Subarctic, alpine nesting by Baird's Sandpiper *Calidris bairdii*. *Wader Study Group Bulletin* 104: 39–50.

Gill, R. E., Tomkovich, P. S. & McCaffery, B. J. 2002. Rock Sandpiper (*Calidris ptilocnemis*). In *The Birds of North America*, (A. Poole & F. Gill, eds), no. 686. The Birds of North America, Inc., Philadelphia.

Golley, M. 1990. The Cley *Calidris:* another apparent hybrid. *Birding World* 3: 237–238.

Golley, M. & Stoddart, A. 1991. Identification of American and Pacific Golden Plovers. *Birding World* 4: 195–204.

Goodwin, C. E. 1981. The Autumn Migration, August 1–November 30, 1980: Ontario Regional Report. *American Birds* 35: 176–179.

Gosbell, K. & Minton, C. 2001. The biometrics and moult of Sanderling *Calidris alba* in Australia. *Stilt* 40: 7–22.

Grant, P. J. 1981. Identification of Semipalmated Sandpiper. *British Birds* 74: 505–509.

Granit, B., Lindroos, R. & Perlman, Y. 1999. Pintail Snipe in Israel in November 1998. *Dutch Birding* 21: 329–333.

Gratto-Trevor, C. L. 1992. Semipalmated Sandpiper (*Calidris pusilla*). The Birds of North America Online (A. Poole, ed.). Cornell Laboratory of Ornithology, Ithaca. http://bna.birds.cornell.edu/bna

Gratto-Trevor, C. L. 2000. Marbled Godwit (*Limosa fedoa*). In *The Birds of North America*, (A. Poole & F. Gill, eds), no. 492. The Birds of North America, Inc., Philadelphia.

Green, I & Overfield, J. 1995. Birding the Polar Ural – Western Palearctic Siberia. *Birding World* 8: 191–197.

Green, R. E. & Bowden, C. G. R. 1986. Field characters for ageing and sexing Stone-curlews. *British Birds* 79: 419–422.

Greenwood, J. G. 1984. Migration of Dunlin *Calidris alpina*: a worldwide overview. *Ringing and Migration* 5: 35–39.

Greenwood, J. G. 1986. Geographical variation and taxonomy of the Dunlin *Calidris alpina* (L.). *Bulletin of the British Ornithologists' Club* 106: 43–56.

Gretton, A. 1991. The ecology and conservation of the Slender-billed Curlew *(Numenius tenuirostris)*. ICBP Monograph No. 6: Cambridge.

Gretton, A., Yurlov, A. K. & Boere, G. C. 2002. Where does the Slender-billed Curlew nest, and what future does it have? *British Birds* 95: 334–344.

Groen, N., Mes, R., Fefelov, I. & Tupitsyn, I. 2006. Eastern Black-tailed Godwit *Limosa limosa melanuroides* in the Selenga Delta, Lake Baikal, Siberia. *Wader Study Group Bulletin* 110: 48–53.

Grieve, A. 1987. Hudsonian Godwit: new to the Western Palearctic. *British Birds* 80: 466–473.

Grønningsaeter, E. 2005. White-rumped Sandpipers in Arctic Norway. *Birding World* 18: 349–350.

Grønningsaeter, E. 2007. Displaying Pectoral Sandpipers in Spitzbergen. *Birding World* 20: 334–335.

Gunnarsson, T., Gill, J. A., Goodacre, S. L., Gélinaud, G., Atkinson, P. W., Hewitt, G. M., Potts, P. W. & Sutherland, W. J. 2006. Sexing of Black-tailed Godwits *Limosa limosa islandica*: a comparison of behavioural, molecular, biometric and field-based techniques. *Bird Study* 53: 193–198.

Gutiérrez, R. 2001. The first breeding record of Cream-coloured Courser in Europe. *Birding World* 14: 323–325.

Guzzetti, B.M., Talbot, S.L., Tessler, D.F., Gill, V.A. & Murphy, E.C. 2008. Secrets in the eyes of Black Oystercatchers: a new sexing technique. *Journal of Field Ornithology* 79: 215–223.

Haig, S. M. & Elliott-Smith, E. 2004. Piping Plover. The Birds of North America Online (A. Poole, ed.). Cornell Laboratory of Ornithology, Ithaca. http://bna.birds.cornell.edu/bna

Hale, W.G. 1980. *Waders*. London, Collins.

Handel, C. M. & Gill, R. E. 1992. Breeding distribution of the Black Turnstone. *Wilson Bulletin* 104: 122–135.

Handel, C. M. & Gill, R. E. 2001. Black Turnstone (*Arenaria melanocephala*). In *The Birds of North America*, (A. Poole & F. Gill, eds), no. 585. The Birds of North America, Inc., Philadelphia.

Harrington, B. A. 2001. Red Knot (*Calidris canutus*). The Birds of North America Online (A. Poole, ed.). Cornell Laboratory of Ornithology, Ithaca. http://bna.birds.cornell.edu/bna

Harrington, B. & Flowers, C. 1996. *The Flight of the Red Knot*. W. W. Norton & Co., New York.

Harrington, B. A., Picone, C., Resende, S. L. & Leeuwenberg, F. 1993. Hudsonian Godwit *Limosa haemastica* migration in southern Argentina. *Wader Study Group Bulletin* 67: 41–44.

▲ **American Avocet**. Florida, late March. RJC.

Harrison, J. M. & Harrison, J. G. 1965. The juvenile plumage of the Icelandic Black-tailed Godwit and further occurrences of this race in England. *British Birds* 58: 10–14.

Harrison, J. M. & Harrison, J. G. 1971. The occurrence of *Calidris alpina sakhalina* (Vieillot) in Britain. *Bull. Brit. Orn. Club* 91: 39–40.

Harrison, J. M. & Harrison, J. G. 1972. Further notes on American and Schioler's Dunlin from Britain. *Bulletin of the British Ornithologists' Club* 92: 38–40.

Hayman, P., Marchant, J.H. & Prater, A.J. 1986. *Shorebirds: An Identification Guide to the Waders of the World*. London and Sydney.

Hazevoet, C. J. 1988. Cape Verde Islands [Cream-coloured] Courser. *Dutch Birding* 10: 30–32.

Heard, C. 1998a. That Curlew: an alternative identification? *Birdwatch*: October 12–13.

Heard, C. 1998b. Identification of Slender-billed Curlew. *Birding World* 11: 275–276.

Helbig, A. J. 1985. Occurrence of White-tailed Plover in Europe. *Dutch Birding* 7: 79–84.

Hendenström, A. 2004. Migration and morphometrics of Temminck's Stint *Calidris temminckii* at Ottenby, southern Sweden. *Ringing and Migration* 22: 51–58.

Heppleston, P. B. 1982. Oystercatcher bills. *Ringing and Migration* 4: 8.

Herbert, C. 2006. Grey-headed Lapwing – First Australian occurrence. *Tattler* 3 (new series), 7.

Higgins, P. J. & Davies, S. J. J. F. (eds), 1996. *Handbook of Australian, New Zealand and Antarctic Birds, Volume 3: Snipe to Pigeons*. Oxford University Press, Melbourne.

Hilty, S. L. & Brown, W. L. 1986. *A Guide to the Birds of Colombia*. Princeton University Press, Princeton.

Hirshfeld, E. 1994. Migration patterns of some regularly occurring waders in Bahrain 1990–1992. *Wader Study Group Bulletin* 73: 36–49.

Hirshfeld, E., Roselaar, C. S, & Shirihai, H. 2000. Identification, taxonomy and distribution of Greater and Lesser Sand Plovers. *British Birds* 93: 162–189.

Hirshfeld, E. & Stawarczyk, T. 1993. Feeding jizz identification of sand plovers. *Birding World* 6: 454–455.

Hirst, P. & Proctor, B. 1995. Identification of Wandering and Grey-tailed Tattlers. *Birding World* 8: 91–97.

Hjort, C. 2005. Siberian Pectoral Sandpipers seen migrating towards the southwest. *British Birds* 98: 261.

Hobson, D., Lassey, A. & Wallace, I. 1984. Dark rumped Whimbrels in Yorkshire. *The Yorkshire Birder* 9/10: 3.

Hockey, P. 1993. Identification forum: jizz identification of sand plovers. *Birding World* 6: 369–372, 454–455.

Hockey, P. A. R. & Aspinall, S. J. 1997. Why do Crab Plovers *Dromas ardeola* breed in colonies? *Wader Study Group Bulletin* 82: 38–42.

Hockey, P. & Douie, C. 1995. *Waders of Southern Africa*. Struik, Cape Town.

Hofland, R. 2007. Sociable Lapwings in Syria and Turkey. *Dutch Birding* 29: 137–138. In Dutch with English summary.

Holden, P. 1985. Measurements of wingspan. *British Birds* 78: 403–404.

Holland, P. K. & Yalden, D. W. 1994. An estimate of lifetime reproductive success for the Common Sandpiper *Actitis hypoleucos*. *Bird Study* 41: 110–119.

Holling, M. and the Rare Breeding Birds Panel. 2007. Rare breeding birds in the United Kingdom in 2003 and 2004. *British Birds* 100: 321–367.

Hollyer, J. N. 1984. Camouflage postures of Jack Snipe at day roost. *British Birds* 77: 319–340.

Holmes, R. T. & Pitelka, F. A. 1998. Pectoral Sandpiper (*Calidris melanotos*). The Birds of North America Online (A. Poole, ed.). Cornell Laboratory of Ornithology, Ithaca. http://bna.birds.cornell.edu/bna

Holt, P. 1999. Long-billed Dowitcher *Limnodromus scolopaceus* at Bharatpur, Rajastan, India: a new species for the Indian subcontinent. *Forktail* 15: 95–96.

Hoodless, A. N. 1995. Eurasian Woodcock. *British Birds* 88: 578–591.

Hoodless, A. N. & Coulson, J. C. 1998. Breeding biology of the Woodcock *Scolopax rusticola* in Britain. *Bird Study* 45: 195–204.

Houston, C. S. & Bowen, D. E. Jr. 2001. Upland Sandpiper (*Bartramia longicauda*). In *The Birds of North America*, (A. Poole & F. Gill, eds), no. 580. The Birds of North America, Inc., Philadelphia.

Howell, S. N. G. & Pyle, P. 2002. Ageing and molt in nonbreeding Black-bellied Plovers. *Western Birds* 33: 268–270.

Howell, S. N. G & Webb, S. 1995. Noteworthy bird observations from Chile. *Bulletin of the British Ornithologists' Club* 115: 57–66.

Howes, J. & Lambert, F. 1987. Some notes on the status, field identification and foraging characteristics of Nordmann's Greenshank *Tringa guttifer*. *Wader Study Group Bulletin* 49: 14–17.

Humphrey, P. S. & Parkes, K. C. 1959. An approach to the study of moults and plumages. *Auk* 76: 1–31.

Iliff, M. J., Sullivan, B. L., Leukering, T. & Gibbons, B. P. 2004. The first record of the Little Stint for Mexico. *Western Birds* 35: 77–87.

Jackson, B. J. S. & Jackson, J. A. 2000. Killdeer (*Charadrius vociferus*). In *The Birds of North America*, (A. Poole & F. Gill, eds), no. 517. The Birds of North America, Inc., Philadelphia.

Jaramillo, A. 2004. Identification of adult Pacific and American Golden Plovers in their southbound migration. *Western Birds* 35: 120–123.

Jaramillo, A. & Henshaw, B. 1995. Identification of breeding plumaged Long- and Short-billed Dowitchers. *Birding World* 8: 221–228.

Jaramillo, A., Pittaway, R. & Burke, P. 1991. The identification and migration of breeding plum-aged dowitchers in southern Ontario. *Birders Journal* 1: 8–25.

Jehl, J. R. Jr., Klima, J. & Harris, R. E,. 2001. Short-billed Dowitcher (*Limnodromus griseus*). In *The Birds of North America*, (A. Poole & F. Gill, eds), no. 564. The Birds of North America, Inc., Philadelphia.

Jenni, D. A. & Mace, T. R. 1999. Northern Jacana (*Jacana spinosa*). *I*In *The Birds of North America*, (A. Poole & F. Gill, eds), no. 467. The Birds of North America, Inc., Philadelphia.

Johnson, O. W. & Connors, P. G. 1996. American Golden-Plover (*Pluvialis dominica*), Pacific Golden-Plover (*Pluvialis fulva*). In *The Birds of North America*, (A. Poole & F. Gill, eds), nos. 201–202. The Birds of North America, Inc., Philadelphia.

Johnson, O. W. & Johnson, P. M. 1983. Plumage-molt-age relationships in "over-summering" and migratory Lesser Golden-Plovers. *Condor* 85: 406–419.

Johnson, O. W. & Johnson, P. M. 2004. Biometrics and field identification of Pacific and American Golden Plovers. *British Birds* 97: 434–443, 553.

Jonsson, L. 1996. Mystery stint at Groot Keeten: first known hybrid between Little and Temminck's Stint? *Dutch Birding* 18: 24–28; also *Dutch Birding* 17: 214–215 & 224.

Jonsson, L. & Grant, P. J. 1984. Identification of stints and peeps. *British Birds* 77: 293–315.

Jukema, J. & Piersma, T. 1987. Special moult of breast and belly feathers during breeding in Golden Plovers *Pluvialis apricaria*. *Ornis Scandinavica* 18: 157–162.

Jukema, J. & Piersma, T. 2000. Contour feather moult of Ruffs *Philomachus pugnax* during north-wards migration, with notes on homology of nuptial plumages in scolopacid waders. *Ibis* 142: 289–296.

Jukema, J. & Piersma, T. 2006. Permanent female mimics in a lekking shorebird. *Biology Letters* 2: 161–164.

Kalchreuter, H. 2000. On the evening flights of fledged Woodcock (*Scolopax rusticola*). *Fifth European Woodcock and Snipe Workshop – Proceedings of an International Symposium of the Wetlands International Woodcock and Snipe Specialist Group 3–5 May 1998*. Wetlands International Global Series No. 4: International Wader Studies 11: Wageningen, The Netherlands, 25.

Karlionova, N., Pinchuk, P., Meissner, W. & Verkuil, Y. 2007. Biometrics of Ruffs *Philomachus pugnax* migrating in spring through southern Belarus with special emphasis on the occurrence of 'faed-ers'. *Ringing and Migration* 23: 134–140.

Kaufman, K. 1990. *A field guide to Advanced Birding*. Boston, Houghton Mifflin.

Kennerley, P. R. & Bakewell, D. N. 1991. Identification and status of Nordmann's Greenshank. *Dutch Birding* 13: 1–8.

Kennerley, P. R., Bakewell, D. N. & Round, P. D. 2008. Rediscovery of a long-lost *Charadrius* plover from South-East Asia. *Forktail* 24: 63–79.

Keppie, D. M. & Whiting, R. M., Jr. 1994. American Woodcock (*Scolopax minor*). The Birds of North America Online (A. Poole, ed.). Cornell Laboratory of Ornithology, Ithaca. http://bna. birds.cornell.edu/bna

Kirwan, G. M. 2007. Two specimens of Red-necked Stint from Iran collected in July 1941. *Dutch Birding* 29: 92–93.

Kiss, J. B. & Szabó, L. 2000. First breeding record of White-tailed Lapwing in Romania (and Europe outside Russia). *British Birds* 93: 400–401.

Klima, J. & Jehl, J. R. (Jr). 1998. Stilt Sandpiper (*Calidris himantopus*). The Birds of North America Online (A. Poole, ed.). Cornell Laboratory of Ornithology, Ithaca. http://bna.birds.cornell.edu/bna

Knopf, F. L. & Wunder, M. 2006. Mountain Plover (*Charadrius montanus*). The Birds of North America Online (A. Poole, ed.). Cornell Laboratory of Ornithology, Ithaca. http://bna. birds.cornell.edu/bna

Koch, M. & Hazevoet, C. J. 2000. Breeding of Cream-coloured Courser in Cape Verde Islands. *Dutch Birding* 22: 18–21.

Konrad, V. 2005. First record of Long-billed Plover *Charadrius placidus* in Singapore. *Forktail* 21: 181.

Konyukhov, N. B. & McCafferty, B. J. 1993. Second record of a Bristled-thighed Curlew from Asia and first record for the former Soviet Union. *Wader Study Group Bulletin* 70, 22–23.

Lambeck, R. H. D., Bianki, V. V., Schekkerman, E. J. C., Wessel, E. G. J., Herman, P. M. J. & Koryakin, A. S. 1995. Biometrics and migration of Oystercatchers (*Haematopus ostralegus*) from the White Sea region (NW Russia). *Ringing and Migration* 16: 140–158.

Lanctot, R. B. & Laredo, C. D.. 1994. Buff-breasted Sandpiper (*Tryngites subruficollis*). The Birds of North America Online (A. Poole, ed.). Cornell Laboratory of Ornithology, Ithaca. http://bna. birds.cornell.edu/bna

Lane, B. A. & Rogers, D. I. 2000. The Australian Painted Snipe *Rostratula (benghalensis) australis*: an endangered species? *Stilt* 36: 26–34.

Laux, E. V. 1994. Mystery sandpiper. *Birding* 26: 66–68. See also *Birding* 27: 306–309, 32: 535–539.

Lawrence, D. 1993. Spotted Sandpiper displaying to and mating with Common Sandpiper. *British Birds* 86: 628.

Leader, P. J. & Carey, G. J. 2003. Identification of Pintail Snipe and Swinhoe's Snipe. *British Birds* 96: 178–198.

▲ **Black-tailed Godwit of race *islandica*.** Iceland, late June. RJC.

Lees, A. C. & Gilroy, J. J. 2004. Pectoral Sandpipers in Europe: vagrancy patterns and the influx of 2003. *British Birds* 97: 638–646.

Legrand, V. 2005. Identification of a Wilson's Snipe on Ouessant, Finistère. *Birding World* 18: 482–484.

Lehman, P. 2006. Autumn plumages from the Bering Sea region, Alaska. *Birding* 38 (5): 26–33.

Lethaby, N. 1995. Undertail-coverts of Solitary Sandpiper. *Birding World* 8: 426–427.

Lethaby, N. & Gilligan, J. 1992. Great Knot in Oregon. *American Birds* 46: 46–47,

Lottin, H. 1962. A study of boreal shorebirds summering on Apalachee Bay, Florida. *Bird Banding* 33: 21–41.

Ludvigsen, S. 1991. Kittlitz's Plovers in Europe. *Birding World* 4: 258.

Mactavish, B. 1996. First ABA-Area Records: Common Redshank in Newfoundland. *Birding* 28: 302–307.

MacWhirter, B., Austin-Smith, P. (Jr) & Kroodsma, D. 2002. Sanderling (*Calidris alba*). In *The Birds of North America*, (A. Poole & F. Gill, eds), no. 653. The Birds of North America, Inc., Philadelphia.

Madge, S. C. 1989. Swinhoe's Snipe *Gallinago megala*: a new species for Nepal. *Forktail* 4: 121–123.

Marks, J. S. 1993. Molt of Bristle-thighed Curlews in the northwestern Hawaiian Islands. *Auk* 110: 573–587.

Marks, J. S., Tibbitts, T. L., Gill, R. E. & McCaffery, B. J. 2002. Bristle-thighed Curlew (*Numenius tahitiensis*). In *The Birds of North America*, (A. Poole & F. Gill, eds), no. 705. The Birds of North America, Inc., Philadelphia.

McCaffery, B. & Gill, R. 2001. Bar-tailed Godwit (*Limosa lapponica*). In *The Birds of North America*, (A. Poole & F. Gill, eds), no. 581. The Birds of North America, Inc., Philadelphia.

McCaffery, B. J. & Harwood, C. M. 2000. Status of Hudsonian Godwits on the Yukon-Kuskokwim Delta. *Western Birds* 31: 165–177.

McWhirter, D. W. 1987. Feeding methods and other notes on the Spoon-billed Sandpiper *Eurynorhynchus pygmaeus* in Okinawa. *Forktail* 3: 60–61.

Marchant, S. & Higgins, P. J. (Eds), 1993. *Handbook of Australian, New Zealand and Antarctic Birds. Volume 2: Raptors to Lapwings*. Oxford University Press, Melbourne.

McGeehan, A. & Meininger, P. L. 2000. American Golden Plover in Tunisia in December 1998. *Dutch Birding* 22: 25–27.

McGowan, R. Y. & Weir, D. N. 2002. Racial identification of Fair Isle Solitary Sandpiper. *British Birds* 95: 313–314.

McLaughlin, K. A. & Wormington, A. 2000. An apparent Dunlin x White-rumped Sandpiper hybrid. *Ontario Birds* 18: 8–12.

Meadows, B. S. 2003. Additional distributional records from the central Hejaz, western Arabia – an addendum to Baldwin & Meadows (1988). *Bulletin of the British Ornithologists' Club* 123: 154–177.

Meeth, P. & Meeth, K. 1989. Long-billed Plover on Bali in November 1973. *Dutch Birding* 11: 114–115.

Meininger, P. L. 1993. Breeding Black-winged Stilts in the Netherlands in 1989–93: including one paired with Black-necked Stilt. *Dutch Birding* 15: 193–197.

Meissner, W. & Scebba, S. 2005. Intermediate stages of age characters create dilemmas in ageing female Ruffs *Philomachus pugnax* in spring. *Wader Study Bulletin* 106: 30–33.

Mengel, R. M. 1948. *Limnodromus semipalmatus* in Arabia. *Auk* 65: 146.

Millington, R. 1994. A mystery *Calidris* at Cley. *Birding World* 7: 61–63. (Reprinted in *Birding* 8: 310–311).

Minton, C. D. T. 2000. Curlew Sandpipers *Calidris ferruginea* swimming while feeding. *Stilt* 37: 22–23.

Miskelly, C. M., Sagar, P. M., Tennyson, A. J. D. & Schofield, R. P. 2001. Birds of the Snares Islands, New Zealand. *Notornis* 48: 1–40.

Mlodinow, S. G, 1999. Spotted Redshank and Common Greenshank in North America. *North American Birds* 53: 124–130.

Mlodinow, S. 2001. Possible anywhere: Sharp-tailed Sandpiper. *Birding* 33: 330–341, 506–510.

Mlodinow, S. G, Feldstein, S. & Tweit, B. 1999. The Bristle-thighed Curlew landfall of 1998: climatic factors and notes on identification. *Western Birds* 30, 133–155.

Moon, S. J. 1983. Little Whimbrel: new to Britain and Ireland. *British Birds* 76: 438–445.

Morlan, J., Dakin, R. E. & Rosso, J. 2004. Apparent hybrids between the American Avocet and Black-necked Stilt in California. *Western Birds* 35: 57–59.

Morony, J. J., Bock, W. J. & Farrand, J. 1975. *Reference List of the Birds of the World*, American Museum of Natural History, New York.

Morozov, V. 1999. Recent status of southern subspecies of the Whimbrel *Numenius phaeopus alboaxillaris* in Russia and Kazakhstan. *Wader Study Group Bulletin* 90, 11.

Morozov, V. V. 2000. Current status of the southern subspecies of the Whimbrel *Numenius phaeopus alboaxillaris* (Lowe 1921) in Russia and Kazakstan. *Wader Study Group Bulletin* 92: 30–37.

Morozov, V. V. 2004. Displaying Swinhoe's Snipe in eastern European Russia: a new species for Europe. *British Birds* 97: 134–138.

Moskoff, W. 1995. Solitary Sandpiper (*Tringa solitaria*). The Birds of North America Online (A. Poole, ed.). Cornell Laboratory of Ornithology, Ithaca. http://bna.birds.cornell.edu/bna

Moskoff, W. & Montgomerie, R. 2002. Baird's Sandpiper (*Calidris bairdii*). The Birds of North America Online (A. Poole, ed.). Cornell Laboratory of Ornithology, Ithaca. http://bna.birds.cornell.edu/bna

Mudge, G. P. & Dennis, R. H. 1995. History of breeding by Temminck's Stints in Britain. *British Birds* 88: 573–577.

Mueller, H. 2005. Wilson's Snipe (*Gallinago delicata*). The Birds of North America Online (A. Poole, ed.). Cornell Laboratory of Ornithology, Ithaca. http://bna.birds.cornell.edu/bna

Mulhauser, B. & Diaz, J-C. 2003. Bill malformation and de-pigmentation amongst Woodcock *Scolopax rusticola* in French-speaking Switzerland. *Nos Oiseaux* 50: 81–87. In French.

Mullarney, K. 1991. Identification of Semipalmated Plover: a new feature. *Birding World* 4: 254–258.

Nethersole-Thompson, D. & Nethersole-Thompson, M. 1979. *Greenshanks*. T & AD Poyser, Berkhamsted.

Nettleship, D. N. 2000. Ruddy Turnstone (*Arenaria interpres*). The Birds of North America Online (A. Poole, ed.). Cornell Laboratory of Ornithology, Ithaca. http://bna.birds.cornell.edu/bna

Nieboer, E., Cronau, J., Letschert, J. & van der Have, T. 1985. Axillary feathers colour patterns as indicators of the breeding origin of Bar-tailed Godwits. *Wader Study Group Bulletin* 45: 34; reprinted *Wader Study Group Bulletin* 69: 73.

Nielsen, B. P. & Colston, P. R. 1984. Breeding plumage of female Caspian Plover. *British Birds* 77: 356–357.

Nol, E. & Blanken, M. S. 1999. Semipalmated Plover (*Charadrius semipalmatus*). In *The Birds of North America*, (A. Poole & F. Gill, eds), no. 444. The Birds of North America, Inc., Philadelphia.

Nol, E. & Humphrey, R. C. 1994. American Oystercatcher (*Haematopus palliatus*). In *The Birds of North America*, (A. Poole & F. Gill, eds), no. 82. The Birds of North America, Inc., Philadelphia.

O'Brien, M. 2006. Subspecific identification of the Willet *Catoptrophorus semipalmatus*. *Birding* 38: 40–47.

O'Brien, M., Crossley, R. & Karlson, K. 2006. *The Shorebird Guide*. Houghton Mifflin, Boston.

Odin, N. 1985. Aberrant Curlew in Gwent and South Glamorgan. *British Birds* 78: 44–45.

Ogle, D. 1992. Oriental Plover *Charadrius veredus*: a new species for Thailand. *Forktail* 7: 156–157.

Olah, J,. Jr. & Tar, J. 2003. The first record of Semipalmated Sandpiper (*Calidris pusilla*) in Hungary. *Aquila* 109–110: 87–90.

Olsson, U. 1987. Separation of Pintail Snipe from Snipe. *British Birds* 80: 248–249.

Oring, L.W., Gray, E.M. & Reed, J.M. 1997. Spotted Sandpiper (*Actitis macularius*). The Birds of North America Online (A. Poole, ed.). Cornell Laboratory of Ornithology, Ithaca. http://bna.birds.cornell.edu/bna

Ozerskaya, T. & Zabelin, V. 2006. Breeding of the Oriental Plover *Charadrius veredus* in southern Tuva, Russia. *Wader Study Group Bulletin* 110: 36–42.

Page, G. W., Warriner, J. S., Warriner, J. C. & Paton, P. W. C. 1995. Snowy Plover (*Charadrius alexandrinus*). In *The Birds of North America*, (A. Poole & F. Gill, eds), no. 154. The Birds of North America, Inc., Philadelphia.

Paige, J. P. 1965. Field identification and winter range of the Asiatic Dowitcher *Limnodromus semipalmatus*. *Ibis* 107: 95–97.

Park, J-Y & Kim, S-W. 1994. First records of Asiatic Dowitcher, Greater Yellowlegs and Gull-billed Tern in Korea. *The Korean Journal of Ornithology* 1: 127–128. In Korean.

Parmelee, D. F. 1992. White-rumped Sandpiper (*Calidris fuscicollis*). The Birds of North America Online (A. Poole, ed.). Cornell Laboratory of Ornithology, Ithaca. http://bna.birds.cornell.edu/bna

Patten, M. A., McCaskie, G. & Morlan, J. 1999. First record of the American Woodcock for California, with a summary of its status in western North America. *Western Birds* 30: 156–166.

Paulson, D. R. 1986. Identification of juvenile tattlers, and a Grey-tailed Tattler record from Washington. *Western Birds* 17: 33–36.

Paulson, D. R. 1993. *Shorebirds of the Pacific Northwest*. University of Washington Press, Seattle and London.

Paulson, D. R. 1995. Black-bellied Plover (*Pluvialis squatarola*). In *The Birds of North America*, (A. Poole & F. Gill, eds), no. 186. The Birds of North America, Inc., Philadelphia.

Payne, L. X. & Pierce, E. P. 2002. Purple Sandpiper (*Calidris maritima*). In *The Birds of North America*, (A. Poole & F. Gill, eds), no. 706. The Birds of North America, Inc., Philadelphia.

Pearson, D. J. 1983. A Blacksmith/Spur-winged Plover *Vanellus armatus* x *V. spinosus* hybrid? *Scopus* 7: 93–94.

Pearson, D. J. & Ash, J. S. 1996. The taxonomic position of the Somali courser *Cursorius (cursor) somalensis*. *Bulletin of the British Ornithologists' Club* 116: 225–229.

Pereira, S. L. & Baker, A. J. 2005. Multiple gene evidence for parallel evolution and retention of ancestral morphological states in the Shanks (*Charadriiformes: Scolopacidae*). *Condor* 107: 514–526.

Pettet, A. 1980. Marsh Sandpipers with yellowish or orange legs. *British Birds* 73: 184–185.

Pettet, A. 1982. Feeding behaviour of White-tailed Plover. *British Birds* 75: 182.

Piersma, T. & Jukema, J. 1993. Red breasts as honest signals of migratory quality in a long-distance migrant, the Bar-tailed Godwit. *Condor* 95: 163–177.

Piersma, T., van Aelst, R., Kurk, K., Berkhouldt, H. & Maas, L. R. M. 1998. A new pressure sensory mechanism for prey detection in birds: the use of principles of seabed dynamics? *Proceedings of the Royal Society, London B*, 265: 1377–1383

Phillips, A. R. 1975. Semipalmated Sandpiper: identification, migrations, summer and winter ranges. *American Birds* 29: 799–807.

Piston, A. W. & Heinl, S. C. 2001. First record of the European Golden-Plover (*Pluvialis apricaria*) from the Pacific. *Western Birds* 32: 179–181.

Porter, R. 2004. An encounter with a Slender-billed Curlew in Yemen. *Birding World* 17: 437–439.

Prater, A. J., Marchant, J. & Vuorinen, J. 1977. *Guide to the identification and ageing of Holarctic waders.* BTO Guide 17: Tring.

Prince, P. A. & Croxall, J. P. 1996. The birds of South Georgia. *Bulletin of the British Ornithologists' Club* 116: 81–104.

Principe, W. L. 1977. A hybrid American Avocet x Black-necked Stilt. *Condor* 79: 128–129.

Putnam, C. 2005. Fall molts of adult dowitchers. *Birding* 37: 380–390; also *Birding* 38: (2), 10–11.

Pyle, P. 1999. Molts by age in the Bristle-thighed Curlew and other shorebirds. *Western Birds* 30: 181–183.

Pym, A. 1982. Identification of Lesser Golden Plover and status in Britain and Ireland. *British Birds* 75: 112–124.

Rasmussen, P. C. & Anderton, J. C. 2005. *Birds of South Asia: The Ripley Guide* Vols. 1 and 2. Smithsonian Institution and Lynx Edicions, Washington D.C. and Barcelona.

Robbins, M. B. & Alderfer, J. K. 2000. "Mystery Sandpiper" revisited: Buff-breasted or Hybrid? *Birding* 32: 535–539.

Robinson, J. A., Oring, L. W., Skorupa, J. P., & Boettcher, R. 1997. American Avocet (*Recurvirostra americana*). In *The Birds of North America*, (A. Poole & F. Gill, eds), no. 725. The Birds of North America, Inc., Philadelphia.

Robinson, J. A., Reed, J. M., Skorupa, J. P. & Oring, L. W. 1999. Black-necked Stilt (*Himantopus mexicanus*). In *The Birds of North America*, (A. Poole & F. Gill, eds), no. 449. The Birds of North America, Inc., Philadelphia.

Rogers, D. I. 2006. Hidden costs: challenges faced by migratory shorebirds living on intertidal flats. Unpublished Ph. D thesis, Charles Sturt University, Australia.

Rogers, D., Styles, G., Moores, N. & Ju Yung-Ki 2008. It's a sensitive Dutch hoe: A new hypothesis on bill-function of the Spoon-billed Sandpiper. In prep.

Rohde, J. 1998. *Kampfläufer* [Ruff *Philomachus pugnax*] *in Schleswig-Holstein: Ein Tagebuch aus den Jahren 1986-97.* Privately published: Bad Bramstedt. In German.

Roos, M. T. & Schrijvershof, P. G. 1993. Red-wattled Lapwing in Israel in winter of 1991/92. *Dutch Birding* 15: 13–15.

Rosair, D. & Gilbert, D. 1988. Oriental Pratincole in Kent – The second British record. *Birding World* 1: 359–360.

Roselaar, C. S. & Gerritsen, G. J. 1991. Recognition of Icelandic Black-tailed Godwit and its occurrence in the Netherlands. *Dutch Birding* 13: 128–135.

Rosenband, G. 1982. Eye-ring of Sharp-tailed Sandpiper. *British Birds* 75: 128.

Rubega, M. A. & Obst, B. S. 1993. Surface-tension feeding in phalaropes: discovery of a novel feeding mechanism. *The Auk*, 110, 169–178.

Rusticali, R., Scarton, F. & Valle, R. 2002. The taxonomic status of the Oystercatcher *Haematopus ostralegus* breeding in Italy. *Bird Study* 49: 310–313.

Ruttledge, R. F. 1978. Greenshank nesting in Ireland. *Irish Birds* 1: 236–238.

Salvig, J. C. 1995. Migratory movements and mortality of Danish Avocets. *Ringing and Migration* 16: 79–90.

Sangster, G. 1996. Hybrid origin of Cox's Sandpiper confirmed by molecular analysis. *Dutch Birding* 18: 255–256.

Savage, S. E. & Johnson, O. W. 2005. Breeding range extensions for the Pacific Golden-Plover and Black-bellied Plover on the Alaska Peninsula. *Wader Study Group Bulletin.* 108: 63–65.

Schekkerman, H. & van Wetten, J. C. J. 1988. Palaearctic Cream-coloured Coursers in Kenya in January-February 1987. *Dutch Birding* 10(1): 26.

Schepers, F., Pineau, O., Geene, R. & Abdelsamad, A.I. 1991. Pectoral Sandpiper in Egypt in May 1990. *Dutch Birding* 13: 95.

Schenk, C. & Ebels, E. B. 2004. Birds of Chukotka and Yakutia. *Dutch Birding* 26: 241–257.

Schols, R. 2005. Birds from northeast China. *Birding World* 18: 297–303.

Schrijvershof, P. J. & Steedman, C. 1986. Terek Sandpiper on Lanzarote in August 1985. *Dutch Birding* 8(1), 28.

Senner, S. E. & McCaffery, B. J. 1997. Surfbird (*Aphriza virgata*). In *The Birds of North America*, (A. Poole & F. Gill, eds), no. 266. The Birds of North America, Inc., Philadelphia.

Serra, L., Baccetti, N. & Zenatello, M. 1995. Slender-billed Curlews wintering in Italy in1995. *Birding World* 8: 295–299; [also photo, vol. 8: 90].

Shama, A. 2003. First records of Spoon-billed Sandpiper *Calidris pygmeus* in the Indian Sunderbans delta, West Bengal. *Forktail* 19: 136–137.

Sharrock, J. T. R. 1980. Kentish Plovers with pale brown legs. *British Birds* 73: 184 (see also *British Birds* 74: 147–149).

Shirihai, H. 1994. Field characters of Senegal Thick-knee. *British Birds* 87: 183–186.

Shirihai, H. 1996. *The birds of Israel.* Academic Press, London.

Shirihai, H. 1988. Pintail Snipe in Israel in November 1984 and its identification. *Dutch Birding* 10: 1–11.

▲ **Adult male breeding Lesser Yellowlegs**. Alaska, mid June. René Pop.

Shirihai, H. & van den Berg, A. B. 1987. Influx of Kittlitz's Sand Plover in Israel in 1986–87. *Dutch Birding* 9: 85–88.

Shrubb, M. 2007. *The Lapwing.* T & AD Poyser, London.

Siblet, J-P. & Spanneut, L. 1998. The Willet in Vendée, France. *Birding World* 11: 386.

Silvius, M. J. 1988. On the importance of Sumatra's east coast for waterbirds, with notes on the Asian Dowitcher *Limnodromus semipalmatus.* *Kukila* 3: 117–137.

Silvius, M. J. & Erftemeijer, P. L. A. 1989. A further revision of the main wintering range of the Asian Dowitcher *Limnodromus semipalmatus.* *Kukila* 4: 49–50.

Sitters, H., Minton, C., Collins, P., Etheridge, B., Hassell, C. & O'Connor, F. 2004. Extraordinary numbers of Oriental Pratincoles in NW Australia. *Wader Study Group Bulletin* 103: 26–31.

Skeel, M. A. & Mallory, E. P. 1996. Whimbrel (*Numenius phaeopus*). In *The Birds of North America*, (A. Poole & F. Gill, eds), no. 219. The Birds of North America, Inc., Philadelphia.

Skewes, J., Minton, C. & Rogers, K. 2004. Primary moult of the Ruddy Turnstone *Arenaria interpres* in Australia. *Stilt* 45: 20–32.

Smart, J. B. & Forbes-Watson, A. 1971. Occurrence of the Asiatic Dowitcher in Kenya. *Bulletin of the East African Natural History Society* 1971: 74–75.

Smiddy, P. & O'Sullivan, O. 1995. Forty-second Irish Bird Report, 1994. *Irish Birds* 5: 325–351.

Smith, R. D. & Summers, R. W. 2005. Population size, breeding biology and origins of Scottish Purple Sandpipers. *British Birds* 98: 579–588.

Snow, D. W. & Perrins, C. M. 1998. *The Birds of the Western Palearctic, Concise Edition.* Volume 1. Oxford University Press, Oxford.

Sondebø, S. 1992. The Willet in Norway. *Birding World* 5: 458–460.

Steele, J. & Vangeluwe, D. 2002. From the Rarities Committee's files: The Slender-billed Curlew at Druridge Bay, Northumberland, in 1998. *British Birds* 95: 279–299.

Stemple, D., Moore, J., Giriunas, I. & Paine, M. 1991. Terek Sandpiper in Massachusetts: first record for eastern North America. *American Birds* 45: 397–398.

Stenning, J. & Hirst, P. 1994. The Grey-tailed Tattler in Grampian – the second Western Palearctic record. *Birding World* 7: 469–472.

Stiles, F. G. & Skutch, A. F. 1989. *A guide to the birds of Costa Rica.* Cornell Univ. Press, Ithaca, NY.

Stoddart, A. 2007. An apparent Hudsonian Dunlin on the Isles of Scilly. *Birding World* 20, 464–466.

Svažas, S., Mongin, E., Grishanov, G., Kuresoo, A. & Meissner, W. 2002. *Snipes of the eastern Baltic region and Belarus.* OMPO Special Publication, Vilnius.

Taldenkov, I. A. 2006. Record of mixed nesting of American Plover and Gray Plover at northern Chukchi Peninsula. *Information Materials of the Working Group on Waders,* 19: 39–41. In Russian.

Takano, T., Hamaguchi, T., Morioka, T., Kanouchi, T. & Kabaya, T. 1985. *Wild Birds of Japan.* Yama-Kai Publishers, Tokyo. In Japanese.

In *The Birds of North America*, (A. Poole & F. Gill, eds), no. 493. The Birds of North America, Inc., Philadelphia.

Taverner, J. H. 1982. Feeding behaviour of Spotted Redshank flocks. *British Birds* 75: 333–334.

Taylor, D. W. 1977. Jack Snipe calling in winter. *Kent Bird Report* (1975) 24: 101.

Taylor, P. B. 1983. Field identification of sand plovers in East Africa. *Dutch Birding* 5: 37–66.

Taylor, P. W. 1984. Field identification of Pintail Snipe and recent records in Kenya. *Dutch Birding* 6: 77–90.

Thorpe, R. I. 1995. Grey-tailed Tattler in Wales: new to Britain and Ireland. *British Birds* 88: 255–262.

Thorup, O. (Compiler). 2006. *Breeding waders in Europe 2000. International Wader Studies* 14. Wader Study Group, U.K.

Tibbitts, T. L. & Moskoff, W. 1999. Lesser Yellowlegs (*Tringa flavipes*). The Birds of North America Online (A. Poole, ed.). Cornell Laboratory of Ornithology, Ithaca. http://bna.birds.cornell.edu/bna

Tomkovich, P. S. 1992a. An analysis of the geographic variability in Knots *Calidris canutus* based on museum skins. *Wader Study Group Bulletin* 64: The Migration of Knots, Supplement.

Tomkovich, P. S. 1992b. Spoon-billed Sandpiper in north-eastern Siberia. *Dutch Birding* 14: 37–41.

Tomkovich, P. S. 1996a. A third report on the biology of the Great Knot *Calidris tenuirostris* on the breeding grounds. *Wader Study Group Bulletin* 81: 88–90.

Tomkovich, P. S. 1996b. *Calidris* sandpipers of north-eastern Siberia. *Dutch Birding* 18: 11–12.

Tomkovich, P. S. 1998. Breeding schedule and primary moult in Dunlins *Calidris alpina* of the Far East. *Wader Study Group Bulletin* 85: 29–34.

Tomkovich, P. S. 2001. A new subspecies of Red Knot *Calidris canutus* from the New Siberian islands. *Bulletin of the British Ornithologists' Club* 121: 257–263.

Tomkovich, P. S., Gill, R. E. & Dementiev, N. 1998. Surfbird in its non-surfing habitats. *Dutch Birding* 20, 233–237.

Tuck, L. M. 1972. *The Snipes; A Study of the Genus Capella.* Canadian Wildlife Service, Ottawa.

Ueki, Y. 1986. *The Painted Snipe.* Tokyo. In Japanese.

Undeland, P. & Sangha, H. S. 2002. Pectoral Sandpiper *Calidris melanotos*: a new species for the Indian subcontinent. *Forktail* 18: 157.

Ura, T., Azuma, N., Hayama, S. & Higashi, S. 2005.

Sexual dimorphism of Latham's Snipe (*Gallinago hardwicki*). *Emu* 105: 259–262.

Urban, E. K., Fry, C. H. & Keith, S. (Eds). 1986. *The Birds of Africa*. Vol. 2. Academic Press, London.

Urfi, A. J. 2002. Waders and other wetland birds on Byet Dwarka Island, Gulf of Kutch, western India. *Wader Study Group Bulletin* 99: 31–34.

Valle, R. & Scarton, F. 1996. Status and distribution of Redshanks *Tringa totanus* breeding along the Mediterranean coasts. *Wader Study Group Bulletin* 81: 66–70.

van den Berg, A. B. 1985. Juvenile plumage of Black-winged Pratincole. *Dutch Birding* 7: 143–144.

van den Berg, A. B. 1988. Identification of Slender-billed Curlew and its occurrence in Morocco in winter of 1987/88. *Dutch Birding* 10: 45–53.

van den Berg, A. B. 1992. Mystery photographs: Green Sandpiper. *Dutch Birding* 14: 223–225.

van Rhijn, J. G. 1991. *The Ruff*. T & AD Poyser, Calton.

van Scheepen, P. & Oreel, G. J. 1995. Herkenning en voorkomen van IJslandse Grutto in Nederland [Identification and status of Islandic Black-tailed Godwits in the Netherlands]. *Dutch Birding* 17: 54–64.

Veit, R. R. & Jonsson, L. 1984. Field identification of the small sandpipers within the genus *Calidris*. *American Birds* 38: 853–876.

Vinicombe, K. E. 1988. Unspecific Golden Plover in Avon. *Birding World* 1: 54–56.

Vlachos, C. 2006. The first Kittlitz's Plover for Greece. *Birding World* 19: 18.

Votier, S. 1997. Ageing of Collared Pratincole. *Birding World* 10: 460.

Wallace, D. I. M. 1970. Identification of Spotted Sandpipers out of breeding plumage. *British Birds* 63: 168–173.

Wallace, D. I. M. 1976. Distinguishing Great Snipe from Snipe. *British Birds* 69: 377–383.

Wallace, D. I. M. 1977. Further definition of Great Snipe characters. *British Birds* 70: 283–289.

Wallace, D. I. M. 1989. Field characters and voice of Swinhoe's Snipe. *British Birds* 82: 269–271.

Wallace, D. I. M. 2003. Shortish or long shot at snipe? *British Birds* 96: 406–407.

Wallace, D. I. M., McGeehan, A. & Allen, D. 2001. Autumn migration in westernmost Donegal. *British Birds* 92: 103–120.

Walsh, T. A. 1985. Yellow orbital ring of Semipalmated Plover. *British Birds* 78: 661.

Warnock, N. D. & Gill, R. E. 1996. Dunlin (*Calidris alpina*). In *The Birds of North America*, (A. Poole &

F. Gill, eds), no. 203. The Birds of North America, Inc., Philadelphia.

Webster, J. D. 1942. Notes on the growth and plumages of the Black Oyster-catcher. *Condor* 44: 205–213.

Weir, J. 1997. The breeding biology of a British Columbia American Avocet colony. *British Columbia Birds* 7: 3–7.

Wenink, P. W. & Baker, A. J. 1996. Mitochondrial DNA lineages in composite flocks of migratory and wintering Dunlins (*Calidris alpina*). *Auk* 113: 744–756.

Wenink, P. W., Baker, A. J., Rosner, H. U. & Tilanus, M. G. J. 1996. Global mitochondrial DNA phylogeography of Holarctic breeding Dunlins (*Calidris alpina*). *Evolution* **50**: 318–330.

Wilson, E. M. 1989. First record of the Terek Sandpiper in California. *Western Birds* 20:63–69.

Wilson, G. E. 1976. Spotted Sandpipers nesting in Scotland. *British Birds* 69: 288–292.

Wilson, J. R. 2000. The northward movement of immature Eastern Curlews in the Austral winter as demonstrated by the population monitoring project. *Stilt* 36: 16–19.

Wilson, W. H. 1994. Western Sandpiper (*Calidris mauri*). The Birds of North America Online (A. Poole, ed.). Cornell Laboratory of Ornithology, Ithaca. http://bna.birds.cornell.edu/bna

Wright, G. 1987. Hudsonian Godwit in Devon. *British Birds* 80: 492–494.

Xeira, A. 1987. The head pattern of Black-winged Stilts. *Wader Study Group Bulletin* 50: 29. (Reprinted *Wader Study Group Bulletin* 69: 74.)

Yésou, P. 1982. Semipalmated Plover in the Western Palearctic [Azores]. *British Birds* 75: 336.

Yésou, P., Chupin, I. I. & Grabovski, V. I. 1992. Notes on the breeding biology of the Bar-tailed Godwit *Limosa lapponica* in Taimyr. *Wader Study Group Bulletin* 66: 45–47.

Yovanovich, G. D. L. 1995. First ABA-Area Records: Collared Plover in Uvalde, Texas. *Birding* 27: 102–104.

Zeillemaker, C. F., Eltzroth, M. S. & Hamerick, J. E. 1985. First North American record of the Black-winged Stilt. *American Birds* 39: 241.

Zielinski, M. 1992. Egyptian Plover in Poland in October-November 1991. *Dutch Birding* 14: 99.

Zink, R. M., Rohwer, S., Andreev, A. V. & Dittmann, D. L. 1995. Trans-Beringia comparisons of mitochondrial DNA differentiation in birds. *Condor* 97: 639–649.

Zöckler, C. 1998. The rare Steppe Whimbrel. *Birding World* 11: 273–274.

▲ **Eurasian Curlew.** Norfolk, mid September. RJC.

INDEX